植物精油

（中英文版）

朱亮锋

李翰祥

杨安坪

李泽贤

编 著

中国出版集团有限公司

世界图书出版公司
广州·上海·西安·北京

图书在版编目（CIP）数据

植物精油：汉文、英文 / 朱亮锋等编著 . -- 广州：世界图书出版广东有限公司, 2025.5. -- ISBN 978-7-5232-2137-2

I . TQ654

中国国家版本馆 CIP 数据核字第 2025ZR8218 号

书　　名	植物精油（中英文版）	
	ZHIWU JINGYOU（ZHONG YING WEN BAN）	
编　　著	朱亮锋　李翰祥　杨安坪　李泽贤	
责任编辑	刘　旭	
装帧设计	柏桐文化	
出版发行	世界图书出版有限公司　世界图书出版广东有限公司	
地　　址	广州市海珠区新港西路大江冲 25 号	
邮　　编	510300	
电　　话	（020）84460408	
网　　址	http://www.gdst.com.cn/	
邮　　箱	wpc_gdst@163.com	
经　　销	新华书店	
印　　刷	广州清粤彩印有限公司	
开　　本	710 mm × 1 000 mm　1/16	
印　　张	36.5	
字　　数	804 千字	
版　　次	2025 年 5 月第 1 版　2025 年 5 月第 1 次印刷	
国际书号	ISBN 978-7-5232-2137-2	
定　　价	268.00 元（USD 53）	

编委会

编　　著：朱亮锋　李翰祥　杨安坪　李泽贤

编　　委：吴　萍　贾永霞　张　贺　王东栋

编　　审：夏念和

参编单位：中国科学院华南植物园
　　　　　广州天赐高新材料股份有限公司
　　　　　广州市立颖贸易有限公司
　　　　　广州归然生态科技有限公司

作者简介
ABOUT THE AUTHOR

朱亮锋（Zhu Liangfeng）

　　教授，研究员，1960年毕业于中山大学，自1964年起在中国科学院华南植物研究所从事植物化学和植物资源学的研究，曾任华南植物研究所植物资源研究室副主任、主任，从1987年开始从事中药材研究开发工作，曾任联合国教科文组织（UNESCO）亚洲及太平洋地区药用植物和芳香植物情报网络（APINMAP）联络员。1989年获国务院有突出贡献科学家津贴；1999年获国际精油和香料行业联盟（IFEAT）成就奖章。

李翰祥（Li Hanxiang）

　　中国科学院华南植物园公共实验室工程师，毕业于中山大学，理学博士。主要研究领域是天然产物化学及有机质谱分析。相关研究成果发表于Journal of Natural Products, Tetrahedron, Tetrahedron Letters, Fitoterapia, Journal of Agricultural and Food Chemistry等期刊。

杨安坪（Winnie Yeung）

　　国际香精香料协会（Exective member of IFEAT Board）董事会成员，亚洲精油及合成香料协会（AAIC）副理事长，中国香料香精化妆品工业协会（CAFFCI）副理事长，中国土畜商会香精香料分会（CHINAEASA）副理事长。从20世纪80年代初起从事香精香料生产和进出口，拓展产品销往欧、美、亚等国际市场及国内市场。30多年来积累了丰富的专业知识及行政管理经验，作为中国香料行业在国际组织的代表，积极进行行业国内国际交流，10多年来多次在中国、美国、英国、加拿大、匈牙利、印度、印度尼西亚的香精香料行业大会、技术及市场高峰论坛等发言，在国内外行业专业杂志发表文章。多次参与筹备在各国举办的国际香精香料协会年会。

李泽贤（Li Zexian）

　　自1960年起在中国科学院华南植物研究所植物分类研究室工作。曾参加西沙群岛、海南岛及其邻近岛屿、广东沿海岛屿及香港等地区植物调查。于2001年开始被香港泰山环境建设有限公司聘任为湿地植物顾问。在各项工作过程中，先后发表论文45篇。参编专著《中国本草图录》卷九、《原色中草药图典》（一）（三）（五）册及《香料植物》等7部。

序言
PREFACE

　　有关精油的书籍近百年来已经出版过不少，较早引入我国的一部是1951年日本人藤田安二著的《植物精油的基础研究》，早期有不少关于精油的文章和专著引用了它。接下来是焦启源1953年著的《芳香植物及其利用》，黄有识1957年编著的《芳香油化学》，江希张、洪慰慈、屠伯范在1960年译的E.坤斯所著的《精油》，刘铸晋等在1964年译的A.R.品德尔所著的《萜类化学》，林启寿1977年编著的《中草药成分化学》，中国科学院上海药物研究所1981年编著的《中草药有效成分提取与分离》，刘树文、胡宗蓍1982年译的《精油手册》，龙康侯等1984年编著的《萜类化学》，朱亮锋等1993年编著的《芳香植物及其化学成分》，以及1993年和1995年Zhu Liangfeng，Li Yonghua，Li Baoling，Lu Biyao，Xia Nianhe and Zhang Weilian编著的 *Aromatic Plants and Essential Constituents* 和 *Aromatic Plants and Essential Constituents*（Supplement I），徐任生1993年主编的《天然产物化学》，杨峻山2005年编著的《萜类化合物》，王锋鹏2009年主编的《现代天然产物化学》，周敏2020年编著的《天然二萜类化合物的波谱分析》和王羽梅2020年主编的《中国芳香植物资源(1—6卷)》等。

　　笔者长期从事植物精油的资源、生产工艺、化学成分开发及应用研究，前后算起来超过50年，并邀请从事于这项工作的同事收集、整理与植物精油相关的资料，初步构建出《植物精油（中英文版）》一书。本书既涵盖了植物精油相关知识，又引用了大量植物精油研究开发的内容。

　　全书一共分为四章，包括我国植物精油资源概况、植物精油的化学成分、植物精油的提取工艺与设备、植物精油和植物精油成分的用途。第一章关于芳香植物概况及我国植物精油资源开发的编排，遵循了生产实际需求与战略应用重要性相结合的分类原则。本书英文版的第一、二章在中文版基础上，结合行业发展进行了内容更新与提升。

目 录
CONTENTS

第一章　我国植物精油资源概况

第一节　芳香植物概况

　　单纯从植物学的角度来说，植物精油资源应该是芳香植物资源，因为植物精油全是从芳香植物中得到的。直观地讲，芳香植物某一部位通过水蒸气蒸馏方法或其他方法，得到一种不溶于水、有气味的挥发性油状液体为精油（essential oil）。精油广泛分布于芳香植物的不同部位（器官）。

　　我国幅员广阔，地处温带、亚热带和热带，气候、土壤、植被多样，为各种芳香植物提供了极为有利的自然生长环境。因此，我国芳香植物十分丰富。据不完全统计，芳香植物包括引种多达1000余种，分属77个科192个属。我国开发利用芳香植物历史非常久远，早在距今2000多年前的秦、汉年代就利用菊科艾蒿类植物沐浴、焚熏以洁身去秽和防病治病，这一古老的方法沿用至今。唐宋时期已有利用茉莉花（*Jasminum sambac*）熏制花茶的记载，这类花茶（现称香片）至今也是深受人们喜爱的茶品种。现今植物精油已涉及化工、轻工、食品、化妆美容、药品、医疗等行业，并且也是我国一类传统出口商品，如松节油、薄荷油、山苍子油、姜油、各类樟油以及一些香花类浸膏，如小花茉莉浸膏、桂花浸膏油、苦水玫瑰油等。当然随着国际市场变化，上述植物精油产品出口情况也在不断调整、变化。

　　我国相当大一部分土地处于南亚热带地区，长江流域和长江流域以南地区是芳香植物最佳的生长之地，故这一广阔地方蕴藏着大量资源。其中已被开发利用和待开发的芳香植物有松科植物马尾松（*Pinus massoniana*）、湿地松（*P. elliottii*），柏科植物刺柏（*Juniperus formosana*）、柏木（*Cupressus funebris*），樟科植物肉桂（*Cinnamomum aromaticum*）、阴香（*C. burmannii*）、樟（*C. camphora*）、黄樟（*C. parthenoxylon*），以及桃金娘科各种桉树如柠檬桉（*Eucalyptus citriodora*）、蓝桉（*E. globulus*）等，其中樟科樟属（*Cinnamomum*）精油具有较大开发潜力。现鼓励改变以往伐木挖根取油的破坏性生产方式，多利用枝叶为原料生产精油。在栽培方面，选择优良植株为母树进行无性繁殖，建立种质资源，并扩大发展成种植基地。

一、樟科樟属植物精油资源

研究樟科樟属种内化学类型时发现了一些重要的植物精油资源，在第二节详细介绍。樟科樟属植物精油资源参看表1-1、表1-2。

表1-1　樟科樟属精油资源一览表

植物名称	主要化学成分（质量占比）	部位（出油率）	备注
1. 毛桂 *Cinnamomum appelianum*	1,8-桉叶油素37.6% 乙酸龙脑酯16.2%	枝叶1.05%	我国特有种，分布于湖南、江西、广东、广西、贵州、四川、云南
2. 滇南桂 *C. austro-yunnanense*	γ-榄香烯20.8%	枝叶0.12%	分布于云南
3. 钝叶桂 *C. bejolghota*	γ-榄香烯28.6% 芳樟醇14.2% β-丁香烯10.1%	枝叶0.31%	分布于云南、四川、贵州
	t-桂醛82.6%	皮0.12%～0.2%	
4. 猴樟 *C. bodinieri*	黄樟油素84.0% 橙花叔醇68.4%	枝叶0.3% 枝叶0.4%	分布于云南、贵州
5. 湖北樟 *C. bodinieri* var. *hupehanum*	樟脑88.5% 柠檬醛95.0%	叶1.5% 叶1.4%	分布于湖北、四川
6. 阴香 *C. burmannii*	d-龙脑51.3% 乙酸龙脑酯7%	枝叶0.3%～0.4%	梅片树型，分布于广东、福建、云南、广西

（续表）

植物名称	主要化学成分（质量占比）	部位（出油率）	备注
	1,8-桉叶油素65.5%	枝叶0.3%～0.4%	计树型，分布于广东
6. 阴香 *C. burmannii*	对伞花烃27.5% 1,8-桉叶油素27.6%	枝叶0.25%～0.3%	油计树型，分布于广东
	芳樟醇57.0%	枝叶	云南产
	柠檬醛76.8%	枝叶	云南产
7. 狭叶阴香 （狭叶桂） *C. burmannii f. heyneanum*	黄樟油素97.5%（枝叶） 黄樟油素98.8%（树皮）	枝叶0.54% 树皮0.81%	分布于湖北、四川、贵州、广西、云南
	樟脑83.9%	枝叶1.0%	脑樟型，分布于江西
	芳樟醇90.6%	枝叶0.3%～0.8%	芳樟型，分布于江西、广东
8. 樟（本樟） *C. camphora*	1,8-桉叶油素50.0%	枝叶0.75%	油樟型，分布于江西
	d-龙脑81.8%	枝叶0.8%	龙脑樟型，分布于江西、湖南
	异橙花叔醇57.7%	枝叶0.4%	花香型，分布于江西
	柠檬醛69.9%	鲜叶1.6%～2.0%	分布于云南西双版纳
9. 芳樟 *C. camphora* var. *linaloolifera*	芳樟醇92.7% 芳樟醇90.1%	鲜叶1.95% 枝叶1.3%	分布于云南西双版纳、福建浦城
10. 肉桂 *C. aromaticum*	桂醛74.1% t-桂醛97.1%	枝叶0.35% 干树皮 0.8%～2.2%	分布于云南、广东肇庆
11. 尾叶樟 *C. caudiferum*	1,8-桉叶油素53.9% 樟脑60.4%	枝叶2.5% 枝叶1.2%	分布于云南西畴
12. 聚花桂 *C. contractum*	t-桂醛	枝叶0.47%	分布于云南

（续表）

植物名称	主要化学成分（质量占比）	部位（出油率）	备注
	1,8-桉叶油素45.8%	枝叶1.10%	
13. 云南樟 *C. glanduliferum*	柠檬醛54.4%	枝叶0.4%～0.5%	分布于云南
	α-水芹烯65.7%	枝叶0.5%～0.8%	
	黄樟油素90%	果实皮2.70%	
14. 天竺桂 *C. japonicum*	丁香酚15.3% 1,8-桉叶油素13.2% 芳樟醇11.4%	枝叶0.25%～1.19%	分布于整个华东地区
	丁香酚29.5% 1,8-桉叶油素16.1%	树皮0.50%~1.26%	
	樟脑30.3% 黄樟油素45.9%	根	
15. 八角樟 *C. ilicioides*	黄樟油素82.1% 黄樟油素82.7%	枝叶0.83% 干侧根0.21%	分布于云南、广西上思
16. 油樟 *C. longepaniculatum*	1,8-桉叶油素58.6% α-松油醇15.4%	枝叶1.2%	分布于四川
	β-桉叶醇41.1%	枝叶1.2%	分布于湖北
	樟脑90.5%	枝叶2.3%	分布于云南西双版纳
17. 沉水樟 *C. micranthum*	癸醛13.8% 癸酸15.9% 月桂烯14.3%	鲜叶0.12%	分布于福建、广西、广东、湖南、江西
	黄樟油素61.3%	树干0.09%	分布于福建
	黄樟油素97.7%	根1.5%	分布于江西吉安

（续表）

植物名称	主要化学成分（质量占比）	部位（出油率）	备注
18. 土肉桂 *C. osmophloeum*	t-桂醛79.5% 芳樟醇83.3%	叶0.28%～1.45%	
	芳樟醇43.9% 苯甲醛35.7%	枝叶	分布于台湾
	芳樟醇40.4% t-桂醛32.9%	枝叶	
19. 少花桂 *C. pauciflorum*	黄樟油素69.7%	枝叶2.4%	分布于四川筠连
	黄樟油素99.3%	枝叶2.5%	分布于云南西双版纳
	黄樟油素99.4%	木材0.44%	
	黄樟油素99.0% 黄樟油素80.4%	鲜树皮1.6% 侧根0.71%	分布于广东、广西、云南、四川、湖北
20. 屏边桂 *C. pingbienense*	芳樟醇38.4% 1,8-桉叶油素18.5%	干叶0.28%	分布于云南屏边
21. 阔叶樟 *C. platyphyllum*	t-甲基异丁香酚94.1%	鲜叶0.57%	分布于陕西、甘肃、四川、湖北
22. 黄樟 （大叶樟） *C. parthenoxylon*	α-蒎烯22.4% 桧烯12.7% 松油醇21.2%	枝叶0.6%～0.8%	分布于广西、广东、福建、江西、湖南、贵州、云南
	1,8-桉叶油素62.4%	枝叶2.0%	大叶油樟
	d-芳樟醇82.8%	枝叶1.1%～1.4%	大叶芳樟
	d-樟脑86.7%	枝叶0.8%～1.0%	大叶脑樟
	柠檬醛72.1%	枝叶0.5%～0.8%	姜樟
	9-氧化橙花叔醇24.2%	枝叶0.6%～0.8%	倍半萜型
	橙花叔醇54.8%	枝叶0.3%～0.4%	倍半萜型
23. 卵叶桂 *C. rigidissimum*	丁香酚甲醚28.6%	根1.4%	分布于海南尖峰岭、广东、广西、台湾
	4-松油醇18.7%	枝叶0.04%	
	苯甲酸苄酯82.8%	叶1.03%	分布于云南西双版纳
	黄樟油素61.7%	根1.44%	

（续表）

植物名称	主要化学成分（质量占比）	部位（出油率）	备注
24. 岩樟 *C. saxatile*	橙花叔醇30.7% 乙酸金合欢酯12.1% 金合欢醇7.6%	干枝叶0.19%	分布于云南麻栗坡
	黄樟油素94.4%	干根1.46%	
25. 银木 *C. septentrionale*	t-甲基丁香酚甲醚85.7%	枝叶1.10% 枝叶1.05%	分布于湖北利川、四川、陕西、甘肃、广西
	1,8-桉叶油素76.1% 丁香酚26.4%	枝叶0.24%	分布于云南西双版纳
26. 香桂 *C. subavenium*	4-松油醇21.4% 1,8-桉叶油素15.3%	枝叶0.1%～0.4%	
	丁香酚67.4%	皮0.89%	分布于云南、四川、湖北、广东、广西、安徽、浙江、江西、福建
27. 柴桂 *C. temala*	d-香茅醛69.1%	枝叶1.1%	分布于云南南部
	黄樟油素44.3% 黄樟油素98.8%	枝叶0.47% 干皮5.42%	分布于云南西双版纳
	l-芳樟醇97.51%	枝叶1.4%	栽培种
	香叶醇92.5%	枝叶1.55%～2.04%	栽培种，香叶醇型
	金合欢醇70.03%	枝叶	金合欢型
	丁香酚甲醚84.3%	枝叶	丁香酚型
28. 细毛樟 *C. tenuipilum*	樟脑85.7%	枝叶	樟脑型
	1,8-桉叶油素50%～57%	枝叶1.4%～1.5%	油樟型
	龙脑45% 柠檬醛59%～75%	枝叶1.43%	龙脑型，分布于云南南部和西部
	榄香脂素84.0%	枝叶1.0%～1.5%	
	榄香烯56.4%	枝叶	倍半萜型
	榄香脂素41.3%	鲜枝叶1.10%	倍半萜型
	肉豆蔻醚33.1%	根0.44%	倍半萜型

（续表）

植物名称	主要化学成分（质量占比）	部位（出油率）	备注
	香叶醇37.3% 芳樟醇21.6% 丁香酚57.5%	枝叶0.47%	
29. 假桂皮树 *C. tonkinense*	黄樟油素36.9% 丁香酚73.4%	皮1.04% 枝叶0.13%	分布于云南麻栗坡
	柠檬醛39.6% 芳樟醇13.3%	鲜叶0.31% 树皮 0.43%～0.53%	
	丁香酚67.7%	鲜叶0.31%	
30. 川桂（柴桂） *C. wilsonii*	芳樟醇24.1% 1,8-桉叶油素18.2%	枝叶	分布于陕西、四川、湖北、湖南、江西、广东、广西
	乙酸龙脑酯26.1% 龙脑16.8%	枝叶	
31. 粗脉桂 *C. validinerve*	芳樟醇43.8% 香叶醇17.4%	叶	分布于云南
32. 锈毛桂 *C. villosulum*	芳樟醇59.5%	鲜叶 0.25%～0.30%	分布于广西
33. 锡兰玉桂 *C. zeylanicum*	丁香酚81.3% 苯甲酸苄酯8.10%	鲜叶 1.5%～2.13% 鲜叶1.53%	分布于云南西双版纳

二、樟科樟属种内化学类型

表1-2 樟科樟属种内化学类型（枝叶精油）一览表

名称	主要化学成分（质量占比）	类型
1. 钝叶桂 *Cinnamomum bejolghota*	t-桂醛82.6% 榄香烯、芳樟醇、丁香烯	桂醛型 混合型
2. 猴樟 *C. bodinieri*	橙花叔醇68.4% 黄樟油素68.0%	橙花樟型 黄樟型
3. 阴香 *C. burmannii*	d-龙脑 51.3% 1,8-桉叶油素65.5% 1,8-桉叶油素27.6%、对伞花烃27.5%	梅片树（龙脑型） 油计树（油樟型） 混合过渡型
4. 樟 *C. camphora*	樟脑88.9% 1-芳樟醇90.6% 1,8-桉叶油素50.0% d-龙脑81.8% 异橙花叔醇57.7%	脑樟型 芳樟型 油樟型 龙脑樟型 橙花樟型
5. 尾叶樟 *C. caudiferum*	1,8-桉叶油素53.9% 樟脑60.4%	油樟型 脑樟型
6. 云南樟 *C. glanduliferum*	柠檬醛54.4% 黄樟油素90.0% 1,8-桉叶油素45.8% α-水芹烯65.7%	姜樟型 黄樟型 油樟型 萜烯型
7. 狭叶阴香（狭叶桂） *C. burmannii f. heyneanum*	黄樟油素98.5% 樟脑48.7% 1,8-桉叶油素33.6%、樟脑35.0%	黄樟型 脑樟型 混合过渡型
8. 油樟 *C. longepaniculatum*	1,8-桉叶油素58.6% 樟脑90.5% β-桉叶醇41.1%	油樟型 脑樟型 倍半萜醇型
9. 长柄樟 *C. longipetiolatum*	柠檬醛71.7% 1,8-桉叶油素、黄樟油素	姜樟型 混合型

（续表）

名称	主要化学成分（质量占比）	类型
10. 土肉桂 *C. osmophloeum*	t-桂醛79.5%	桂醛型
	芳樟醇83.3%	芳樟型
	芳樟醇43.9%、苯甲醛35.7%	混合过渡型
11. 少花桂 *C. pauciflorum*	黄樟油素97.5%	黄樟型
	柠檬醛、α-蒎烯、1,8-桉叶油素	混合型
12. 黄樟（大叶樟） *C. parthenoxylon*	1,8-桉叶油素62.4%	油樟型
	d-芳樟醇94.3%	芳樟型
	d-樟脑86.7%	脑樟型
	α，β-柠檬醛72.1%	姜樟型
	丁香酚甲醚71.5%	丁香酚型
13. 卵叶桂 *C. rigidissimum*	苯甲酸苄酯82.8%	芳香族型
	丁香酚甲醚28.6%、4-松油醇18.7%	混合型
14. 岩樟 *C. saxatile*	樟脑40.5%	脑樟型
	t-异丁香酚甲醚85.7%	丁香酚型
15. 银木 *C. septentrionale*	t-丁香酚甲醚85.7%	丁香酚型
	1,8-桉叶油素76.8%	油樟型
16. 香桂 *C. subavenium*	樟脑85.0%	脑樟型
	1,8-桉叶油素76.1%	油樟型
	丁香酚甲醚84%～89%	丁香酚型
17. 柴桂 *C. tamala*	香茅醛69.1%	香茅醛型
	黄樟油素44.3%	黄樟型
18. 细毛樟 *C. tenuipilum*	金合欢醇64%～70%	倍半萜型
	香叶醇92.5%	香叶醇型
	γ-榄香烯56.4%	萜烯型
	榄香脂素	倍半萜型
19. 假桂皮树 *C. tonkinense*	α，β-柠檬醛53.2%	姜樟型
	芳樟醇37.0%；香叶醇22.1%	花香型
20. 锡兰玉桂 *C. zeylanicum*	丁香酚81.3%	丁香酚型
	苯甲酸苄酯8.1%	芳香族型
	t-桂醛31.2%；芳樟醇34.1%	混合型

三、乔木、灌木精油资源

我国木本精油资源除上述松科、柏科、樟科外，还有其他非常丰富的精油资源，绝大多数都是利用其枝叶、果实、种子来获取精油，对资源再生和生态环境起到重要作用，也确保了精油资源可持续生产，详情见表1–3。

表1–3 我国乔木、灌木精油资源一览表

植物名称	主要化学成分	分布	备注
白豆杉 *Pseudotaxus chienii* 叶含精油 红豆杉科	α-蒎烯、莰烯、β-蒎烯、δ-3-蒈烯、月桂烯、柠檬烯、β-侧柏烯、t-石竹烯、对伞花-8-醇、t-金合欢烯、α-依兰油烯、β-荜澄茄烯、t-β-金合欢烯、δ-杜松烯、去氢白菖烯、橙花叔醇等	江西、浙江、湖南、广东、广西等省区。生长于常绿阔叶林及落叶阔叶林中	为庭园观赏树种，木材可供雕刻
香榧 *Torreya grandis* 种子含精油 红豆杉科	α-蒎烯、月桂烯、2-蒈烯、4-松油醇、α-松油醇、芳樟醇、α-荜澄茄烯、β-荜澄茄烯、β-金合欢烯、β-荜澄茄异构体、β-杜松烯、δ-杜松烯等	浙江、江苏、江西、福建、湖南、贵州等省有栽种	种子为著名"干果"，供食用，精油可用于日化工业，如牙膏、香皂，种子入药作驱虫用
竹柏 *Podocarpus nagi* 枝叶含精油0.10% 罗汉松科	3-己烯醇、α-蒎烯、7-辛烯-4-醇、1,8-桉叶油素、α-松油醇、α-石竹烯、依兰油烯、β-荜澄茄烯、γ-杜松烯、γ-榄香烯、γ-杜松烯异构体、δ-杜松烯、橙花叔醇等	台湾、福建、浙江、江西、湖南、广东、四川等省区。生长于低海拔常绿阔叶林中	为优质用材，亦是庭园观赏树种，种子富含脂肪油，供工业用和食用
黄果冷杉 *Abies ernestii* 枝叶含精油 0.15%～0.28% 松科	3-乙烯醇、柠檬烯、2,3,3-三甲基二环[2,2,1]庚-2-醇、龙脑、4-松油醇、α-松油醇、辣薄荷醇、辣薄荷酮、乙酸龙脑酯、环丙乙酸2-己酯、δ-杜松醇、β-桉叶醇等	我国特有种，产于四川西部和北部、西藏东部。生长于海拔2500～3500米山地次生林中	主要建筑用材，枝叶精油用于天然香料原料

（续表）

植物名称	主要化学成分	分布	备注
广西九里香 *Murraya kwangsiensis* 叶含精油0.27% 芸香科	α-水芹烯、β-蒎烯、邻伞花烃、柠檬烯、3-蒈烯、γ-松油烯、芳樟醇、香茅醛、异胡薄荷酮、胡薄荷酮、α-松油醇、2-（3,3-二甲基）亚环己基乙醇、β-柠檬醛、橙花醇、α-柠檬醛、香醇、乙酸香叶酯等	广西。生长于海拔800米的山地林中	果能通经止痛，根能健胃，叶能治麻疹骨折，枝叶能祛风止痛、行气止咳
巴山冷杉 *Abies fargesii* 枝叶含精油0.2%～0.3% 松科	3-乙烯醇、柠檬烯、2,3,3-三甲基二环[2,2,1]庚-2-醇、龙脑、4-松油醇、α-松油醇、环丙乙酸2-己酯、δ-杜松醇、β-桉叶醇等	甘肃、陕西、河南、湖北、四川等省。组成纯林或生长于杂木林中，海拔2500～3700米	枝叶精油可开发成皂用香精原料，木材质轻软可作建筑用材
岷江冷杉 *A. faxoniana* 枝叶含精油0.4%～0.7% 松科	檀烯、三环烯、α-蒎烯、莰烯、β-蒎烯、月桂烯、柠檬烯、柠檬醛、丁香酚甲醚、α-依兰油烯、橙花叔醇、喇叭茶醇等	甘肃、四川等省。生长于海拔2700～3900米的冷湿气候、排水良好、酸性棕色灰化土地和草甸	枝叶精油可用作涂料的溶剂
杉松（辽东冷杉） *A. holophylla* 枝叶和松脂含精油 松科	枝叶精油：三环烯、α-蒎烯、莰烯、β-蒎烯、月桂烯、3-蒈烯、柠檬烯、樟脑、龙脑、对伞花烃、马鞭草烯酮、2-异丙基-5-甲基-茴香醚、乙酸龙脑酯、乙酸香叶酯、γ-依兰油烯、β-甜没药烯等 松脂精油：α-蒎烯、莰烯、δ-3-蒈烯、对伞花烃、α-柠檬烯、γ-松油烯、α-异松油烯、樟脑、龙脑、4-松油醇、月桂烯、乙酸龙脑酯、δ-杜松烯等	东北牡丹江流域、长白山区及辽河东部山区。俄罗斯、朝鲜亦有分布。生长于海拔500～1200米的冷湿环境灰化棕色森林土地带	民间用精油防治感冒

（续表）

植物名称	主要化学成分	分布	备注
臭冷杉（东陵冷杉） *A. nephrolepis* 叶（鲜）含精油2.2% 松科	松烯、三环烯、α-蒎烯、莰烯、β-蒎烯、月桂烯、环小茴香烯、柠檬烯、龙脑、芳樟醇、乙酸龙脑酯、乙酸松油酯、α-柏木醇、α-蛇麻烯、β-甜没药烯、橙花叔醇、α-甜没药醇等	小兴安岭南坡、长白山，山西和河北亦有分布。生长于海拔300～2100米的冷湿环境、排水良好的山坡	
雪松（喜马拉雅杉） *Cedrus deodara* 枝叶含精油0.75% 松科	α-蒎烯、莰烯、β-蒎烯、β-月桂烯、柠檬烯、2-蒈烯、小茴香醇、樟脑、龙脑、4-松油醇、蒽、α-松油醇、乙酸龙脑酯、β-石竹烯、α-石竹烯、β-荜澄茄烯、δ-杜松醇等	西藏西南部，北京有栽种。常见庭园观赏树种	世界著名庭园观赏树种
鱼鳞云杉 *Picea jezoensis* var. *microsperma* 松脂含精油 松科	α-蒎烯、莰烯、月桂烯、δ-3-蒈烯、α-柠檬烯、α-异松油烯、樟脑、龙脑、4-松油醇、月桂烯醇、t-石竹烯、蛇麻烯、α-依兰油烯、δ-杜松烯等	黑龙江、吉林、辽宁。常生长于海拔300～800米的丘陵和缓坡地带	民间用松脂精油防治感冒
白杆 *P. meyeri* 枝叶含精油0.35% 松科	三环烯、α-蒎烯、莰烯、β-月桂烯、3-蒈烯、柠檬烯、2-蒈烯、芳樟醇、樟脑、龙脑、2,3,3-三甲基二环[2,2,1]庚-2-醇、4-松油醇、α-松油醇、乙酸龙脑酯、β-石竹烯、δ-杜松醇、δ-杜松醇异构体等	山西、河北、内蒙古。喜生长于较冷湿灰化棕色土森林上	轻质用材，为华北地区主要造林树种，枝叶精油可用作涂料溶剂
青杆 *P. wilsonii* 叶含精油0.24% 松科	三环烯、α-蒎烯、莰烯、桧烯、β-蒎烯、β-月桂烯、柠檬烯、2-蒈烯、樟脑、龙脑、4-松油醇、α-松油醇、香茅醇、乙酸龙脑酯、δ-杜松烯、β-金合欢烯等	内蒙古、河北、山西、甘肃、青海、四川等省区。适应性强，耐干冷气候，生长于海拔700～2800米处	枝叶精油可用于一般涂料的溶剂

（续表）

植物名称	主要化学成分	分布	备注
华山松（五针松） *Pinus armandi* 叶含精油0.60% 松科	三环烯、α-蒎烯、莰烯、β-蒎烯、β-月桂烯、β-水芹烯、2-莰烯、α-松油醇、樟脑、龙脑、4-松油醇、乙酸龙脑酯、β-石竹烯、α-石竹烯、γ-杜松烯、δ-杜松烯、δ-杜松醇、δ-杜松醇异构体等	山西、陕西、甘肃、河南、湖北、四川、云南、贵州、西藏。生长于海拔1600～3600米的山坡和沟谷纯林、混交林	可采树脂提取松节油，木材为优良造纸原料，树皮可提取栲胶，种子可提取脂肪油供食用
白皮松 *P. bungeana* 叶含精油0.90% 松科	三环烯、α-蒎烯、莰烯、β-蒎烯、β-月桂烯、柠檬烯、3,6,6-三甲基二环[3,1,1]庚-2-烯、樟脑、α-松油醇、β-石竹烯、α-石竹烯、γ-杜松烯、β-荜澄茄烯、δ-杜松烯等	山西、河南、陕西、甘肃、四川、湖北等省。适于干冷气候，能耐-30℃低温，生长在酸性土和石灰岩中	材用树种和庭园观赏树种，种子可供食用，精油可作松节油用
加勒比松（古巴松） *P. caribaea* 松脂含精油10%～12% 松科	三环烯、α-侧柏烯、α-蒎烯、莰烯、β-蒎烯、月桂烯、对伞花烃、β-水芹烯、大茴香醚、马鞭草烯酮、枯茗醛等	原产于加勒比海地区。广东有引种	近年引种优良树种，松脂可生产松节油和松香
赤松 *P. densiflora* 叶含精油0.6%～0.8% 松科	三环烯、α-蒎烯、莰烯、β-蒎烯、β-月桂烯、β-水芹烯、2-莰烯、α-松油醇、对叔丁基苯甲醇、乙酸龙脑酯、β-石竹烯、α-石竹烯、β-荜澄茄烯、γ-榄香烯、δ-杜松烯等	黑龙江东部、吉林长白山、辽宁中部辽东半岛、山东胶州、江苏北部沿海地区均有生长	为优良材用树种，松针精油可用作松节油
湿地松 *P. elliottii* 树脂含精油19%～28.6% 松科	α-蒎烯、莰烯、β-蒎烯、柠檬烯、α-龙脑烯醛、蒎葛缕醇、加州月桂酮、桃金娘烯醇、马鞭草烯酮、枯茗醛、乙酸葛缕酯等	原产于美国东部湿暖湿地区。我国华南、华东有引种	为优良速生树种，树脂产量高，松节油、蒎烯含量高

（续表）

植物名称	主要化学成分	分布	备注
乔松 *P. griffithii* 叶含精油 0.7%～0.8% 松科	三环烯、α-蒎烯、莰烯、β-蒎烯、β-月桂烯、α-水芹烯、柠檬烯、2-蒈烯、α-松油醇、β-石竹烯、γ-依兰油烯、δ-杜松烯、δ-杜松烯异构体等	云南、西藏。阿富汗、缅甸、尼泊尔亦有分布。生长于海拔1200～3000米的山坡和沟谷	为材用树种，松脂枝叶精油均可作松节油用
思茅松 *P. kesiya* var. *langbianensis* 松脂含精油 松科	α-蒎烯、莰烯、β-蒎烯、月桂烯、3-蒈烯、β-水芹烯、异松油烯、爱草脑、α-柏木烯、长叶烯、α-石竹烯等	云南南部思茅、麻栗坡、普洱、景东及西部潞西。其经济价值较马尾松高	松节油中β-蒎烯含量较高
红松 *P. koraiensis* 叶含精油0.5%， 皮含精油 松科	叶精油：三环烯、α-蒎烯、莰烯、β-蒎烯、月桂烯、3-蒈烯、柠檬烯、α-异松油烯、樟脑、龙脑、4-松油醇、对伞花-α-醇、α-松油醇、乙酸龙脑酯、乙酸松油醇酯、cis-石竹烯、蛇麻烯、β-荜澄茄烯、γ-榄香烯、γ-杜松烯、δ-杜松烯等 皮精油：檀烯、三环烯、α-蒎烯、莰烯、β-蒎烯、月桂烯、3-蒈烯、异松油烯、龙脑烯醛、樟脑、龙脑、4-松油醇、α-松油醇、桃金娘烯醇、马鞭草烯酮、乙酸龙脑酯、古巴烯、长叶烯、cis-石竹烯、蛇麻烯、β-甜没药烯等	我国东北长白山区、吉林山区及小兴安岭。多生长于多湿寒冷棕色林地，海拔150～1800米	为材用树种，叶富含精油，可用作松节油，叶、树皮均可入药
华南五针松 *P. kwangtungensis* 松脂含松节油 松科	α-蒎烯、莰烯、月桂烯、β-蒎烯、3-蒈烯、α-松油烯、柠檬烯、樟脑、龙脑、4-松油醇、月桂烯醇、t-石竹烯、[α]-β-金合欢烯、蛇麻烯等	我国特有树种，分布于贵州、广西、广东、湖南等省区。喜温湿气候土深厚酸性土岩石山坡上	为材用树种，用于建筑矿坑、制作家具等，松脂生产松节油

（续表）

植物名称	主要化学成分	分布	备注
马尾松 *P. massoniana* 松脂含精油， 叶含精油0.2% 松科	松脂精油（松节油）：α-蒎烯、莰烯、月桂烯、β-蒎烯、长叶烯、柠檬烯、β-石竹烯、α-蛇麻烯 叶精油：α-蒎烯、莰烯、β-蒎烯、月桂烯、柠檬烯、4-蒈烯、蒎葛缕醇、龙脑、4-松油醇、α-松油醇、乙酸龙脑酯、β-榄香烯、β-石竹烯、α-石竹烯、β-荜澄茄烯、α-石竹烯、雅槛蓝烯、γ-榄香烯、γ-杜松烯、橙花叔醇等	安徽、陕西，河南南部汉水流域以下，广东、广西亦有分布。耐干旱，能生于瘠薄红土	树脂是生产松节油的主要原料，松针和松果精油可配制日用化妆品和日化香精
偃松 *P. pumila* 叶含精油1.0%， 树皮含精油4.1% 松科	叶精油：α-侧柏烯、α-蒎烯、莰烯、桧烯、月桂烯、3-蒈烯、对伞花烃、柠檬烯、α-异松油烯、二甲基苏合香烯、4-松油醇、对伞花-α-醇、α-松油烯、爱草脑、乙酸松油酯等 树皮精油：α-蒎烯、莰烯、β-蒎烯、月桂烯、3-蒈烯、β-水芹烯、1,8-桉叶油素、松油烯、樟脑、乙酸龙脑酯、长叶烯、蛇麻烯、母菊薁等	东北大兴安岭、小兴安岭，海拔1000米以上	松针精油和树皮精油均可作松节油用
新疆五针松 *P. sibirica* 叶含精油2.8%， 一年生树皮含精油4.16% 松科	叶精油：α-蒎烯、莰烯、β-蒎烯、月桂烯、3-蒈烯、柠檬烯、β-水芹烯、1,8-桉叶油素、蛇麻烯、松油烯、松油醇、ε-杜松烯、γ-杜松烯、母菊薁等 树皮精油：α-蒎烯、莰烯、β-蒎烯、月桂烯、3-蒈烯、松油醇、ε-杜松烯、γ-杜松烯、母菊薁等	新疆阿尔泰山、西北部喀纳斯河和霍姆河流域。生长于海拔1600～2300米处，与落叶松混生	为一优质材用树种，松针富含精油，可用作松节油

（续表）

植物名称	主要化学成分	分布	备注
樟子松（海拉尔松） *P. sylvestris* var. *mongolica* 叶含精油0.67% 松科	三环烯、α-蒎烯、莰烯、β-蒎烯、月桂烯、3-蒈烯、α-松油醇、马鞭草烯酮、乙酸龙脑酯、柠檬烯、t-葛缕醇、β-榄香烯、γ-依兰油烯、α-依兰油烯、γ-杜松烯、去氢白菖烯、喇叭茶醇等	黑龙江、内蒙古有分布。生长于海拔400～800米的山地和沙丘	树干可采松脂，生产松香和松节油，木材防腐力强，可作建筑用材
长白松 *P. sylvestris* var. *sylvestriformis* 叶含精油0.78% 松科	三环烯、α-蒎烯、莰烯、β-蒎烯、月桂烯、对伞花烃、γ-松油烯、α-异松油烯、古巴烯、β-榄香烯、cis-石竹烯、蛇麻烯、β-荜澄茄烯、β-芹子烯、γ-榄香烯、α-依兰油烯、γ-杜松烯、δ-杜松烯、去氢白菖烯等	吉林省。生长于海拔800～1600米的山地上	树干产松脂，用途与马尾松相同
油松（巨果松） *P. tabuliformis* 叶含精油0.50% 松科	三环烯、α-蒎烯、莰烯、桧烯、β-蒎烯、β-月桂烯、柠檬烯、2-蒈烯、龙脑、4-松油醇、α-松油醇、乙酸龙脑酯、β-石竹烯、α-石竹烯、γ-榄香烯、δ-杜松烯、γ-依兰油烯、δ-杜松醇、δ-杜松醇异构体等	吉林、辽宁、河北、河南、山东、山西、内蒙古、陕西、甘肃、宁夏、青海、四川等省区。生长于海拔100～2600米深厚和排水良好的土地	为我国主要材用树种，树干可采脂，树皮可提栲胶
火炬松 *P. taeda* 叶含精油 松科	三环烯、α-蒎烯、莰烯、β-蒎烯、月桂烯、α-侧柏烯、γ-松油烯、α-松油烯、4-松油醇、α-松油醇、乙酸龙脑酯、β-榄香烯、t-石竹烯、蛇麻烯、β-荜澄茄烯、α-依兰油烯、γ-杜松烯、δ-杜松烯等	原产于北美洲。我国华东中南各省区有引种，生长良好	是近期引种优良松树种
黑松（日本黑松） *P. thunbergii* 叶含精油1.0% 松科	α-蒎烯、莰烯、β-蒎烯、β-月桂烯、β-水芹烯、2-蒈烯、α-松油醇、对叔丁基苯甲醇、乙酸龙脑酯、β-石竹烯、α-石竹烯、β-荜澄茄烯、δ-杜松烯等	原产于日本、朝鲜。辽东半岛、山东、江苏、浙江、福建、台湾有栽种。喜生长于温带海洋性气候	树干可产松脂，松针精油富含β-蒎烯

（续表）

植物名称	主要化学成分	分布	备注
兴凯湖松 *P. takahasii* 叶含精油0.39% 松科	三环烯、α-蒎烯、莰烯、β-蒎烯、月桂烯、α-侧柏烯、α-松油烯、4-松油醇、α-松油醇、乙酸龙脑酯、β-榄香烯、t-石竹烯、蛇麻烯、β-荜澄茄烯、δ-杜松烯等	黑龙江。俄罗斯也有分布。生长于湖边沙丘及山顶石砾上	为材用树种，轻质耐腐蚀，可加工为建筑、家具、车辆用材，枝叶精油可用于涂料溶剂
杉木（沙杉）*Cunninghamia lanceolata* 0.9%～1.1% 柏科	三环烯、α-蒎烯、莰烯、β-蒎烯、β-月桂烯、柠檬烯、对伞花烃、α-松油醇、乙酸龙脑酯、桉醇、乙酸香叶酯、β-石竹烯、α-石竹烯等	长江流域及秦岭以南广泛栽种	根部可作救生圈、瓶塞，是水网地区和庭园绿化树种
柳杉 *Cryptomeria japonica* var. *sinensis* 枝叶含精油0.62% 柏科	桉烯、1,8-间盏二烯、γ-松油烯、2-蒈烯、4-松油醇、α-松油醇、乙酸龙脑酯、δ-杜松烯、β-桉叶醇、罗汉松烯等	广东、广西、云南、贵州、四川等省区，江苏、安徽、山东、河南也有栽种	庭园观赏绿化树种
水松 *Glyptostrobus pensilis* 0.4%～0.6% 柏科	三环烯、α-蒎烯、莰烯、β-蒎烯、β-月桂烯、柠檬烯、对伞花烃、α-松油醇、乙酸龙脑酯、桉醇、乙酸香叶酯、β-石竹烯、α-石竹烯等	我国特有树种，分布于广东、广西、云南。生长于低海拔地区湿地，耐水不耐寒	
池杉 *Taxodium distichum* var. *imbricarium* 枝叶含精油 0.3%～0.5% 柏科	2-己烯醛、3-己烯醇、丙基环丙烷、三环烯、α-蒎烯、莰烯、β-蒎烯、盏烯、β-月桂烯、间伞花烃、柠檬烯、3-蒈烯、4-松油醇、α-松油醇、乙酸龙脑酯、β-荜澄茄烯等	江苏、浙江、河南、湖北等省有栽种。喜生长于低海拔平原湖泊地区，抗风力强	为水网堤岸庭园绿化树种，也是一种材用树种。枝叶精油可用作松节油

（续表）

植物名称	主要化学成分	分布	备注
柏木 *Cupressus funebris* 根茎材含精油 3%～5% 柏科	α-蒎烯、莰烯、β-蒎烯、月桂烯、柠檬烯、α-龙脑烯醛、桃金娘烯醇、t-葛缕醇、乙酸松油酯、马鞭草烯酮、乙酸橙花酯、乙酸香叶酯、异辣薄荷烯酮、去氢白菖烯、橙花叔醇、榄香烯、石竹烯氧化物、β-桉叶醇、花柏烯、柏木烯醇等	浙江、福建、江西、湖南、广东、贵州、云南、四川等省。越南亦有分布	木材为优质用材，耐腐蚀，精油为主要天然香料
福建柏 *Chamaecyparis hodginsii* 叶含精油 柏科	α-蒎烯、莰烯、β-蒎烯、月桂烯、柠檬烯、α-龙脑烯醛、t-葛缕醇、4-松油醇、龙脑、α-松油醇、乙酸龙脑酯、桃金娘烯醛、桃金娘烯醇、乙酸香叶酯、异辣薄荷烯酮、β-桉叶醇、芮木烯、脱氢松香烯等	浙江、福建、广东、贵州、云南、四川等省。越南亦有分布	我国特有种，叶精油具特殊香气，可供日化香精使用
侧柏 *Platycladus orientalis* 叶含精油0.8%，树皮含精油0.31%，果壳含精油（干）0.4%，木材含精油（干）1.1%，柏科	叶精油：α-侧柏烯、α-蒎烯、桧烯、β-蒎烯、月桂烯、α-水芹烯、3-蒈烯、水合桧烯、γ-松油烯、2,4-对蓋二烯、4-松油醇、乙酸龙脑酯、乙酸松油酯、t-石竹烯、罗汉柏烯、β-芹子烯、β-荜澄茄烯、δ-杜松烯、榄香醇、柏木烯醇等 树皮精油：α-侧柏烯、α-蒎烯、小茴香烯、β-水芹烯、月桂烯、3-蒈烯、对伞花烃、柠檬烯、γ-松油烯、α-松油烯、4-松油醇、乙酸松油酯、t-石竹烯、罗汉柏烯、β-芹子烯、β-柏木烯、柏木烯醇等	我国南北各地均有分布	果实入药，可祛痰止咳，精油香气清爽，特别是木材精油，持久宜人，可用作化妆品和香皂香精，也可用作室内消毒剂和杀虫剂

（续表）

植物名称	主要化学成分	分布	备注
侧柏 *P. orientalis* 叶含精油0.8%， 树皮含精油0.31%， 果壳含精油（干）0.4%， 木材含精油（干）1.1%， 柏科	果壳精油：α-柏木烯、α-蒎烯、桧烯、β-蒎烯、月桂烯、3-蒈烯、对伞花烃、水合桧烯、γ-松油烯、乙酸龙脑酯、2,4（8）-对蓋二烯、4-松油醇、α-松油醇、β-芹子烯、柏木烯醇等 木材精油：α-柏木烯、罗汉柏烯、β-花柏烯、愈创木醇、柏木烯醇、霍香醇、斯佩舒尔醇、d-[2（12）]罗汉柏烯-3-α-醇、金钟柏醇等	我国南北各地均有分布	果实入药，可祛痰止咳，精油香气清爽，特别是木材精油，持久宜人，可用作化妆品和香皂香精，也可用作室内消毒剂和杀虫剂
祁连圆柏 *Juniperus przewalskii* 叶含精油（干）6% 柏科	α-蒎烯、β-蒎烯、α-松油烯、柠檬烯、3-蒈烯、侧柏酮、4-松油醇、α-荜澄茄烯、γ-杜松烯、古巴烯、t-石竹烯、雪松烯、δ-荜澄茄烯、δ-杜松烯、古巴烯醇等	青海、甘肃、四川。生长于海拔2600～4000米的向阳坡	叶药用，可止血、镇咳，主治咳血、吐血、血尿
夜合花 *Magnolia coco* 叶含精油0.24% 木兰科	α-蒎烯、β-松油烯、β-月桂烯、4-蒈烯、1,8-桉叶油素、三环烯、3-蒈烯、芳樟醇、4-松油醇、α-松油醇、α-佛手烯、α-姜黄烯、β-雪松烯、去氢白菖烯、β-金合欢烯、α-姜黄烯等	我国南部广泛栽种，亦广植于东南亚各地	为常见庭园观赏树种，花芳香，可用于熏茶
玉兰 *M. denudata* 叶含精油0.04%～0.05% 木兰科	α-蒎烯、桧烯、β-月桂烯、柠檬烯、1,8-桉叶油素、对伞花烃、cis-3-己烯醇、芳樟醇、β-波旁烯、α-松油醇、乙酸龙脑酯、β-石竹烯、α-蛇麻烯、γ-依兰油烯、t-橙花叔醇、ε-依兰油醇、β-桉叶醇等		

植物名称	主要化学成分	分布	备注
武当木兰 *M. sprengeri* 枝叶含精油， 花蕾含精油0.20% 木兰科	枝叶精油：α-蒎烯、莰烯、β-蒎烯、1,8-桉叶油素、对伞花烃、樟脑、芳樟醇、乙酸龙脑酯、t-石竹烯、蛇麻烯、γ-杜松烯、香茅醇、芳姜黄烯、去氢白菖烯、石竹烯氧化物、丁香酚甲醚、香榧醇、γ-桉叶醇、β-桉叶醇、α-桉叶醇等	生长于山地杂木林中	花蕾和枝叶精油可开发成天然香料
莽草 *Illicium tashiroi* 果实含精油0.83% 八角科	α-蒎烯、柠檬烯、β-水芹烯、茴香脑、龙脑、黄樟油素、丁香酚甲醚、细辛醚、t-石竹烯、三甲氧基苯丙烯、δ-杜松烯等	四川、湖南、广东、广西、台湾等省区。日本亦有分布。生长于山沟、水边或阳光充足的灌木丛中	为有待开发的大茴香脑资源
红花八角 *I. dunnianum* 果皮含精油0.42% 八角科	α-蒎烯、β-蒎烯、月桂烯、柠檬烯、β-水芹烯、α-异松油烯、1,8-桉叶油素、对伞花烃、芳樟醇、4-松油醇、古巴烯、α-松油醇、十一酮、龙脑、t-葛缕醇、t-β-金合欢烯、δ-杜松烯、橙花叔醇、β-桉叶醇、刺柏脑等	福建、湖南、广东、广西、贵州等省区。生长于海拔500～700米的山谷溪旁、河流山地密林中	果皮精油为有待开发的天然香料资源
红茴香 *I. henryi* 果实含精油2.25% 八角科	3-蒈烯、柠檬烯、β-水芹烯、β-蒎烯、桧烯、葛缕酮、橙金娘烯醛、cis-葛缕醇、蒎葛缕醇、1,8-桉叶油素、芳樟醇、4-松油醇、α-松油醇、细辛醚、肉豆蔻醚、乙酸龙脑酯、乙酸松油酯、花柏烯、古巴烯、芹菜脑等	陕西、安徽、江苏、江西、福建、河南、湖北、四川、贵州、云南等省。生长于海拔750～1500米山坡、沟旁的林下或灌丛中	果叶均入药，有舒筋活血、止血、止痛、行气功效

（续表）

植物名称	主要化学成分	分布	备注
披针叶八角 *I. lanceolatum* 果实含精油 八角科	二甲基苯乙烯、对伞花烃、甲基苯乙酮、莰烯、δ-3-蒈烯、柠檬烯、α-蒎烯、β-蒎烯、葛缕酮、桃金娘烯醛、紫苏醛、樟脑、胡椒酮、桃金娘醇、t-葛缕醇、蒎-葛缕醇、4-松油醇、丁香酚甲醚、乙酸龙脑酯、乙酸松油酯、δ-杜松烯、t-石竹烯、榄香脂素等	江苏、安徽、浙江、江西、福建、广东、广西、云南等省区。生长于阴湿沟边或杂木林两旁	果实精油为有待开发的天然香料
大八角 *I. majus* 果实含精油 八角科	二甲基苯乙烯、对伞花烃、甲基苯乙酮、2-蒎烯、莰烯、δ-3-蒈烯、柠檬烯、α-蒎烯、β-蒎烯、枯茗醛、葛缕酮、桃金娘烯醛、α-龙脑烯醛、桃金娘烯醇、t-葛缕醇、α-松油醇、蒎葛缕醇、龙脑、芳樟醇、1,8-桉叶油素、4-松油醇、丁香酚甲醚、异丁香酚甲醚、乙酸松油酯、榄香脂素、异榄香脂素等	四川、贵州、湖南、广西等省区。生长于海拔200～2000米的混交林中	果实精油为有待开发的天然香料
闽皖八角 *I. minwanense* 果实含精油 八角科	α-蒎烯、β-蒎烯、桧烯、月桂烯、侧柏烯、柠檬烯、β-水芹烯、α-异松油烯、1,8-桉叶油素、对伞花烃、芳樟醇、4-松油醇、α-松油醇、乙酸松油醇、乙酸香叶酯、γ-杜松烯、β-桉叶醇等	福建。生长于海拔1000～1850米的沟谷两旁和林缘	果实精油为有待开发的天然香料
野八角 *I. simonsii* 果实含精油1% 八角科	α-蒎烯、β-蒎烯、月桂烯、柠檬烯、芳樟醇、樟脑、桧醇、爱草脑、d-葛缕酮、乙酸龙脑酯、乙酸香叶酯、β-香柠檬烯等	贵州。生长于林中	民间用果实代替八角作辛香调料用

（续表）

植物名称	主要化学成分	分布	备注
厚皮香八角 *I. ternstroemioides* 果实含精油 八角科	α-蒎烯、β-蒎烯、柠檬烯、β-水芹烯、α-异松油烯、1,8-桉叶油素、对伞花烃、芳樟醇、古巴烯、α-松油醇、cis-石竹烯、β-榄香烯、蛇麻烯、α-依兰烯、γ-古芸烯、乙酸香叶酯、γ-杜松烯等	广东。生长于高海拔的密林中或溪涧旁	果实精油为有待开发的天然香料资源
八角（大茴香） *I. verum* 果实含精油 8%～12%， 鲜叶含精油 0.3%～0.4% 八角科	果实精油：α-蒎烯、月桂烯、柠檬烯、芳樟醇、4-松油醇、爱草脑、大茴香醛、t-大茴香醚、β-香柠檬烯、丁香酚甲醚等 鲜叶精油：柠檬烯、芳樟醇、黑椒酚甲醚、cis-大茴香醚、对-大茴香醛、β-石竹烯、t-大茴香醚、α-佛手烯等	广西、广东、福建、云南、贵州，很多为栽培。生长于湿暖的山谷中	富含大茴香醚的果实和鲜叶是提取大茴香脑的最佳原料，调味料主要原料，入药具开胃下气、暖胃散寒、止痛等功效，是合成雌性激素己烷雌酚的主要原料
刺果番荔枝 *Annona muricata* 果实含精油 番荔枝科	3-甲基-3-丁烯醇、3-甲基-2-丁烯醇、丁酸乙酯、乙酸-3-甲基丁酯、乙酸-3-甲基-2-丁烯酯、α-蒎烯、β-蒎烯、2-羟基-4-甲基戊酸甲酯、β-月桂烯、己酸乙酯、7-甲基-4-癸烯、1,8-桉叶油素、柠檬烯、辛酸甲酯、4-松油醇、α-松油醇、辛酸乙酯等	原产于热带美洲。现我国广东、广西、福建、云南、台湾等省区有栽培	果实可作水果食用，木材可作造船用材

（续表）

植物名称	主要化学成分	分布	备注
紫玉盘 *Uvaria macrophylla* 枝叶含精油 番荔枝科	1,8-桉叶油素、3,6,6-三甲基二环[3,1,1]庚-2-烯、α-罗勒烯、古巴烯、β-榄香烯、β-石竹烯、α-石竹烯、β-荜澄茄烯、γ-榄香烯、δ-杜松烯等	我国南部。越南也有分布。生长于低海拔灌木丛中	为庭园观赏植物
肉豆蔻 *Myristica fragrans* 干花含精油 肉豆蔻科	α-侧柏烯、α-蒎烯、桧烯、β-蒎烯、β-月桂烯、β-水芹烯、3-蒈烯、α-松油烯、α-水芹烯、γ-松油烯、2-蒈烯、4-松油醇、黄樟油素、异丁香酚、肉豆蔻醚等	原产于马来西亚。现广东、云南、台湾有栽种	种仁可药用，精油用作食品调料
番木瓜 *Carica papaya* 果肉含精油	甲基环己烷、2-甲基丁烷、丁酸甲酯、甲苯、己酸甲酯、cis-β-罗勒烯、cis-氧化芳樟醇（吡喃型）、t-氧化芳樟醇（吡喃型）、芳樟醇、癸酸甲酯、苯乙醛、丁酸苯甲酯、苯乙酯、异硫氢酸苯甲酯等	原产于亚洲和美洲的热带、亚热带地区。我国引种为常见水果	
中华猕猴桃 *Actinidia chinensis* 果实含精油	乙酸乙酯、丙酸乙酯、丁酸甲酯、丁酸乙酯、t-2-己烯醛、t-2-己烯醇、己酸甲酯、己酸乙酯、苯甲酸甲酯、芳樟醇、苯甲酸乙酯、4-松油醇、β-达马烯等	陕西、湖北、河南、江苏、安徽、浙江、江西、福建、广西、广东。生长于林内或灌木丛中	果实富含维生素，可作水果食用，也可药用
岗松 *Baeckea frutescens* 枝叶含精油1.4% 桃金娘科	1,8-桉叶油素、γ-松油烯、芳樟醇、4-松油醇、α-松油醇、丁香酚、β-石竹烯、α-石竹烯、δ-杜松烯、桂醇等	江西、福建、广东、广西等省区。多生长于低丘荒山草坡灌丛	民间用其治肠炎、腹泻、脚癣和皮肤瘙痒，治蚊虫咬

植物名称	主要化学成分	分布	备注
美叶桉 *Eucalyptus calophylla* 0.4%～0.5% 桃金娘科	α-蒎烯、莰烯、β-蒎烯、对伞花烃、1,8-桉叶油素、γ-松油烯、小茴香醇、蒎葛缕醇、4-松油醇、乙酸桧酯、香芹酚、麝香草酚、β-石竹烯、香树烯、γ-榄香烯、喇叭茶醇等	原产于澳大利亚。广东有栽种	精油为医药和香料工业原料，花为蜜源植物
赤桉 *E. camaldulensis* 枝叶含精油 0.14%～0.28% 桃金娘科	α-蒎烯、β-蒎烯、α-水芹烯、柠檬烯、1,8-桉叶油素、γ-松油烯、对伞花烃、芳樟醇、4-松油醇、香树烯、α-松油醇、辣薄荷酮、枯茗醇、丁香酚等	原产于澳大利亚。华南地区至西南地区有栽种	木材纹理致密，易打磨、耐腐，对白蚁菌类抵抗能力强，可用于枕木桩等，精油为医药香料工业原料
柠檬桉 *E. citriodora* 枝叶含精油 0.5%～2.0% 桃金娘科	α-蒎烯、β-蒎烯、柠檬烯、1,8-桉叶油素、异胡薄荷醇、香茅醛、香茅醇、乙酸香茅酯、β-石竹烯、β-桉叶醇等	原产于澳大利亚。广东、广西、福建有栽种	精油为香料工业重要原料之一，可进一步提取香茅醛，亦可供药用如配十滴水等
窿缘桉 *E. exserta* 叶含精油 0.8%～1.0% 桃金娘科	α-蒎烯、莰烯、β-蒎烯、间伞花烃、1,8-桉叶油素、小茴香醇、蒎葛缕醇、龙脑、α-松油醇、桃金娘烯酮、乙酸龙脑酯等	原产于澳大利亚。华南地区有栽种	为造林树种，叶精油是提取1,8-桉叶油素的原料
蓝桉 *E. globulus* 枝叶含精油 0.7%～0.9% 桃金娘科	α-蒎烯、β-蒎烯、β-月桂烯、1,8-桉叶油素、γ-松油烯等	原产于澳大利亚。广西、云南、四川有栽种	提取1,8-桉叶油素的主要原料，亦可配置香精原料，有杀菌等功能

（续表）

植物名称	主要化学成分	分布	备注
直杆蓝桉 *E. globulus* subsp. *maidenii* 叶含精油 1.5%～2.3% 桃金娘科	α-蒎烯、β-蒎烯、月桂烯、对伞花烃、1,8-桉叶油素、γ-松油烯、α-松油醇、香叶醇、别香树烯、刺柏脑、α-石竹烯、β-桉叶醇等	原产于澳大利亚。云南有栽种	为造林树种，叶精油是提取1,8-桉叶油素原料
桉树（大叶桉） *E. robusta* 叶含精油 0.8%～1.0% 桃金娘科	α-蒎烯、β-蒎烯、月桂烯、α-水芹烯、对伞花烃、柠檬烯、1,8-桉叶油素、2-蒈烯、α-松油醇、辣薄荷酮、β-石竹烯、α-松油醇、别香树烯、金合欢（E.Z）等	原产于澳大利亚。广东、广西、云南、四川有栽种	枝叶精油可作为香料工业原料，亦可作为行道树种
细叶桉 *E. tereticornis* 枝叶含精油 0.7%～0.9% 桃金娘科	α-蒎烯、莰烯、β-蒎烯、间伞花烃、1,8-桉叶油素、小茴香醇、α-龙脑烯醛、蒎葛缕醇、龙脑、α-松油醇、马鞭草烯酮、甲酸龙脑酯、葛缕酮、乙酸龙脑酯等	原产于澳大利亚。广东、广西、福建、云南有栽种	为造林树种之一，精油可用于香料工业和医药工业
白千层 *Melaleuca leucadendron* 枝叶含精油 1.0%～1.5% 桃金娘科	α-蒎烯、β-蒎烯、β-月桂烯、1,8-桉叶油素、γ-杜松烯、4-松油醇、α-松油醇、β-石竹烯、十氢-1,1,4,7-四甲基-4aH环丙[e]薁-4a-醇、喇叭茶醇异构体、喇叭茶醇、β-桉叶醇等	原产于澳大利亚。广西、广东、福建、台湾均有栽种	叶精油可代白树油，精油有镇痛、驱虫及防腐等功效，可治牙痛、耳痛、风湿痛、神经痛等
白树 *M. leucadendron* var.*cajaputi* 枝叶含精油0.14% 桃金娘科	α-蒎烯、β-蒎烯、1,8-桉叶油素、芳樟醇、α-松油烯、乙酸松油酯、β-石竹烯、β-榄香烯、喇叭茶醇、β-桉叶醇等	原产于澳大利亚。我国华南植物园从印尼引种	精油可作风油精用，也可作杀菌剂用

植物名称	主要化学成分	分布	备注
番石榴 *Psidium guajava* 果实含精油 桃金娘科	己醛、2-己烯醛、3-己烯醇、α-蒎烯、乙酸己烯酯、乙酸己酯、柠檬烯、1,8-桉叶油素、辛酸甲酯、α-松油烯、乙酸-3-苯丙酯、古巴烯、β-石竹烯等	原产于南美洲。广东、云南均有栽种	果实是广东南部常见水果，也可制成饮料，叶有止泻收敛作用
丁子香 *Syzygium aromaticum* 花蕾含精油1.5%～30%，叶含精油6.6% 桃金娘科	花蕾精油：丁香酚、1,8-桉叶油素、苯甲醛、苯甲醇、水杨酸甲酯、香荆芥酚、β-石竹烯、α-蛇麻烯、环氧石竹烯等 叶精油：丁香酚、1,8-桉叶油素、香荆芥酚、β-石竹烯、α-蛇麻烯、环氧石竹烯等	马来群岛和非洲盛产，东非桑及巴尔主产丁香花蕾。华南地区有引种	为主要芳香药用植物，具抗菌驱虫、健胃止痛作用，治胃寒腹痛、呃逆、吐泻、牙痛，亦可用于配制风油精，花蕾可作食品调料
蒲桃 *S. jambos* 果肉含精油 桃金娘科	己醛、乙酸-1-乙氧基乙酯、2-己烯醛、3-己烯醇、二环[4,2,0]辛-1,3,5-三烯、苯甲醛、苯甲醇、芳樟醇、苯乙醇、枯茗醇、α-松油醇、苯骈噻唑、橙花醇、香叶醇、桂醛、桂醇、α-依兰油烯等	云南、广东、广西、福建、台湾等省区栽培或野生。中印半岛、南洋有分布。生长于肥湿润土中	果实可用作水果和蜜饯等
红枝蒲桃 *S. rehderianum* 枝叶含精油0.13% 桃金娘科	α-蒎烯、β-蒎烯、β-月桂烯、枞油烯（R）、3,6,6-三甲基-二环[3,1,1]庚-2-烯、β-罗勒烯、芳樟醇、α-荜澄茄烯、古巴烯、β-波旁烯、α-石竹烯、β-荜澄茄烯、α-愈创木烯、γ-杜松烯、β-荜澄茄烯异构体、δ-杜松烯、γ-依兰油烯、γ-芹子烯等	福建、广东、广西等省区。生长于低海拔杂木林中	枝叶精油可作为化妆品和皂用香料原料

（续表）

植物名称	主要化学成分	分布	备注
黄牛木 *Cratoxylum cochinchinense* 枝叶含精油0.20% 藤黄科	α-罗勒烯、γ-松油烯、4-松油醇、α-松油醇、δ-榄香烯、古巴烯、姜烯、β-石竹烯、β-金合欢烯、α-石竹烯、β-荜澄茄烯、γ-榄香烯、δ-杜松烯、喇叭茶醇、δ-杜松醇等	我国南部。东南亚亦有分布。常生长于丘陵地	精油可作日化和医药原料，根和树枝民间药用，可治感冒咳嗽，外用可治跌打损伤
金丝桃 *Hypericum monogynum* 枝叶含精油0.26～0.30% 藤黄科	桧烯、β-月桂烯、β-罗勒烯、α-罗勒烯、cis-乙酸菊酯、β-榄香烯、β-荜澄茄烯、γ-榄香烯、γ-榄香烯异构体等	河北、河南、陕西、江苏、浙江、台湾、福建、江西、湖北、四川、广东等省区	为庭园观赏植物，药用有清凉解毒、祛风消肿功效
榅桲 *Cydonia oblonga* 果实含精油0.41% 蔷薇科	t-β-金合欢烯、己酸乙酯、癸酸乙酯、cis-3-己烯醇、己醇、t-紫花前胡内酯、丁香酚甲醚等	原产于中亚西亚。新疆、陕西、江西、福建有栽种	果实药用，治呕吐酸水、食积胸闷
油楠（火水树） *Sindora glabra* 树干含大量精油 云实科	乙酸龙脑酯、α-荜澄茄烯、古巴烯、β-荜澄茄烯、β-石竹烯、α-石竹烯、γ-依兰油烯、α-依兰油烯、γ-杜松烯、δ-杜松烯、δ-杜松醇等	海南。生长于山地林中	树干内贮大量轻质油，其中精油含80%，可用于制作香料和治疗皮肤病
海南黄檀（花梨公） *Dalbergia hainanensis* 花含精油 蝶形花科	α-蒎烯、苯甲醛、7-辛烯-4-醇、5-甲基-3-庚烯醇、柠檬烯、3-甲基-3-庚酮、苯甲酸甲酯、芳樟醇、苯乙醇、苯甲酸、对-羟基苯甲酸甲酯、N-苯基-1-萘胺等	海南省特有种。生长于山林地带	海南常见材用树种，花精油有待开发

（续表）

植物名称	主要化学成分	分布	备注
枫香（三角枫） *Liquidambar* *formosana* 树脂含精油 10%～12%， 叶含精油 金缕梅科	树脂精油：甲酸乙酯、2-甲基-3-戊醇、乙酸乙酯、4-甲基己醇、莰烯、3-甲基-2-丁醇、戊二烯基环戊烷、对伞花烃、4-松油醇、樟脑、龙脑、异龙脑、对-蓋-1,4-二烯醇、桃金娘烯醇、α-金合欢烯、马鞭草烯酮、丙基苯甲醛等 叶精油：(z)-3-己烯醇、β-侧柏烯、γ-松油烯、1,8-蓋二烯、乙酸松-4-酯、α-异松油烯、伞花酮、4-松油醇、α-松油醇、驱蛔素、2-壬烯-2-酮、β-波旁烯、β-石竹烯、γ-石竹烯、γ-依兰油烯、β-荜澄茄烯、脱氢白菖蒲烯、茉莉酮、菲等	秦岭及淮河以西各省区，北至台湾，西至四川、云南，南至海南	枫香树脂是一种较好的定香剂，可配制烟用和皂用香精，添于牙膏中有止血、止痛功效
半枫荷 *Semiliquidambar* *cathayensis* 枝叶含精油 0.5%～0.6% 金缕梅科	己醛、2-己烯醛、3-己烯醇、丙基环丙烷、α-蒎烯、苯甲醛、β-蒎烯、β-月桂烯、对伞花烃、柠檬烯、α-罗勒烯、4-松油醇、α-松油醇等	广东、广西、湖南。生长于低海拔天然林中	根及叶药用，可治跌打风湿、产后风
亮叶桦 *Betula luminifera* 树皮含精油 0.2%～0.5% 桦木科	水杨酸甲酯（97%）	云南、贵州、四川、陕西、甘肃、湖北、江西、浙江、广东、广西等省区。生长于海拔500～2500米的山地阳坡杂木林中	树皮入药可除湿、消食、解毒，外用精油治关节痛，可配制"沙示"饮料

（续表）

植物名称	主要化学成分	分布	备注
沙针 *Osyris wightiana* 根含精油 檀香科	l-乙基丙基苯、对异丁基甲苯、枞油烯、1,4-二丙基苯、α-檀香烯、β-香柠檬烯、β-甜没药烯、α-檀香醇、α-甜橙醛、β-檀香醇、棕榈酸乙酯、9-十八烯醛、蒿素、6,9-十八碳二烯酸甲酯等	西藏、四川、云南、广西。东南亚亦有分布。生长于海拔500～2700米的灌丛中	根药用可消肿、止痛、祛风，精油可用于调香
檀香 *Santalum album* 木材含精油 2.5%～6.5% 檀香科	α-檀香烯、β-檀香烯、β-檀香烯立体异构体、α-檀香烯立体异构体、α-檀香醇立体异构体、β-檀香醇立体异构体、α-檀香醇	原产于南洋群岛、印度、缅甸。广东、台湾、云南有引种	檀香木为重要香熏原料，精油为名贵香料，果实可食用，木材可制各种名贵工艺品
贡甲（白山柑） *Acronychia oligophlebia* 叶含精油 0.8～1.0% 芸香科	α-蒎烯、β-蒎烯、月桂烯、枞油烯、β-罗勒烯、芳樟醇、3-环己烯基甲醇、2,2,3-三甲基-3-环戊基乙醛、龙脑、樟脑、丁香酚甲醚、α-古巴烯、β-榄香烯、cis-石竹烯、香树烯、蛇麻烯、β-杜松烯、橙花叔醇、β-杜松醇等	海南。生长于低海拔的湿润次生林或雨林中	精油可调配较低档香精，去萜烯后香精可提高档次
山油柑（降真香） *A. pedunculata* 叶含精油 0.8%～0.9% 芸香科	α-侧柏烯、α-蒎烯、β-侧柏烯、对伞花烃、柠檬烯、芳樟醇、水合莰烯、乙酸松油酯、乙酸β-侧柏酯、β-松油醇、α-松油醇、水合桧烯、t-氧化芳樟醇（吡喃型）、3-癸烯-2-酮、cis-氧化芳樟醇（吡喃型）等	广东、广西、云南。生长于常绿阔叶林中	精油可配制香精，去萜烯后其香气更优雅
酸橙（枸头橙） *Citrus aurantium* 叶含精油0.18% 芸香科	月桂烯、δ-柠檬烯、γ-松油烯、氧化芳樟醇、柠檬醛等	秦岭以南各地均有栽种，野生不多见	嫩果入药，加工后可代"积实"

（续表）

植物名称	主要化学成分	分布	备注
巴柑檬（香柠檬） *C. aurantifolia* 叶含精油0.52%， 果皮含精油0.41% 芸香科	叶精油：月桂烯、β-蒎烯、柠檬烯、β-罗勒烯、芳樟醇、柠檬醛、α-松油醇、乙酸芳樟酯、乙酸香叶酯、柠檬醛、β-橙花叔醇等 果皮精油：α-蒎烯、桧烯、β-蒎烯、月桂烯、3-蒈烯、γ-松油烯、芳樟醇、α-松油醇、柠檬醛、乙酸芳樟酯、乙酸橙花酯、乙酸香叶酯等	原产于意大利，欧洲南部有栽种。云南有引种	叶和果皮均可调配香精，果皮精油尤为优质，可配制化妆品、食品等香精
柚（胡柑） *C. grandis* 花含精油， 果皮含精油 0.06%～0.2% 芸香科	果皮精油：δ-己烯醇、乙酸-4-己烯酯、柠檬烯、cis-氧化芳樟醇（呋喃型）、t-氧化芳樟醇（呋喃型）、α-松油醇、橙花醇、cis-葛缕醇、香叶醇、乙酸香叶酯、吲哚、紫苏醇等 花精油：α-蒎烯、β-水芹烯、β-蒎烯、柠檬烯、α-罗勒烯、芳樟醇、玫瑰呋喃、3-甲基-4-庚酮、邻氨基苯甲酸甲酯等	长江以南各省区均有栽种，品种不少	柚花浸膏为名贵天然香料，可配制花香型化妆品和食品香精，果皮亦可药用和提取果胶
柠檬（洋柠檬） *C. limon* 果皮含精油1.50% 芸香科	2-己烯醛、2-己烯醇、α-蒎烯、β-蒎烯、月桂烯、对伞花烃、柠檬烯、γ-松油烯、芳樟醇、4-松油醇、α-松油醇、乙酸香叶酯、乙酸异香叶酯、α-佛手烯、甜没药烯等	广东、四川等省有栽种（外来种）	果皮精油主要用于调配食用香精和皂用化妆品香精，叶精油亦可配制香精，也是日常水果

（续表）

植物名称	主要化学成分	分布	备注
黎檬 *C. limonia* 0.14%～0.26% 芸香科	α-蒎烯、β-蒎烯、甲基庚酮、柠檬烯、罗勒烯、芳樟醇、香茅醇、香茅醛、橙花醇、α-柠檬醛、香草醇、柠檬醛、乙酸香茅酯、乙酸橙花酯、乙酸香叶酯、丙酸芳樟酯等	云南、广西、广东、福建、台湾、贵州和四川等省区。中南半岛有分布	果实可制凉果和配制饮料
香橼 *C. medica* 叶含精油2.4% 芸香科	α-蒎烯、β-蒎烯、桧烯、甲基庚烯酮、月桂烯、α-水芹烯、罗勒烯、芳樟醇、香茅醛、橙花醇、β-柠檬醛、香叶醇、柠檬醛、乙酸橙花酯、乙酸香叶酯、石竹烯等	长江以南各省区均有栽种	精油可用于配制食用香精、化妆品香精，果实入药，药用有理气化痰功效
木里柠檬 *C. medica* var. *muliensis* 叶含精油0.21～0.32% 芸香科	甲基庚烯酮、柠檬烯、芳樟醇、松油醇、香茅醛、薰衣草醇、4-松油醇、香茅醇、橙花醇、柠檬醛、乙酸香茅酯、乙酸橙花酯、乙酸香叶酯等	四川。生长于海拔1600～2200米的山区	民间以果入药，有理气、舒郁消痰等功效
佛手柑（佛手） *C. medica* var. *sarcodactylis* 果皮含精油1.8% 芸香科	β-月桂烯、间伞花烃、柠檬烯、γ-松油烯、2-蒈烯、芳樟醇、4-松油醇、α-松油醇、橙花醇、β-柠檬醛、香叶醇、α-柠檬醛、α-佛手烯、β-甜没药烯等	我国南部和西南部多有栽种	果皮入药用于镇咳祛痰，精油可配制食用香精
柑橘 *C. reticulata* 果皮含精油 芸香科	芳樟醇、α-松油醇、对伞花烃、柠檬烯、月桂烯、α-蒎烯、β-蒎烯、α-松油烯、异松油烯、α-侧柏烯、香茅醛、水合桧烯等	秦岭以南各省区广泛栽种	是著名水果，品种多，果皮精油是调配食品的香料，果皮入药有健胃助消化作用

（续表）

植物名称	主要化学成分	分布	备注
蕉柑 *C. reticulata* 'Jiangan' 果皮含精油1.2% 芸香科	月桂烯、柠檬烯、芳樟醇、α-松油醇、甲酸香叶酯、β-柠檬醛、别二氢香芹酮、α-柠檬醛、紫苏醛等	广东、广西、福建、台湾有栽种	为著名水果之一，果皮入药为芳香健胃剂，精油为食品香精原料
椪柑 *C. reticulata* 'Ponkan' 果皮含精油0.90%～1.20% 芸香科	β-月桂烯、辛醛、对伞花烃、柠檬烯、γ-松油烯、戊基环丙烯、芳樟醇、β-柠檬醛、香叶醇、α-柠檬醛、香芹酚、香叶醇、紫苏醛、柏木烯醇等	广东、广西、福建、江西、浙江、湖南、四川等省区均有栽种	为著名水果，精油可配制食品和化妆品香精
甜橙 *C. sinensis* 果皮含精油0.70%～0.90% 芸香科	3-蒈烯、异松油烯、桧烯、月桂烯、柠檬烯、氧化二戊烯、乙酸二氢葛缕酯、香茅醛、壬醛、芳樟醇、2-辛烯、α-紫罗兰酮、α-松油醇、石竹烯、葛缕酮、t-葛缕醇、香叶醇、cis-葛缕醇等	我国华南各省区均有栽种	为著名水果，橙皮精油用于调配食品、牙膏用香精
齿叶黄皮 *Clausena dunniana* 枝叶含精油，黄皮叶味类型含0.7% 芸香科	八角类型精油：柠檬烯、β-罗勒烯、γ-松油烯、1-甲基-1，2-二乙烯基-5-环己烯、芳樟醇、异大茴香醚、大茴香醚、丁香酚甲醚等 黄皮类型精油：4-蒈烯、δ-蒎烯、β-水芹烯、β-罗勒烯、3-蒈烯、对伞花烃、1-甲基-2,3-二亚乙基-4-环己烯、1-甲基-2,3-二亚乙基-5-环己烯、4-松油醇、β-石竹烯、蛇麻烯、β-荜澄茄烯、橙花叔醇等	湖南、广东、广西、贵州、云南。生长于山坡密林或疏林中	异大茴香醚可直接合成大茴香脑，进一步合成大茴香醛，用于牙膏、食品、香皂和化妆品香精，并有较强杀菌功效

（续表）

植物名称	主要化学成分	分布	备注
黄皮 *C. lansium* 果实含精油 芸香科	2-己烯醛、β-水芹烯、苯乙醛、γ-松油烯、芳樟醇、4-松油醇、α-松油醇、辣薄荷醇、β-石竹烯、β-甜没药烯等	广西、广东、福建等省区。亚洲、美洲均有栽种	为我国南方著名水果，叶可药用治感冒、肠炎、胃炎，根可治胃痛等
三叉苦（三桠苦） *Melicope pteleifolia* 叶含精油 芸香科	α-蒎烯、6-甲基庚烯二酮、柠檬烯、对伞花烃、β-罗勒烯（E）、cis-氧化芳樟醇（呋喃型）、芳樟醇、α-松油醇、古巴烯、γ-依兰油烯、γ-杜松烯、柏木脑、δ-松油醇等	台湾、福建、广东、广西、贵州、云南南部。生长于平原、溪边、林缘或灌丛中	叶、茎皮、根药用，可治胃痛、咽炎、流感、流脑、关节炎等
翼叶九里香 *Murraya alata* 叶含精油0.08% 芸香科	δ-榄香烯、β-榄香烯、t-石竹烯、α-愈创木烯、β-愈创木烯、香树烯、β-芹子烯、α-依兰油烯、β-荜澄茄烯、γ-广藿香烯、γ-榄香烯、α-古芸烯、愈木醇、花柏酮、香榧醇、马兜铃烯、胡萝卜醇、β-桉叶醇、异愈创木醇等	广东西南部。生长于干旱草地次生林或海滨灌林中	民间用其煮水洗身，防治疮疥
豆叶九里香 *M. euchrestifolia* 枝叶含精油0.4% 芸香科	柠檬烯、苯甲醛、α-蒎烯、β-蒎烯、胡薄荷酮、紫苏醇、乙酸二氢葛缕酯、香树烯等	台湾、广西、云南等省区。多生长于山地疏林或密林中	精油抗菌力强，对滴虫有杀灭作用，可配制软膏治感冒，入药具祛风活血功效
九里香 *M. exotica* 叶含精油 1.4%～1.6% 芸香科	α-荜澄茄烯、α-杜松烯、t-石竹烯、蛇麻烯、α-姜黄烯、γ-柏木烯、t-β-金合欢烯、δ-杜松烯、cis-t-金合欢醇等	台湾、福建、广东和广西。生长于干旱的空旷地或灌丛中	精油可配制日用化工业香精和皂用化妆品

植物名称	主要化学成分	分布	备注
调味九里香 *M. koenigii* 枝叶含精油0.15% 芸香科	柏木烯、莰烯、α-水芹烯、β-蒎烯、邻伞花烃、柠檬烯、3-蒈烯、γ-松油烯、芳樟醇、香茅醛、异胡薄荷酮、胡薄荷酮、α-松油醇、2-（3,3-二甲基）亚环己基乙醇、β-柠檬醛、橙花醇、α-柠檬醛、乙酸橙花酯、香叶醇、乙酸香叶酯、β-石竹烯等	海南和云南南部。生长于路旁和林中	果能通经止痛，根能健胃，叶能治麻疹骨折，枝叶祛风
小叶九里香 *M. microphylla* 叶含精油0.36% 芸香科	α-水芹烯、α-蒎烯、β-水芹烯、α-松油烯、3-蒈烯、4-松油烯、α-松油醇、乙酸芳樟酯、榄香烯、石竹烯等	海南省。生长于沿海群落附近	小叶九里香适应性强，生长于贫瘠的环境，是海南有利用潜力的植物，精油可调配化妆品和皂用香料
千里香 *M. paniculata* 枝叶含精油0.3% 芸香科	紫苏醛、δ-榄香烯、β-榄香烯、t-石竹烯、t-β-金合欢烯、雅槛兰烯、蛇麻烯、别香树烯、β-荜澄茄烯、橙花叔醇、花柏烯等	台湾、福建、广东、广西和湖南等省区，贵州和云南南部亦有分布。生长于较干旱疏林中	精油可用于调配香精，叶入药可行气止痛、活血散瘀以及麻醉和镇痛
满天香 *M. tetramera* 枝叶含精油3.0%～3.3% 芸香科	α-蒎烯、桧烯、月桂烯、α-松油烯、罗勒烯、对伞花烃、枞油烯、β-水芹烯、芳樟醇、薄荷酮、异薄荷酮、（+）新薄荷醇、薄荷醇、α-松油醇、胡椒酮、乙酸薄荷酯等	云南省。生长于干热河谷疏林中	叶和根入药，有祛风解毒、行气止痛、活血止痛功效，精油可调日化用香精

（续表）

植物名称	主要化学成分	分布	备注
乔木茴芋 *Skimmia arborescens* 叶含精油 芸香科	2-己烯醛、3-己烯醇、柠檬烯、芳樟醇、苯并噻唑、乙酸香叶酯、橙花叔醇等	我国东南部，广西西南部，云南南部和西藏东南部。生长于山地杂木林中较湿润处	精油的香气幽雅，可调配较高档香精
飞龙掌血 *Toddalia asiatica* 枝叶含精油0.32% 芸香科	α-侧柏烯、α-蒎烯、莰烯、β-松油烯、β-月桂烯、d-水芹烯、α-松油烯、对伞花烃、柠檬烯、γ-松油烯、2-蒈烯、4-松油醇、α-松油醇、乙酸龙脑酯、β-榄香烯、β-石竹烯、α-石竹烯、β-荜澄茄烯、γ-榄香烯、δ-杜松烯、桂醇等	广东、广西、湖南、湖北、贵州、云南、四川、陕西、福建和浙江等省区。生长于丛林中	根药用能散瘀消肿、祛风镇痛
竹叶花椒 *Zanthoxylum armatum* 叶含精油 0.4%～0.6% 芸香科	2-己烯醛、3-己烯醇、己醇、1,8-桉叶油素、芳樟醇、4-松油醇、β-柠檬醛、α-柠檬醛、紫苏醛、桂酸甲酯异构体、桂酸甲酯、橙花叔醇、植醇等	山东、河南、陕西、甘肃等省以南各地。生长于海拔低至300米的山地杂林，石灰岩地较常见	根皮、茎皮和果实药用可祛风、散寒、行气、镇痛，干果可作驱虫药，亦可作调味剂
簕欓花椒 *Z. avicennae* 果实含精油0.5% 芸香科	α-蒎烯、β-侧柏烯、月桂烯、罗勒烯、α-侧柏烯、枞油烯、β-水芹烯、辛醛、庚醇、辛醇、乙酸辛酯、4-甲基-6-乙基己醛、芳樟醇、十二醛、龙脑、依兰油烯、β-榄香烯、蛇麻烯等	台湾、福建、海南、广东、广西、云南等省区。生长于疏林中	精油可直接调配香精，亦具较强杀菌能力，根入药可治咽喉痛等，亦可治风湿骨痛
花椒（萘椒） *Z. bungeanum* 果实含精油 0.2%～0.4% 芸香科	月桂烯、柠檬烯、1,8-桉叶油素、芳樟醇、α-松油醇、辣薄荷酮、t-β-金合欢烯、α-荜澄茄烯、γ-杜松烯等	我国除东北及新疆外各地均野生或栽种	精油作调香原料，果实亦为麻辣型调味料，并有杀菌消毒作用，入药治牙痛

（续表）

植物名称	主要化学成分	分布	备注
川陕花椒 Z. piasezkii 果实含精油 0.2%～0.4% 芸香科	β-月桂烯、柠檬烯、1,8-桉叶油素、β-罗勒烯、芳樟醇、4-松油醇、α-松油醇、乙酸松油酯、β-荜澄茄烯、橙花叔醇等	四川、陕西、甘肃等省。主要生长于干燥的山坡上和路旁	精油可用于化妆品及皂用香精，亦可作麻辣型调味料
大叶臭椒 Z. myriacanthum 果实含精油0.32% 芸香科	桧烯、柠檬烯、1,8-桉叶油素、α-侧柏烯、4-松油醇、α-松油醇、cis-辣薄荷醇、t-葛缕醇、橙花叔醇、β-桉叶醇等	福建、广东、广西、湖南、贵州。生长于疏林中或路旁湿地	精油可用于配制香精
青花椒 Z. schinifolium 果皮（干）含精油 1.70% 芸香科	桧烯、月桂烯、α-水芹烯、柠檬烯、β-水芹烯、β-罗勒烯-X、β-罗勒烯-Y、1,8-桉叶油素、芳樟醇、4-松油醇、爱草脑、α-松油醇、β-榄香烯、t-石竹烯、乙酸松油酯、蛇麻烯、大茴香醚、β-荜澄茄烯、橙花叔醇异构体等	黄河南北各省区多数有产	果皮入药，有温中助阳、散寒祛湿、止痒、驱虫功效
三角榄 Canarium bengalense 叶含精油0.23% 橄榄科	α-蒎烯、α-松油烯、α-荜澄茄烯、β-石竹烯、α-石竹烯、异丁香酚甲醚、γ-榄香烯、α-金合欢烯等	云南和广西、广东有栽培。生长于海拔400～1300米的杂木林中	果可生食或糖渍成凉果，精油作皂用和化妆品香精
黄连木 Pistacia chinensis 枝叶含精油0.12% 漆树科	α-蒎烯、莰烯、β-蒎烯、β-月桂烯、辛醛、α-水芹烯、间伞花烃、柠檬烯、β-罗勒烯、α-罗勒烯、β-石竹烯、雅槛兰烯、γ-榄香烯、橙花叔醇、β-桉叶醇等	长江以南各省区及华北、西北。生长于海拔1400～3500米的石山杂木林中	精油可用于调配香精

（续表）

植物名称	主要化学成分	分布	备注
枫杨 *Pterocarya atenoptera* 枝叶含精油0.25% 胡桃科	α-蒎烯、β-蒎烯、枞油烯、1-（1-环己烯基)-2-丙酮、t-葛缕醇、桃金娘烯醇、β-石竹烯、α-佛手烯、cis-β-金合欢烯、α-甜没药烯、β-甜没药烯、α-甜没药烯异构体、α-甜没药醇等	我国南部和中部。多生长于河滩低湿地	为庭园行道绿化树种，根、树皮和叶药用治痛疮及脚癣
刺参 *Oplopanax elatus* 根叶含精油 0.08%～0.1% 五加科	α-蒎烯、β-蒎烯、辛醛、柠檬烯、罗勒烯、紫苏醛、2,6-二甲基庚烯、龙脑、十二醛、乙酸龙脑酯、α-古巴烯、十四醛、异石竹烯、异金合欢烯、γ-杜松烯、δ-杜松烯、橙花叔醇、愈创木醇、柏木烯醇、榧醇、布黎醇、金合欢醇等	吉林长白山。生长于海拔1400～1500米的落叶阔叶林下	根茎有补气、助阳、兴奋中枢神经功能，可治疗神经衰弱、低血压、阳痿、精神分裂、糖尿病
地檀香 *Gaultheria forrestii* 枝叶含精油 0.3%～0.8% 杜鹃花科	水杨酸甲酯等	云南。生长于灌木丛中	精油有消炎止痛功效，亦可调配香精，但主要用于合成水杨酸和水杨醇
滇白珠 *G. leucocarpa* var. *crenulata* 枝叶含精油0.7% 杜鹃花科	t-2-癸烯醛、6-甲基-5-庚烯-2-酮、桧烯、壬醛、水杨酸甲酯、水杨酸乙酯、邻羟苯甲酸苯酯等	四川、云南、贵州、湖北、湖南、广东、广西等省区。生长于灌木丛中	与地檀香精油相同用途
狭叶杜香 *Ledum palustre* var. *angustum* 叶含精油 杜鹃花科	α-侧柏烯、α-蒎烯、莰烯、β-侧柏烯、β-蒎烯、α-水芹烯、对伞花烃、环小茴香烯、1,8-桉叶油素、t-蓋烯、桃金娘烯醛、月桂烯醇、枯茗醛、乙酸龙脑酯、甲酸香叶酯、别香树烯等	我国东北地区及内蒙古。生长于疏林下、石坡上，以及林缘和沼泽地	长白山民间用叶作药用，治月经不调、不孕症，精油可治慢性支气管炎

（续表）

植物名称	主要化学成分	分布	备注
杜香 *L. palustre* 叶含精油0.2% 杜鹃花科	对伞花烃、桧烯、β-蒎烯、α-蒎烯、松油烯、小茴香烯等	我国东北及内蒙古。生长于水甸子或湿草地	精油供药用，祛痰效果良好
烈香杜鹃 *Rhododendron anthopogonoides* 枝叶含精油0.7%～1.9% 杜鹃花科	β-月桂烯、柠檬烯、β-罗勒烯（Z）、3-苯基-2-丁酮、1-甲基-3-苯基丙醇、乙酸龙脑酯、古巴烯、乙酸-1-甲基-3-苯基丙酯、α-佛手烯、γ-榄香烯、α-石竹烯、γ-依兰油烯、雅槛蓝烯、α-金合欢烯、γ-杜松烯、β-杜松烯、α-古芸烯、γ-榄香醇、吉玛酮、别香树烯、α-依兰油烯、芹子-3、7（11）-二烯等	青海、甘肃和四川。生长于高山，自成灌丛	精油对慢性支气管炎效果好
头花杜鹃 *R. capitatum* 枝叶含精油0.5～2.0% 杜鹃花科	α-蒎烯、莰烯、β-蒎烯、β-月桂烯、柠檬烯、β-罗勒烯（Z）、α-罗勒烯、γ-松油烯、乙酸龙脑酯、β-古芸烯、α-石竹烯等	青海、甘肃。生长于海拔2500～3600米的高山草原或灌丛	精油对慢性支气管炎有较好疗效，其毒性较烈香杜鹃低
北方雪层杜鹃 *R. nivale* subsp. *boreale* 枝叶含精油1.2% 杜鹃花科	α-蒎烯、β-蒎烯、α-荜澄茄油烯、古巴烯、β-波旁烯、α-檀香烯、γ-榄香烯、β-古芸烯、β-愈创木烯、杜松烯、α-依兰油烯、δ-杜松烯、α-古巴烯、γ-芹子烯、愈创木醇、β-桉叶醇、香榧醇、吉玛酮、异愈创木醇等	云南、西藏、青海、四川。生长于海拔3000～4400米的高山杜鹃灌丛中或云杉下	精油为有待开发的天然香料
樱叶杜鹃 *R. cerasinum* 枝叶含精油0.4%～0.6% 杜鹃花科	α-蒎烯、莰烯、β-蒎烯、α-松油烯、对伞花烃、柠檬烯、水合桧烯、芳樟醇、4-松油醇、α-松油醇、乙酸龙脑酯、t-石竹烯、香树烯、α-佛手烯、δ-杜松烯、γ-芹子烯、α-愈创木烯、β-桉叶醇等	云南、四川、西藏。生长于海拔3400～4500米的石岩坡或针叶林缘	精油为有待开发的天然香料

（续表）

植物名称	主要化学成分	分布	备注
青海杜鹃 *R. przewalskii* 枝叶含精油0.4% 杜鹃花科	愈创木烯、异卡拉米二醇、α-杜松醇、异愈创木醇、d-斯潘连醇、δ-杜松醇、杜鹃烯、杜鹃次烯、牻牛儿酮、柠檬烯、β-月桂烯、香树烯等	青海、甘肃、陕西和四川北部。生长于海拔4000米左右的高山上，自成灌丛	精油用于治疗慢性支气管炎
腋花杜鹃 *R. racemosum* 枝叶含精油0.3% 杜鹃花科	α-蒎烯、β-蒎烯、对伞花烃、1,8-桉叶油素、乙酸龙脑酯、t-石竹烯、氧化石竹烯等	云南中部和北部，四川西南部。生长于海拔800～2800米的疏灌木丛中	精油有杀菌、祛痰、消毒作用，亦可治疗慢性支气管炎
千里香杜鹃 *R. thymifolium* 枝叶含精油 0.5%～2.0% 杜鹃花科	α-蒎烯、莰烯、β-蒎烯、月桂烯、柠檬烯、壬醛、蛇麻烯、金合欢烯、牻牛儿酮、桧脑等	青海和甘肃。生长于海拔2400～3800米的湿润山坡形成灌丛	精油对慢性支气管炎有较好疗效
紫金牛 *Ardisia japonica* 全株含精油0.2% 紫金牛科	乙酸乙酯、2,4-二甲基-2-戊醇、异硫氢酸丙酯、2-甲基-3-（1-甲基乙基）环氧乙烷、2,3-二羟基-6-甲基呋喃-4-酮、1,3,5-环庚三烯、4-甲基-3-戊烯-2-酮、3-甲基-2-丁烯醇、己酸乙酯、2-辛烯醇、3,4-二甲基-3-己烯-2-酮、2-十一酮、4-松油醇、邻二甲氧基苯、龙脑、水杨酸甲酯、异胡薄荷酮、苯甲醇、2-甲氧基-4-异丙基苯酚、3,4,5-三甲氧基苯甲醛、β-桉叶醇、α-石竹烯醇、1,6-二甲基-4-异丙基萘等	长江以南各省区。朝鲜、日本也有分布。生长于林下、谷地溪旁阴湿处	为民间草药，对慢性支气管炎和肺炎有一定疗效

（续表）

植物名称	主要化学成分	分布	备注
连翘 *Forsythia suspensa* 种子含精油4.0% 木犀科	β-蒎烯、α-蒎烯、芳樟醇、对伞花烃、γ-松油烯、β-水芹烯、月桂烯、莰烯、β-罗勒烯、4-松油醇、3-蒈烯等	云南、江苏、湖北、甘肃、陕西、山西、河南、山东、河北及东北地区。生长于海拔1000米的山地，多为栽培	精油有抑制流行性感冒病毒的活性
黄荆（山黄荆） *Vitex negundo* 枝叶含精油 0.30%～0.60% 马鞭草科	α-石竹烯、α-水芹烯、α-蒎烯、桧烯、β-蒎烯、1,8-桉叶油素、对伞花烃、γ-松油烯、α-松油烯、4-松油醇、α-松油醇、δ-榄香烯、α-荜澄茄烯、β-榄香烯、蛇麻烯、环氧石竹烯等	长江流域以南各省区，北达秦岭淮河。生长于山坡路旁或灌木丛中	精油具祛痰、镇咳、平喘的功效，治慢性气管炎，果实为清凉性镇静药
牡荆 *V. negundo* var. *cannabifolia* 枝叶含精油 0.3%～0.6% 马鞭草科	桧烯、7-辛烯-4-醇、1,8-桉叶油素、γ-松油烯、4-松油醇、α-松油醇、β-波旁烯、β-榄香烯、β-石竹烯、γ-依兰油烯、α-石竹烯、β-荜澄茄烯、δ-杜松烯、γ-榄香烯、β-桉叶醇等	华东各省及河北、湖南、湖北、广东、广西、四川、贵州、云南。生长于山坡路旁	精油具祛痰、镇咳、平喘功效，治疗慢性气管炎，果实为清凉性镇痛药
荆条 *V. negundo* var. *heterophylla* 枝叶含精油0.15% 马鞭草科	桧烯、芳-枞三烯、1,8-桉叶油素、4-松油醇、乙酸-α-松油酯、β-石竹烯、β-金合欢烯（Z）、β-荜澄茄烯、榄香烯、β-桉叶醇等	辽宁、河北、山西、山东、河南、陕西、甘肃、江苏、安徽、江西、湖南、贵州、四川。生长于山坡路旁	精油具祛痰、镇咳、平喘功效
蔓荆 *V. trifolia* 枝叶含精油 0.11%～0.12% 马鞭草科	α-蒎烯、β-石竹烯、α-水芹烯、α-松油烯、桧烯、β-蒎烯、1,8-桉叶油素、对伞花烃、γ-松油烯、4-松油醇、α-松油醇、δ-榄香烯、α-荜澄茄烯、β-榄香烯、雅槛蓝酮等	福建、台湾、广东、广西、云南。生长于平原河滩、疏林及树林附近	精油有祛痰、镇咳和平喘作用，治慢性气管炎，茎叶入药，治跌打损伤、风湿疼痛，果实治感冒、风热神经性头痛

（续表）

植物名称	主要化学成分	分布	备注
木香薷 *Elsholtzia stauntoni* 叶含精油1.30% 唇形科	α-蒎烯、1,8-桉叶油素、苯乙酮、芳樟醇、4-松油醇、α-松油醇、t-石竹烯、cis-石竹烯、δ-杜松烯等	河北、河南、山西、陕西、甘肃。生长于海拔700～1600米的谷地溪旁或河川沿岸草坡石山	精油对痢疾、肠胃炎、感冒等疾病有较好疗效

补充樟属以外种属

植物名称	主要化学成分	分布	备注
厚壳桂 *Cryptocarya chinensis* 枝叶含精油0.36% 樟科	β-水芹烯、乙酸龙脑酯、δ-榄香烯、α-荜澄茄烯、古巴烯、β-荜澄茄烯、β-石竹烯、α-石竹烯、β-荜澄茄烯异构体、γ-榄香烯、δ-杜松烯、榄香醇、愈创木醇、δ-杜松醇等	四川、广西、广东、福建、台湾。生长于海拔300～1000米的山谷常绿阔叶林中	为丰富的倍半萜资源，可开发成理想的定香剂
黄果厚壳桂 *C. concinna* 枝叶含精油0.16% 樟科	β-蒎烯、β-月桂烯、柠檬烯、α-松油醇、芳樟醇、β-石竹烯、γ-石竹烯、β-松油烯、橙花叔醇、γ-榄香烯、δ-杜松醇等	广东和广西	精油可配制日用化工香精和皂用牙膏
鼎湖钓樟 *Lindera chunii* 枝叶含精油0.20%～0.30% 樟科	古巴烯、波旁烯、β-石竹烯、β-古芸烯、α-石竹烯、γ-杜松烯、β-荜澄茄烯、β-马榄烯、γ-榄香烯、β-杜松烯、1,2,3,4,4a,7-六氢-1,6-二甲基4-(1-甲基乙基)萘、δ-杜松醇、香榧醇、布黎醇、刺柏脑等	广东、广西等省区	为丰富倍半萜资源，可开发成定香剂

（续表）

植物名称	主要化学成分	分布	备注
香叶树 *L. communis* 枝叶含精油0.16% 樟科	δ-榄香烯、β-榄香烯、α-榄香烯、α-檀香烯、α-佛手烯、α-依兰油烯、α-石竹烯、γ-杜松烯、β-荜澄茄烯、雅槛蓝烯、古芸烯、γ-依兰油烯、δ-杜松烯、β-马榄烯、δ-杜松醇等	陕西、甘肃、湖北、湖南、江西、浙江、广东、广西、福建、台湾、四川、贵州、云南等省区。中南半岛亦有分布。散生或混生常绿阔叶林中	富含倍半萜，可开发成定香剂
红果山胡椒 *L. erythrocarpa* 枝叶含精油 樟科	α-蒎烯、莰烯、β-蒎烯、柠檬烯、β-月桂烯、对伞花烃、香叶醇、乙酸龙脑酯、石竹烯、乙酸香叶酯、佛手烯、蛇麻烯、γ-杜松烯、金合欢醇、α-侧柏烯、β-水芹烯、桧烯、3-蒈烯等	陕西、河南、山东、江苏、安徽、浙江、江西、湖北、湖南、福建、台湾、广东、广西、四川等省区。生长于海拔1000米以下的山坡山谷溪边林下	精油可用于调配皂和化妆品香精
山胡椒 *L. glauca* 果皮含精油 樟科	β-蒎烯、莰烯、罗勒烯、壬醛、1,8-桉叶油素、龙脑、柠檬醛、对伞花烃、黄樟油素、乙酸龙脑酯、γ-广藿香烯等	山东、河南、陕西、甘肃、山西、江苏、安徽、浙江、江西、福建、台湾、广东、广西、湖南、湖北、四川等省区。生长于海拔400米以下的山坡林缘路旁	果皮精油可配制皂用香精

（续表）

植物名称	主要化学成分	分布	备注
三桠乌药 *L. obtusiloba* 鲜叶含精油 0.9%～1.1% 樟科	α-蒎烯、莰烯、β-蒎烯、β-月桂烯、cis-罗勒烯、γ-松油烯、樟脑、β-榄香烯、石竹烯、β-芹子烯、α-侧柏烯、β-水芹烯、柠檬烯、柠檬醛、葛缕酮、芳樟醇、龙脑、薄荷脑、α-松油醇、4-松油醇、乙酸龙脑酯、乙酸香叶酯等	辽宁、山东、安徽、江苏、河南、陕西、甘肃、浙江、江西、福建、湖南、湖北、四川、西藏等省区。生长于密林灌木丛中	精油可配制化妆品和皂用香精等
山橿（大叶钩樟） *L. reflexa* 叶含精油 樟科	α-蒎烯、莰烯、β-蒎烯、柠檬烯、3-蒈烯、β-月桂烯、γ-松油酯、1,8-桉叶油素、芳樟醇、龙脑、香叶醇、乙酸龙脑酯、乙酸香叶酯等	河南、江苏、安徽、浙江、江西、湖南、湖北、贵州、云南、广西、广东、福建。生长于海拔1000米以下的山谷山坡、林下灌丛中	根入药性温味辛，有止血消肿功效，亦可治胃痛、风疹、疥癣等
山苍子（山鸡椒） *Litsea cubeba* 果实含精油 3.0%～4.0% 花含精油 1.5%～1.6% 樟科	α-蒎烯、莰烯、桉烯、β-蒎烯、6-甲基-5-庚烯-2-酮、柠檬烯、β-月桂烯、对伞花烃、4-松油醇、芳樟醇、β-柠檬醛、α-柠檬醛、香叶醇、β-石竹烯等 α-侧柏烯、α-蒎烯、桉烯、β-月桂烯、α-水芹烯、α-松油烯、间伞花烃、1,8-桉叶油素、γ-松油烯、2-蒈烯、4-松油醇、α-松油醇、榄香烯、β-石竹烯等	广西、广东、福建、台湾、浙江、江苏、安徽、江西、湖南、湖北、四川、贵州、云南、西藏。生长于灌丛路旁、水边等	果皮精油主含柠檬醛，其中60%是合成紫罗兰酮的主要原料，可用作食品、烟草、化学品原料；花精油可开发为天然香料，具有一定杀菌能力

植物名称	主要化学成分	分布	备注
清香木姜子 *L. euosma* 鲜果实含精油 2.5%～3.0% 樟科	α-蒎烯、莰烯、柠檬烯、对伞花烃、甲基庚烯酮、香茅醛、芳樟醇、樟脑、β-柠檬醛、α-柠檬醛、香叶醇、α-蛇麻烯等	分布与山鸡椒相同	果实精油是提取柠檬醛主要原料之一，精油可直接调香等用途或化工原料，可提取柠檬醛
毛叶木姜子 *L. mollis* 果实（皮）含精油 0.08% 樟科	α-蒎烯、莰烯、β-蒎烯、柠檬烯、1,8-桉叶油素、对伞花烃、6-甲基-5-庚烯-2-酮、芳樟醇、香茅醛、樟脑、α-松油醇、香茅醇、β-柠檬醛、α-柠檬醛等		
木姜子（辣姜子） *L. pungens* 叶含精油0.44% 樟科	α-蒎烯、莰烯、β-蒎烯、月桂烯、β-松油烯、1,3,3-三甲基-2-氧杂二环[2,2,2]辛烷、1,8-桉叶油素、小茴香烯、柠檬烯、（−）-异胡薄荷酮、香茅醇、3-甲基-5-（1-甲基乙烯基）环己酮、香茅醛、乙酸橙酯、cis-盖烯-8、γ-杜松烯等	湖北、湖南、广东、广西、贵州、云南、四川、西藏、甘肃、陕西、河南、山西、浙江等省区。生长于溪旁山地阳坡的杂木林缘	精油可用于配制化妆品、食品、皂用等香精
刨花润楠 *Machilus pauhoi* 鲜叶含精油0.05% 樟科	辛烷、壬醇、壬烷、环小茴香烯、β-蒎烯、辛醛、间伞花烃、龙烯、1,8-桉叶油素、β-罗勒烯、侧柏烯、α-罗勒烯、壬醛、癸醛、癸醇、乙酸龙脑酯、2-十一酮、8-十八醛、t-盖二烯、芹子烯、β-榄香烯、十二醛、乙酸二氢葛缕酯、薄荷脑等	浙江、福建、江西、湖南、广东、广西等省区	精油可用于配制化妆品和皂用香精

（续表）

植物名称	主要化学成分	分布	备注
绒楠 *Machilus velutina* 枝叶含精油 0.20%～0.25% 樟科	α-松油醇、乙酸龙脑酯、β-石竹烯、α-石竹烯、β-荜澄茄烯、γ-榄香烯、δ-杜松烯、橙花叔醇、δ-杜松醇等	广西、广东、福建、江西、湖南、浙江等省区	为材用树种，枝叶和树皮为"香粉"优质原料。精油可配香精
新樟 *Neocinnamomum delavayi* 叶含精油0.7% 樟科	α-蒎烯、莰烯、β-蒎烯、月桂烯、α-水芹烯、对伞花烃、1,8-桉叶油素、柠檬烯、t-罗勒烯、γ-松油烯、芳樟醇、桧醇、樟脑、香茅醇、乙酸龙脑酯、乙酸香茅酯、cis-石竹烯、t-石竹烯、α-蛇麻烯、别香树烯、γ-广藿香烯、杜松烯、γ-榄香烯、α-荜澄茄烯、α-金合欢烯等	云南、四川及西藏。生长于海拔1100～2300米的灌丛林缘或密林中	精油可开发成药用和化妆品用香精
长叶新木姜 *Neolitsea oblongifolia* 叶含精油 0.3%～0.4% 樟科	α-侧柏烯、α-蒎烯、桧烯、β-蒎烯、对伞花烃、柠檬烯、1,8-桉叶油素、α-龙脑烯醛、蒎葛缕醇、5-（1-甲基乙基）二环[3,1,0]己-2-酮、（-）-cis-桧醇、4-松油醇、枯茗醇、马鞭草烯酮、乙酸龙脑酯、乙酸香叶酯等	广东、广西。生长于海拔300～900米的山谷密林中或林缘处	精油可调配皂用香精
小新木姜 *N. umbrosa* 叶含精油 樟科	α-蒎烯、莰烯、1,8-桉叶油素、优藏茴香酮、蒎葛缕醇、对伞花烃-8-醇、桃金娘烯醛、马鞭草烯酮、桃金娘烯醇、乙酸龙脑酯、古巴烯、γ-榄香烯、δ-杜松烯、β-桉叶醇等	广东、广西等省区。生长于山谷的混交林中	精油可配制化妆品和皂用香精

（续表）

植物名称	主要化学成分	分布	备注
楠木 *Phoebe zhennan* 果皮含精油1.6% 樟科	α-蒎烯、莰烯、β-蒎烯、α-水芹烯、α-松油烯、对伞花烃、β-水芹烯、γ-松油烯、cis-桉醇、龙脑、乙酸辛酯、α-松油醇、甲酸香叶酯、α-古巴烯、δ-杜松烯、橙花叔醇、榄香醇、愈创木醇、β-桉叶醇、4-（5-甲基-2-呋喃基）丁酮-2等	湖北、湖南、四川等省。野生楠木多见于海拔1500米以下的阔叶林中	为优良材用树种，木材坚实，为建筑家具、造船用材。精油可作化妆品原料

四、草本（含藤本）精油资源

我国草本精油资源更为丰富。草本精油资源包括一年生和多年生草本，其生长周期短，生产更加灵活，更容易适应市场需求。草本植物还有几大优点：再生潜力强，如香叶天竺葵（*Pelargonium graveolens*）、香根鸢尾（*Iris pallida*）；可供食品调味用，如胡椒（*Piper nigrum*）、茴香（*Foeniculum vulgare*）等；可用于轻工原料，如柠檬草（香茅）（*Cymbopogon citratus*）；可供药用，如人参（*Panax ginseng*）、当归（*Angelica sinensis*）等，详见表1-4。

表1-4　我国草本（含藤本）精油资源一览表

植物名称	主要化学成分	分布	备注
尾花细辛 *Asarum caudigerum* 全草含精油0.08% （马兜铃科）	1,8-桉叶油素、(Z)-异丁香酚甲醚、异丁香酚甲醚、榄香脂素、(Z)-异榄香脂素、3,4,5-三甲基苯甲醛、三甲基烯丙基苯(II)等	浙江、江西、福建、台湾、湖南、广东、广西等。生长于阴湿林下溪旁	全株入药，民间用于代细辛

（续表）

植物名称	主要化学成分	分布	备注
牛蹄细辛 *A. delavayi* 全草含精油1.4% （马兜铃科）	α-蒎烯、莰烯、β-蒎烯、1,8-桉叶油素、对伞花烃、龙脑、4-松油醇、樟脑、乙酸龙脑酯、黄樟油素、3,5-二甲氧基甲苯、t-石竹烯、β-古云烯、丁香酚甲醚、2,3,5-三甲氧基甲苯、3,4,5-三甲氧基甲苯、橙花叔醇、细辛醚、肉豆蔻醚、榄香脂素等	四川、云南北部。生长于海拔800~1600米的林下阴湿岩坡上	全草在四川峨眉作土细辛入药
金耳环 *A. insigne* 全草含精油1.05% （马兜铃科）	α-蒎烯、β-小茴香烯、β-蒎烯、龙脑、乙酸龙脑酯、黄樟油素、丁香酚甲醚、t-石竹烯、马兜铃烯、β-金合欢烯、细辛醚、橙花叔醇、金合欢醇（Z）等	广西。生长于深山溪边、林下阴湿处	当地用作细辛入药
单叶细辛 *A. himalaicum* 全草含精油0.4% （马兜铃科）	氧化芳樟醇、芳樟醇、龙脑、萘、异丙基茴香醚、t-葛缕醇、乙酸龙脑酯、古巴烯、3,5-二甲氧基甲苯、乙酸松油酯、β-古芸烯、丁香酚甲醚、去氢白菖烯、榄香脂素等	湖北、四川、云南、贵州、西藏、甘肃、陕西。生长于海拔1300~3100米的溪边林下阴湿处	作为细辛入药
小叶爬崖香 *Piper arboricola* （藤本） 全草含精油 0.22%~0.28% （胡椒科）	β-月桂烯、柠檬烯、芳樟醇、松油醇、癸醛、癸醇、4-松油醇、2-十一酮、β-榄香烯、十二醛、β-石竹烯、α-石竹烯、γ-杜松烯、2-十三酮、β-榄香烯异构体、榄香醇、橙花叔醇等	我国东南至西南均有分布。生长于林中，攀援于岩石和树干上	民间芳香草药，治偏头痛
华南胡椒 *P. austrosinense*（木质藤本） 枝叶含精油 0.4%~0.6% （胡椒科）	α-蒎烯、桧烯、β-蒎烯、β-月桂烯、α-水芹烯、3-蒈烯、α-松油烯、枞油烯、γ-松油烯、芳樟醇、4-松油醇、古巴烯、β-石竹烯、α-石竹烯、β-雪松烯、β-荜澄茄烯、γ-榄香烯等	广东、广西。生长于林中，攀援于树干和岩石上	当地民间用作芳香草药

（续表）

植物名称	主要化学成分	分布	备注
蒟酱（蒌叶） *P. betle*（攀援藤本） 枝叶含精油 （胡椒科）	胡椒酚、蒌叶酚、丙烯基焦儿茶酚、香荆芥酚、丁香酚、对伞花烃、1,8-桉叶油素、丁香酚甲醚、石竹烯、杜松烯等	我国东至西南各省区均有分布。以人工栽种为主	叶入药可祛风祛湿、杀虫、止痒等，亦治胃痛
胡椒 *P. nigrum*（木质攀援藤本） 果实含精油 2.5%～3.0% （胡椒科）	α-蒎烯、β-蒎烯、β-月桂烯、α-水芹烯、3-蒈烯、间伞花烃、柠檬烯、2-蒈烯、芳樟醇、δ-榄香烯、古巴烯、β-石竹烯、α-石竹烯	原产于东南亚。现我国东南、西南各省区均有栽种	果实入药治消化不良、腹泻，可暖肠胃，亦被大量用作食品调料
蕺菜 *Houttuynia cordata* 全草含精油 0.4%～0.6% （三白草科）	3-己烯醇、β-月桂烯、α-罗勒烯、苯乙醛、芳樟醇、4-松油醇、α-松油醇、辣薄荷酮、香叶醇、2-十三酮、肉豆蔻醚、去氢白菖烯、橙花叔醇、月桂酸等	黄河以南及西藏、台湾、云南、甘肃。生长于溪边和林下湿地	全草入药，有散热、消肿、解毒功效
三白草 *Saururus chinensis* 全草含精油0.32% （三白草科）	3-己烯醇、2-甲基环戊醇、1,8-桉叶油素、芳樟醇、樟脑、黄樟油素、肉豆蔻醚等	长江以南各省区塘边、溪边其他湿地。日本、越南也有	民间用药，有消热利尿功效
芥菜 *Brassica juncea* 种子含精油 0.21%～0.25% （十字花科）	3-异硫氰基-1-丙烯、4-异硫氰基-1-丁烯、对伞花烃等	原产于亚洲，我国各地以蔬菜栽种	常见蔬菜，种子为辛香调味料
土荆芥 *Chenopodium ambrosioides* 全草含精油 0.4%～1.0% （藜科）	α-松油烯、对伞花烃、1,8-桉叶油素、α-松油醇、乙酸4-松油醇酯、香芹酚、麝香草酚、驱蛔脑、喇叭茶醇等	华东、华南以及四川等地区也有栽种	用于驱蛔虫、十二指肠虫、绦虫等

（续表）

植物名称	主要化学成分	分布	备注
香叶天竺葵 *Pelargonium graveolens* 茎叶含精油 1.0%～1.4% （牻牛儿苗科）	香茅醇、甲酸香茅酯、香叶醇、甲酸香叶酯、玫瑰醚、芳樟醇、氧化芳樟醇等	原产于非洲南部。现我国各地均有栽种	精油香气近似玫瑰油，可配制化妆品香精
黄果西番莲 *Passiflora edulis f. flavicarpa* 果肉含精油 （西番莲科）	3-己烯醇、丙基环内烷、β-蒎烯、乙酸乙酯、3-甲基丙酸异丙酯、丁酸己酯、4-羟基苯乙酯、己酸3-己烯酯、己酸己酯、2,6-二叔丁基对甲酚等	原产于南美洲。广东、云南、福建、台湾有栽种	果肉汁可配制饮料和冰激凌，种子含脂肪油
黄海棠 *Hypericum ascyron* 全草含精油 0.32%～0.38% （藤黄科）	乙酸1-乙氧基乙酯、3-己烯醛、3-己烯醇、己醇、3-己烯酸、1,8-桉叶油素、β-石竹烯、α-金合欢烯（ZE）、β-榄香烯、α-金合欢醇等	东北、华北、华中及华南地区。生长于山坡林下灌丛林和草丛中	果为民间用药，治头痛吐血等，种子治胃痛
黄葵 *Abelmoschus moschatus* 种子含精油0.3% （锦葵科）	乙酸癸酯、金合欢醇、月桂酸乙酯、t-2-t-6-金合欢醇、cis-2-t-6-金合欢醇、乙酸月桂烯酯、十六烯醛、乙酸-cis-2-t-6-金合欢酯、黄葵内酯、乙酸-t-6-金合欢酯、亚油酸、亚油酸乙酯等	广东、湖南、江西、台湾。生于山谷、沟傍或草坡	种子烤焙后可代咖啡
臭节草（臭草） *Boenninghausenia albiflora* 全株含精油0.18% （芸香科）	侧柏烯、桧烯、β-蒎烯、3-蒈烯、对伞花烃、1,8-桉叶油素、β-水芹烯、别罗勒烯、4-松油醇、桃金娘醛、α-松油醇、癸醛、乙酸癸酯、β-荜澄茄烯、γ-杜松烯等	长江流域以南各地。生长于石灰岩山地、阴湿林缘或灌木丛	全草入药具清热凉血、舒筋活血、消炎功效

植物名称	主要化学成分	分布	备注
芸香 *Ruta graveolens* 全草含精油0.06% （芸香科）	4-甲基-2-庚酮、2-壬酮、2-壬醇、乙酸1-甲基辛酯、2-十一酮、2-十一醇、2-十二酮、乙酸-1-甲基癸酯、2-十三酮、当归素等	原产于欧洲。现我国南北均有栽种	全草入药可祛风、退热、利尿等，亦用作煮糖水
飞龙血掌 *Toddalia asiatica* 枝叶含精油0.32% （芸香科）	α-蒎烯、莰烯、β-松油烯、β-月桂烯、对伞花烃、柠檬烯、γ-松油烯、β-石竹烯、β-荜澄茄烯、β-榄香烯、乙酸龙脑酯、δ-杜松烯、2-蒈烯等	广东、广西、湖南、贵州、云南、四川、福建、浙江等省区。生于丛林中	根入药可散瘀消肿、祛风镇痛
人参 *Panax ginseng* 根含精油 1.14%～1.16% （五加科）	α-愈创木烯、β-香树烯、β-榄香烯、β-古芸烯、β-石竹烯、t-金合欢醇、α-榄香烯、3,3-二甲基己烷、十七烷、肉豆蔻酸、十五酸-2,7-二甲基辛烷等	吉林东部、辽宁东部、黑龙江东部。朝鲜半岛亦有分布。生长于海拔1400～1500米的落叶阔叶林	为重要强壮剂，花亦含精油
莳萝 *Anethum graveolens* 种子含精油 1.2%～3.5% （伞形科）	柠檬烯、t-二氢葛缕酮、t-葛缕酮、葛缕醇、大茴香醚、芹菜脑、1-甲基乙烯基甲苯等	原产于欧洲。我国东北、甘肃、四川、广东等地区有栽种	茎叶果实有茴香味，可作调味料，果实入药，有祛风健胃等效果，芽可作蔬菜
灰绿叶当归 *Angelica glauca* 根含精油0.48% （伞形科）	6-甲基-二环[3,2,0]庚-6-烯-2-酮、6-丙基-二环[3,2,0]庚-6-烯-2-酮、戊基苯、β-甜没药烯、δ-杜松烯、枯茗醛、γ-榄香烯、榄香醇、去氢喇叭茶醇等	新疆。生长于海拔1000米左右的河谷、林下、林缘的杂草丛中	民间入药代姜活，有祛风湿、发汗等功效

（续表）

植物名称	主要化学成分	分布	备注
当归 *A. sinensis* 鲜叶含精油0.4% 根含精油 0.4%～0.7% （伞形科）	α-蒎烯、月桂烯、柠檬烯、马鞭草烯酮、藏红花醛、对-乙基苯甲醛、3,4二甲基苯甲醛、优葛缕酮、1,1,5-三甲基-2-甲酸基2,5-环己二烯-4-酮、古巴烯、t-β-金合欢烯、2,4,6-三甲基苯甲醛、β-芹子烯、佛手烯、γ-杜松烯、δ-杜松烯等	陕西、甘肃、湖北、四川、云南、贵州。多为栽培，少见野生	叶精油制成面霜治面黄褐斑，根亦含精油，是著名药材，根精油含蒿本酯48%～60%
芹菜 *Apium graveolens* 种子含精油0.2% （伞形科）	β-蒎烯、β-月桂烯、柠檬烯、1-乙烯基-2-己烯基-环丙烷、β-石竹烯、β-芹子烯、α-芹子烯、α-蒎烯等	我国南北各地均有栽种，为一种蔬菜	种子精油和油树脂为重要调味料，全株入药降压利尿
新疆藁本 *Conioselinum tataricum* 根茎含精油1.3% （伞形科）	α-蒎烯、β-月桂烯、α-水芹烯、3-蒈烯、β-水芹烯、β-罗勒烯(*cis*-X)、β-罗勒烯(*trans*-Y)、γ-松油烯、异松油烯、α-玫瑰醚、蛇床酞内酯、α-松油醇、胡椒烯酮、乙酸-α-松油酯、丁香酚甲醚、肉豆蔻醚、榄香脂素、β-金合欢烯等	新疆。生于草地	根茎入药，有散风寒、止痛、燥湿功效
芫荽 *Coriandrum sativum* 果实种子含精油1.0% （伞形科）	α-蒎烯、β-蒎烯、对伞花烃、γ-松油烯、芳樟醇、香叶醇、乙酸香叶酯等	原产于欧洲地中海，我国大部分地区有栽种	种子精油为重要中西餐调味料，为我国南方调味蔬菜，称"香草"，干果入药具健胃、祛风、祛痰功效

植物名称	主要化学成分	分布	备注
孜然芹 *Cuminum cyminum* 种子含精油3.7% （伞形科）	α-蒎烯、β-蒎烯、对伞花烃、柠檬烯、1,8-桉叶油素、γ-松油烯、4-松油醇、α-松油醇、对-异丙基苯酚、枯茗醛、对-异丙基苯甲酸等	外来种，新疆等地有栽种	精油有枯茗醛香味，可调制食品香精，也是咖喱的原料之一
环根芹 *Cyclorhiza waltonii* 根含精油 （伞形科）	α-侧柏烯、α-蒎烯、桧烯、月桂烯、α-松油烯、对伞花烃、β-水芹烯、γ-松油烯、异松油烯、4-松油醇、杜松烯、异杜松烯、古巴烯、cis-石竹烯等	为我国特有种，分布于西藏、云南、四川。生长于海拔3500～4600米的山坡	根入药有清热解毒作用
茴香 *Foeniculum vulgare* 果实含精油 3.0%～6.0% （伞形科）	α-蒎烯、桧烯、β-月桂烯、间伞花烃、柠檬烯、γ-松油烯、小茴香酮、异大茴香醛、大茴香醚等	原产于地中海地区，现我国各省区均有栽种	精油主要用于食品和牙膏、酒类香精，入药有抗菌、温胃、散寒功效
匙叶甘松 *Nardostachys jatamansi* 根茎含精油 5%～6.5% （败酱科）	9-马兜铃烯、1(10)-马兜铃烯、β-马榄烯、1,2,9,10-四脱氢马兜铃烷、甘松醇等	分布于我国西南部。生于山地林下	根茎为我国常用药材，具理气、健胃、止痛功效，亦可配制香精
糙叶败酱 *Patrinia rupestris subsp. scabra* 根茎含精油0.3% （败酱科）	β-石竹烯、α-蛇麻烯、古巴烯、δ-愈创木烯、β-芹子烯、β-愈创木烯、δ-杜松烯、大茴香醚、α-金合欢烯、喇叭茶醇、δ-杜松烯等	我国东北和华北。生于向阳山坡和土坎上	全草和根入药，有清热燥湿、止血止带等功效

（续表）

植物名称	主要化学成分	分布	备注
蜘蛛香 *Valeriana jatamansi* 根含精油 0.5%～0.8% （败酱科）	α-蒎烯、柠檬烯、1,8-桉叶油素、对伞花烃、乙酸龙脑酯、龙脑、橙花叔醇、马鞭草烯醇等	陕西、河南、湖北、湖南、四川、贵州、云南、西藏。生于山坡草地	根入药有消食健胃、理气止痛、祛风解毒功效
缬草（欧缬草） *V. officinalis* 全草含精油 0.5%～0.6% （败酱科）	α-蒎烯、莰烯、β-蒎烯、对伞花烃、柠檬烯、龙脑、乙酸龙脑酯、β-侧柏醇、β-古芸烯、马兜铃-1,9-二烯、缬草酮、对-叔丁基苯甲醇等	我国东北至西南地区。生长于高山山坡、林下和沟边	精油可配制烟用、食用、酒类、化妆品用香精，有安神功效
宽叶缬草（蜘蛛香） *V. officinalis* var. *latifolia* 根含精油2% （败酱科）	异缬草酸、α-蒎烯、莰烯、桧烯、β-蒎烯、月桂烯、柠檬烯、龙脑、乙酸龙脑酯、乙酸葛缕酯、乙酸二氢葛缕酯、β-石竹烯、蛇麻烯、麝香草酚甲醚、4-松油醇等	我国东北、西南、华东等地区。生长于高山山坡、林下或溪边	功效与缬草相同
胜红蓟（咸虾花） *Ageratum conyzoides* 全草含精油0.44% （菊科）	3-己烯醇、莰烯、乙酸龙脑酯、β-石竹烯、香兰素、β-金合欢烯、6-去甲氧基胜红蓟素、β-荜澄茄烯、γ-榄香烯、胜红蓟素等	原产于南美洲。江西、福建、广东、广西、云南、贵州、四川有大量栽种。生长于山坡、林缘、河边	精油可用于调制香精，全草可作绿肥和鱼花草
细裂青亚菊 *Ajania przewalskii* 全草含精油0.26% （菊科）	1,8-桉叶油素、α-侧柏酮、α-侧柏酮异构体、樟脑、龙脑、4-松油醇、α-松油醇、桃金娘烯醇、cis-辣薄荷醇、cis-葛缕醇、cis-桧醇等	分布于西北地区和四川省等	烟熏驱虫

植物名称	主要化学成分	分布	备注
细叶亚菊 *A. tenuifolia* 全草含精油0.5% （菊科）	α-蒎烯、莰烯、桧烯、β-蒎烯、1,8-桉叶油素、樟脑、龙脑、4-松油醇、α-松油醇、t-乙酸菊酯、香芹酚等	西北各省区和西藏	精油可开发成风油精原料
珠光香青（香青） *Anaphalis margaritacea* 全草含精油0.2% （菊科）	α-蒎烯、β-雪松烯、β-石竹烯、α-愈创木烯、α-石竹烯、γ-愈创木烯、α-姜黄烯、十五烷、δ-愈创木烯、γ-依兰油烯、δ-杜松烯、橙花叔醇等	西北各省区和云南、湖北、湖南等省。生长于海拔300～3400米的半高山石砾地、沟边	精油可用于日化、香烟，入药具止痢、止血、驱虫等功效
阿垻蒿 *Artemisia abaensis* 全草含精油0.3% （菊科）	己醛、4-甲基-3-戊烯醛、苯甲醛、2,5,5-三甲基-1,2,6-庚三烯、桧烯、7-辛烯-4-醇、间伞花烃、1,8-桉叶油素、苯乙醛、蒿酮、丁酸-3-己烯酯、α-侧柏酮、2,6-二甲基-7-辛烯-3-酮、樟脑、龙脑、4-松油醇、吲哚、杜松醇、β-桉叶醇、α-芹子醇等	青海东部、甘肃西南部、四川阿坝地区。生长于湖边、沟边、路旁	
阿克赛蒿 *A. aksaiensis* 全草含精油0.36% （菊科）	3-己烯醛、间伞花烃、1,8-桉叶油素、蒿酮、3,6,6-三甲基-2-降蒎醇、芳樟醇、α-侧柏酮、樟脑、龙脑、4-松油醇、辣薄荷酮、香豆素、β-荜澄茄烯、γ-榄香烯等	甘肃西部阿克塞哈萨克族自治县。生长于海拔3000～3800米的坡地	
碱蒿 *A. anethifolia* 全草含精油0.98% （菊科）	7-辛烯-4-醇、1,8-桉叶油素、桧醇、龙脑、4-松油醇、桃金娘烯醇、对-异丙基苯酚、桧醇异构体、β-芹子烯等	华北、西北和黑龙江。生长于干旱山坡	基生叶可用作茵陈精，精油可提取1,8-桉叶油素

（续表）

植物名称	主要化学成分	分布	备注
莳萝蒿 *A. anethoides* 全草含精油0.6% （菊科）	3-蒈烯、间伞花烃、1,8-桉叶油素、γ-松油烯、4-蒈烯、α-侧柏酮、α-侧柏酮异构体、樟脑、龙脑、4-松油醇、枯茗醇、α-松油醇、桃金娘烯醛、乙酸-α-松油酯、β-金合欢烯等	东北、华北、西北各省区及山东、河南、四川北部。可生长于盐碱地	甘肃产精油可提取1,8-桉叶油素，四川产精油可提辣薄荷酮
奇蒿 *A. anomala* 全草含精油0.21% （菊科）	己醛、2-己烯醛、1,8-桉叶油素、t-氧化芳樟醇（呋喃型）、芳樟醇、蒎葛缕醇、樟脑、龙脑、4-松油醇、桃金娘烯醇、β-榄香烯、β-石竹烯、β-雪松烯、α-姜黄烯、δ-杜松烯、橙花叔醇、α-金合欢醇等	秦岭以南各省区。生长于低海拔林缘、沟边荒地	全草入药，有活血、通经、消热解暑、止痛、消食功效
艾蒿 *A. argyi* 全草含精油0.42% （菊科）	1,8-桉叶油素、α-侧柏酮、α-侧柏酮异构体、侧柏醇、樟脑、龙脑、4-松油醇、α-松油醇、丁香酚、荜澄茄烯、α-芹子醇等	除极旱高寒地区外基本全国都有分布。生长于低海拔荒地山坡	全草入药，有温经去湿、散寒、止血、消炎功效，其栽培种为蕲艾
黄花蒿（青蒿） *A. annua* 全草含精油0.51% （菊科）	1,8-桉叶油素、蒿酮、3,3,6-三甲基-2-降蒎酮、樟脑、戊酸苯甲酯、β-石竹烯、香豆素、β-金合欢烯、别-香树烯、β-芹子烯、对-辛基甲氧基苯、δ-松油醇、青蒿素等	全国各地均有分布。东部生长于海拔1500米，西北、西南均在海拔3000～3650米处	全草入药有清热、解暑、凉血利尿、止汗功效，从中提取的青蒿素为治疟疾专用药。不同地区精油成分有一定差异
暗绿蒿 *A. atrovirens* 全草含精油0.14% （菊科）	己醛、2-己烯醛、7-辛烯-醇-4、1,8-桉叶油素、蒿酮、3,6,6-三甲基-2-降蒎酮、芳樟醇、樟脑、龙脑、萘、4-松油醇、桃金娘烯醇、β-石竹烯、金合欢烯、α-石竹烯、β-荜澄茄烯、橙花叔醇、α-甜没药醇、雅槛蓝烯等	秦岭与黄河流域以南，四川、云南等地。泰国也有分布。生长于低海拔山坡草地、路旁	

植物名称	主要化学成分	分布	备注
高岭蒿（长白山蒿） *A. brachyphylla* 全草含精油0.48% （菊科）	7-辛烯-4-醇、3-辛烯、1,8-桉叶油素、γ-松油醇、3,6,6-三甲基-2-降蒎醇、樟脑、龙脑、4-松油醇、α-松油醇、丁香酚、β-石竹烯、雅槛蓝烯、α-芹子醇等	吉林省东部。朝鲜亦有分布。生长于海拔1100米以上亚高山草甸、森林边缘	
美叶蒿 *A. calophylla* 全草含精油 0.13%~0.25% （菊科）	7-辛烯-4-醇、1,8-桉叶油素、蒿酮、丁酸-3-己烯酯、3,6,6-三甲基-2-降蒎醇、壬醛、桧醇、樟脑、龙脑、4-松油醇、辣薄荷酮、桃金娘烯醛等 7-辛烯-4-醇、1,8-桉叶油素、蒿酮、芳樟醇、桧醇、樟脑、龙脑、4-松油醇、桃金娘醛、马鞭草烯酮、葛缕酮、辣薄荷酮、1,4-对蓋二烯醇-7等	青海南部、广西西部、四川、云南、贵州及西藏东部。生长于海拔1600~3000米的林缘路旁	
茵陈蒿（绵茵陈） *A. capillaris* 全草含精油 0.3%~0.6% （菊科）	丁香酚、丁香酚甲醚、β-石竹烯、α-姜黄烯、茵陈炔、橙花叔醇、丙酸香茅酯、2-甲基丙酸香叶酯、茵陈炔酮、2-甲基丙酸丁香酯、2-甲基丙酸异丁香酯、2-甲基丁酸丁香酯、戊酸丁香酯、6,10,14-三甲基-2-十五酮等	我国南北各省均有分布	精油可配制清凉油、喷雾剂，全草入药有发汗、利尿、解热作用
沙蒿 *A. desertorum* 全草含精油0.45% （菊科）	3-甲基苯酚、桧醇、乙酸香茅酯、茉莉酮、α-姜黄烯、丙酸香茅酯、α-甜没药烯氧化物、α-甜没药醇等	华北、西北和东北及西南各省区。朝鲜、日本、印度亦有分布。生长于海拔4000米草原林缘、草甸	精油具较强抗菌和消炎作用

（续表）

植物名称	主要化学成分	分布	备注
牛尾蒿 *A. dubia* 全草含精油0.27% （菊科）	柠檬烯、1,8-桉叶油素、芳樟醇、樟脑、龙脑、对伞花烃-8-醇、α-松油醇、α-姜黄烯、茵陈炔、丁香酚甲醚等	内蒙古、甘肃、四川、云南、西藏。生长于海拔低于3500米的草原疏林	新稍可当作茵陈用
无毛牛尾蒿 *A. dubia* var. *subdigitata* 全草含精油0.5% （菊科）	己醛、苯甲醛、6-甲基-5-庚烯-2-酮、对伞花烃-8-醇、吲哚、丁香酚甲醚、α-姜黄烯、茵陈炔、榄香脂素、3,4,5-三甲基苯甲醛、榄香脂素异构体、十三醛等	内蒙古、河北、山西、陕西、宁夏、甘肃、青海、山东、河南	
直茎蒿 *A. edgeworthii* 全草含精油 0.32%～0.36% （菊科）	间伞花烃、1,8-桉叶油素、3,3,6-三甲基-2-降蒎醇、t-氧化芳樟醇（呋喃型）、芳樟醇、侧柏醇、樟脑、2,3-二氢苯并呋喃、侧柏醇异构体、4-松油醇、对伞花烃-8-醇、α-松油醇、桃金娘烯醇、香豆素、β-金合欢烯、α-姜黄烯、橙花叔醇等	青海、新疆、云南、西藏等省区。生长于海拔2200～4700米的山坡路旁	青海民间入药当茵陈用
海州蒿 *A. fauriei* 全草含精油0.38% （菊科）	间伞花烃、1,8-桉叶油素、芳樟醇、桧醇、樟脑、龙脑、4-松油醇、桃金娘烯醇、桃金娘烯醛	河北、山东、江苏三省沿海地区。日本、朝鲜亦有分布	地区不同精油成分不尽相同
冷蒿 *A. frigida* 全草含精油0.72% （菊科）	1,8-桉叶油素、蒿酮、3,6,6-三甲基-2-降蒎醇、芳樟醇、α-侧柏酮、α-侧柏酮异构体、樟脑、龙脑、4-松油醇、α-松油醇等	华北、东北、西北及西藏。中亚、北美亦有分布。生长于干旱及半干旱地区的山坡、沙丘	全草入药有止痛、消炎、镇咳作用，亦作茵陈代用品，产地不同精油成分不尽相同

植物名称	主要化学成分	分布	备注
华北米蒿 *A. giraldii* 全草含精油 0.5%～0.6% （菊科）	己醛、6-甲基-5-庚烯-2-酮、苯甲醛、己烯醛、α-蒎烯、香叶醇、1,8-桉叶油素、对伞花烃-8-醇、α-松油醇、β-石竹烯、β-金合欢烯、α-姜黄烯、茵陈炔、α-金合欢烯、茵陈炔酮等	内蒙古、河北、山西、陕西、宁夏、甘肃及四川等省区。生长于海拔1000～2300米的黄土高原	全草入药有清热、解毒、利肺功效
江孜蒿 *A. gyangzeensis* 全草含精油0.36% （菊科）	己醛、2-己烯醛、间伞花烃、柠檬烯、γ-松油烯、4-蒈烯、芳樟醇、樟脑、对伞花烃-8-醇、4-松油醇、香茅醇、β-雪松烯、茵陈炔、γ-榄香烯、β-杜松烯、橙花叔醇、柏木醇等	西藏江孜地区、青海甘肃南部。生长于海拔3900米的山坡上	
盐蒿 *A. halodendron* 全草含精油 0.12%～0.42% （菊科）	β-月桂烯、β-水芹烯、氧化芳樟醇（呋喃型）、樟脑、4-松油醇、α-姜黄烯、茵陈炔、2-甲基丙酸香叶酯、橙花叔醇、α-甜没药醇氧化物、α-甜没药醇等	东北西部和华北、西北各省区。生长于低海拔的沙丘草原、荒漠砾质土	嫩枝叶入药，有止咳祛痰、镇喘、消炎、解表之功效
臭蒿 *A. hedinii* 全草含精油0.46% （菊科）	3-甲基丁酸乙酯、乙酸-2-甲基丁酯、cis-氧化芳樟醇（呋喃型）、芳樟醇、苯乙醇、α-松油醇、乙酸芳樟酯、吲哚、乙酸香叶酯、乙酸香叶酯异构体、β-金合欢烯、γ-芹子烯、2,6-二叔丁基对甲酚、橙花叔醇、1-壬烯-3-醇等	内蒙古和西北各省区。印度、尼泊尔、克什米尔亦有分布。生长于海拔2000～5000米的路旁、坡地林缘等	为青海甘肃民间入药，有清热解毒、凉血、消炎、祛湿等功效

（续表）

植物名称	主要化学成分	分布	备注
五月艾 *A. indica* 全草含精油 0.18%～0.22% （菊科）	苯甲醛、2,2-二甲基己醛、1,8-桉叶油素、芳樟醇、α-侧柏酮、α-侧柏酮异构体、侧柏醇、樟脑、龙脑、4-松油醇、桃金娘烯醛、葛缕醇、乙酸龙脑酯、β-石竹烯、雅槛蓝烯、δ-杜松烯、缬草酮等	除新疆、青海和宁夏外，全国各地均有分布。生长于海拔较低的林缘坡地	全草入药，作"艾"的代用品，有清热解毒、止血、消炎作用。地区不同精油成分有差异
牡蒿 *A. japonica* 全草含精油0.23% （菊科）	3-己烯醛、δ-氧杂二环[15,10]辛烷、辛烷、苯甲醛、9-氧杂-二环[6,1,0]壬烷、1,8-桉叶油素、氧化芳樟醇（呋喃型）、芳樟醇、蒎葛缕醇、樟脑、4-松油醇、龙脑、桃金娘烯醇、α-松油醇、3,4,5-三甲基-2-环戊烯酮、2-乙烯基、2,5-二甲基-4-己烯醛、α-姜黄烯、2,6-二叔丁基对-甲酚、2,2-二甲基-3-（二甲基-1-丙烯基)-环丙酸乙酯、异榄香脂素等	除新疆、青海、内蒙古外全国各地区均有分布。生长于海拔3300米以下的林缘、旷野、山坡、路旁	全草入药，有清热解毒、消暑祛湿、止血、消炎、散瘀功效，可作"青蒿"代用品
柳叶蒿 *Artemisia integrifolia* 全草含精油0.34% （菊科）	己醛、7-辛烯-4-醇、1,8-桉叶油素、苯乙醛、壬醛、樟脑、龙脑、萘、茵陈炔、α-芹子醇等	东北及华北北部。朝鲜、俄罗斯东部亦有分布。多生长于低海拔林缘、路旁	精油有待开发成风油精原料
山艾 *A. kawakamii* 全草含精油0.90% （菊科）	6-甲基-5-庚烯-2-酮、芳樟醇、异龙脑、桃金娘烯醇、葛缕醇、cis-桧醇、乙酸葛缕酯、丁香酚、乙酸龙脑酯、α-姜黄烯、愈创木醇、α-松油醇等	台湾特有种，产于台中、宜兰、苗栗、嘉义等	精油有待开发成风油精原料

（续表）

植物名称	主要化学成分	分布	备注
白苞蒿 *A. lactiflora* 全草含精油0.41% （菊科）	β-石竹烯、α-金合欢烯、α-姜黄烯、β-金合欢烯（Z）、姜烯、β-金合欢烯（E）、橙花叔醇等	南岭山脉以南、四川、云南、贵州。越南、老挝亦有分布。多生长于林缘山谷	全草入药，广东、广西民间用作"奇蒿"代用品，有活血通经、散瘀、治疗肝肾疾病功效
矮蒿 *A. lancea* 全草含精油0.41% （菊科）	2-己烯醛、7-辛烯-4-醇、6-甲基-5-庚烯酮-2、1,2-桉叶油素、芳樟醇、马鞭草烯酮、樟脑、龙脑、4-松油醇、α-松油醇、乙酸葛缕酯、乙酸龙脑酯、β-石竹烯、β-金合欢烯、β-荜澄茄烯、β-芹子醇等	除新疆、青海、宁夏及西藏外全国均有分布。日本、朝鲜、印度也有分布。生长于低海拔林缘、荒坡等	民间用作"艾"和茵陈代用品，有散寒、温经止血、安胎、消炎功效
野艾蒿 *A. lavandulaefolia* 全草含精油0.39% （菊科）	7-辛烯-4-醇、1,8-桉叶油素、芳樟醇、α-侧柏酮、樟脑、龙脑、4-松油醇、α-松油醇、桃金娘烯醇、辣薄荷醇、葛缕醇、β-石竹烯、β-金合欢烯、β-荜澄茄烯等	除新疆、青海及东南沿海、台湾外全国各地均有分布。日本、朝鲜、蒙古、俄罗斯亦有分布。生于低海拔林缘、河滨、草地	全草入药，作用与"艾"相同
白叶蒿 *A. leucophylla* 全草含精油 0.32%～0.36% （菊科）	7-辛烯-4-醇、1,8-桉叶油素、蒿酮、3,6,6-三甲基-2-降蒎醇、α-侧柏酮、樟脑、龙脑、异龙脑、4-松油醇、α-松油醇、桃金娘烯醛、辣薄荷醇、吲哚、榄香脂素、α-芹子醇等	东北、华北、西北、西南各省区。蒙古、朝鲜、俄罗斯亦有分布。生长于路旁、林缘、河岸、湖边	作"艾"代用品，有逐湿、止血、消炎作用

（续表）

植物名称	主要化学成分	分布	备注
大花蒿 A. macrocephala 全草含精油0.23%（菊科）	戊酸、3-甲基己酸、1,8-桉叶油素、芳樟醇、壬醛、α-侧柏酮、苯乙醇、樟脑、2,4-己二烯酸甲酯、龙脑、蒽、松油醇、α-松油醇、2,2-二甲基丙酸香叶酯、2-甲基丁酸香叶酯等	西北、西藏。生长于海拔1500～3400米的干旱、半干旱地区	全草入药，牧区也可用作牲畜饲料
东北牡蒿 A. manshurica 全草含精油0.20%（菊科）	己醛、2-己烯醛、8-氧杂二环[5,1,0]辛烷、苯甲醛、6-甲基-5-庚烯-2-酮、9-氧杂二环[6,1,0]壬烷、柠檬烯、1,8-桉叶油素、蒿酮、8-壬烯-2-酮、樟脑、龙脑、4-松油醇、萘、大茴香醚、乙酸龙脑酯、丁香酚、茵陈炔、2,6-二叔丁基对-甲酚等	黑龙江、吉林、辽宁、河北北部。生长于低海拔湿润地区和山坡林缘	
蒙古蒿 A. mongolica 全草含精油0.18%（菊科）	己醛、2-己烯醛、苯甲醛、7-辛-4-醇、1,8-桉叶油素、2,5,5-三甲基-2、6-庚二烯酮-4、α-侧柏酮、α-侧柏酮异构体、樟脑、龙脑、4-松油醇、辣薄荷酮等	除海南、云南、西藏，全国各地均有栽种。多生长于中、低海拔地区	
小球花蒿 A. moorcoftiana 全草含精油0.85%（菊科）	蒿酮、2,6,6-三甲基-2-降蒎醇、cis-桧醇、樟脑、4-松油醇、龙脑、t-辣薄荷醇、香芹酚、橙花叔醇、桧烯等	甘肃、宁夏、青海、四川、云南。巴基斯坦也有栽种。生长于海拔3000～4800米的亚高山、草原、草甸。生产区域不同精油成分有差异	

（续表）

植物名称	主要化学成分	分布	备注
多花蒿 *A. myriantha* 全草含精油0.16% （菊科）	己醛、2-己烯醛、α-侧柏烯、α-蒎烯、β-松油烯、β-蒎烯、月桂烯、6-甲基-5-庚烯酮-2、间伞花烃、柠檬烯、1,8-桉叶油素、γ-松油烯、芳樟醇、松醇、樟脑、龙脑、4-松油醇、萘、α-松油醇、β-石竹烯、β-金合欢烯、α-姜黄烯、愈创木醇等	山西、甘肃、青海、四川、贵州、云南及广西。印度、不丹、尼泊尔亦有分布。生长于海拔1000～2000米的路旁和灌丛中	云南全草入药，用于消炎。产地不同精油成分有差异
白毛多花蒿 *A. myriantha* var. *pleiocephala* 全草含精油0.47% （菊科）	2-己烯醛、α-水芹烯、间伞花烃、1,8-桉叶油素、α-松油烯、芳樟醇、4-松油醇、枯茗醇、α-松油醇、1,3-对蓝二烯醇-7、α-姜黄烯、雅槛蓝烯、茵陈炔、橙花叔醇、愈创木醇、茵陈炔酮等	青海及西南各省区。印度北部、不丹、尼泊尔亦有分布。生长于中高海拔山坡路旁。	由于生产地区不同精油成分有差异
川西腺毛蒿 *A. occidentali-* *sichuanensis* 全草含精油0.56% （菊科）	2-己烯醛、2,5,5-三甲基-1,3,6-庚三烯、桧烯、7-辛烯-4-醇、6-甲基-5-庚烯-2-酮、间伞花烃、1,8-桉叶油素、芳樟醇、樟脑、龙脑、4-松油醇、α-松油醇、辣薄荷醇、β-石竹烯、β-金合欢烯、β-荜澄茄烯、雅槛蓝烯、β-杜松烯、姜烯、γ-杜松烯、愈创木醇、δ-杜松醇等	四川西部	精油可开发为风油精原料
黑沙蒿 *A. ordosica* 全草含精油0.64% （菊科）	2-己烯醛、α-蒎烯、β-蒎烯、月桂烯、柠檬烯、β-罗勒烯（Z）、β-罗勒烯（E）、4-松油醇、姜黄烯、茵陈炔、橙花叔醇、白菖烯、α-甜没药烯氧化物B、α-甜没药醇等	华北北部及西北各省。生长于海拔1500米以下的荒漠、干旱草原坡地	内蒙古地区民间用于消炎、止血、祛风、消热，牧区也用于喂牲畜

（续表）

植物名称	主要化学成分	分布	备注
黑蒿 *A. palustris* 全草含精油1.20% （菊科）	α-蒎烯、6-甲基-5-庚烯-2-酮、1,8-桉叶油素、α-罗勒烯、蒿酮、α-侧柏酮、樟脑、龙脑、4-松油醇等	东北三省西部、内蒙古河北北部。生于低海拔草地	具开发为天然精油潜力
西南牡蒿 *A. parviflora* 全草含精油0.30%～0.42% （菊科）	己醛、2-己烯醛、8-氧杂二环[5,1,0]辛烷、6-甲基-5-庚烯-2-酮、9-氧杂二环[6,1,0]壬烯、1,8-桉叶油素、蒿酮、芳樟醇、樟脑、龙脑、4-松油醇、吲哚、2,6-二叔丁基对甲酚、橙花叔醇等	甘肃、青海、陕西、湖北、四川、云南、西藏。阿富汗、印度、尼泊尔、缅甸亦有分布。生长于海拔2000～3100米的草丛、坡地	全草入药，有清热、解毒、止血、祛湿功效，可代替"青蒿"
叶苞蒿 *A. phyllobotrys* 全草含精油0.21%～0.38% （菊科）	1,8-桉叶油素、丁酸-3-己烯酯、芳樟醇、α-侧柏酮、α-侧柏酮异构体、龙脑、4-松油醇、枯茗醇、α-松油醇、β-榄香烯、丁香酚甲醚、β-石竹烯、α-姜黄烯、茵陈炔等	青海南部、四川西部。生长于海拔3000～3900米的高山草原	
魁蒿 *A. princeps* 全草含精油0.13%～0.20% （菊科）	己醛、2-己烯醛、苯甲醛、苯乙醛、7-辛烯-4-醇、1,8-桉叶油素、蒿酮、芳樟醇、蒎葛缕醇、丁香酚、樟脑、龙脑、4-松油醇、桃金娘烯醇、2,3-二氢苯并呋喃、别香树烯、α-姜黄烯、β-荜澄茄烯、β-芹子烯、杜松醇等	除干旱地区外全国均有分布。日本、朝鲜亦有分布。生长于低中海拔林缘路旁	民间入药作"艾"代用品，具逐寒、理血气、调经、安胎、止血、消炎功效
红足蒿 *A. rubripes* 全草含精油0.40%～0.50% （菊科）	莰烯、7-辛烯-4-醇、6-甲基庚烯酮-2、1,8-桉叶油素、蒿酮、3,6,6-三甲基-2-降蒎醇、桧醇、樟脑、龙脑、4-松油醇、辣薄荷酮、β-荜澄茄烯、α-芹子烯等	东北、华东、华北各省区。俄罗斯东部、日本、朝鲜等也有分布。生长于低海拔荒地、林缘路旁	全草入药可作"艾"代用品，有温经、散寒、止血作用

（续表）

植物名称	主要化学成分	分布	备注
白莲蒿 *A. sacrorum* 全草含精油0.50% （菊科）	5,5-二甲基-2-呋喃酮、7-辛烯-4-醇、1,8-桉叶油素、樟脑、龙脑、4-松油醇、乙酸龙脑酯、β-桉叶醇、α-松油醇等	全国各地均有分布。日本、印度、朝鲜等亦有分布。生长于低海拔地区	可作"茵陈"代用品，民间以其入药有清热、解毒等功效
猪毛蒿 *A. scoparia* 全草含精油0.88% （菊科）	己醛、苯甲醛、1,8-桉叶油素、蒿酮、丁香酚、丁香酚甲醚、β-石竹烯、β-金合欢烯、β-荜澄茄烯、雅槛蓝烯、茵陈炔、橙花叔醇、茵陈炔酮、2-甲基丙酸丁香酯等	全国均有分布。欧亚大陆都有分布。生长于海拔4000米以下的草原、荒漠、林缘路旁	基生叶、幼苞均入药，用法与茵陈蒿相同，称土茵陈或北茵陈。产地不同精油成分有差异
蒌蒿 *A. selengensis* 全草含精油0.52% （菊科）	7-辛烯-4-醇、1,8-桉叶油素、蒿酮、α-侧柏酮、α-侧柏酮异构体、樟脑、龙脑、4-松油醇、α-松油醇、桃金娘烯醇、马鞭草烯酮、δ-杜松醇、β-桉叶醇、α-芹子醇等	除西北地区、东南南部海边、西藏外，全国其他地区均有产。蒙古、朝鲜亦有分布。生长于湖边、沼泽地	全草入药，治肝炎，有止血、消炎、镇咳、化痰功效
大籽蒿 *A. sieversiana* 全草含精油 0.30%～0.40% （菊科）	7-辛烯-4-醇、1,8-桉叶油素、丁酸3-己烯酯、芳樟醇、α-侧柏酮、α-侧柏酮异构体、樟脑、龙脑、4-松油醇、α-松油醇、辣薄荷醇、橙花醇、茉莉酮、2-甲基丁酸香叶酯、α-芹子醇、2,2-二甲基丁酸香叶酯等	东北、华北、西南高山地区。印度、巴基斯坦亦有分布。生长于干旱、半干旱地区	民间全草入药有消炎、止血功效，精油有防晒功效，牧区用作饲料，种子有果胶，作拉面条用。采摘地不同精油成分有差异

（续表）

植物名称	主要化学成分	分布	备注
柔毛蒿 *A. pubescens* 全草含精油 （菊科）	1,8-桉叶油素、芳樟醇、龙脑、4-松油醇、乙酸龙脑酯、香豆素、α-姜黄烯、雅槛蓝烯、茵陈炔、茵陈炔酮、t-葛缕醇等	东北、华北、四川、西北。蒙古、日本亦有分布。生长于低、中海拔草原、草地	春梢可作茵陈，牧区也有用来喂牲畜
秦岭蒿 *A. qinlingensis* 全草含精油 （菊科）	7-辛烯-4-醇、1,8-桉叶油素、樟脑、龙脑、4-松油醇、丁香酚甲醚、茵陈炔、β-桉叶醇、缬草酮等	湖南西南部、陕西南部、甘肃东部。生长于中低海拔山坡、路旁	全株入药可代替"艾"，也可提取天然樟脑
粗茎蒿 *A. robusta* 全草含精油 （菊科）	己醛、2-己烯醛、3,3-二甲基-6-亚甲基-1,4,6-庚三烯、桧烯、辛烯醇、3-辛醇、间伞花烃、1,8-桉叶油素、樟脑、龙脑、8-对伞花醇、乙酸龙脑酯、β-金合欢烯、β-芹子烯、姜烯、2,6-二叔丁基对-甲酚、δ-杜松烯、愈创木醇、β-甜没药烯醇等	四川、云南。印度北部亦有分布。生长于海拔2200～3500米的山坡路旁	
灰苞蒿 *A. roxburghiana* 全草含精油 （菊科）	5,5-二甲基-2-呋喃酮、1,8-桉叶油素、芳樟醇、α-侧柏酮、樟脑、龙脑、4-松油醇、辣薄荷醇、乙酸龙脑酯、枯茗醇、β-石竹烯等	陕西、青海、湖北及西南各省区。阿富汗、印度、尼泊尔亦有分布。生长于中、高海拔地区干河谷路旁	精油可开发成风油精原料
紫苞蒿 *A. roxburghiana* var. *purpurascens* 全草含精油 （菊科）	己醛、苯甲醛、1,8-桉叶油素、α-侧柏酮、α-侧柏醇异构体、桧醇、樟脑、加州月桂酮、龙脑、4-松油醇、枯茗醇、丁香酚甲醚、茵陈炔、榄香脂素、橙花叔醇等	四川西部及西藏。尼泊尔、巴基斯坦、克什米尔、印度北部和泰国亦有分布	精油富含侧柏酮，可开发成侧柏酮资源

（续表）

植物名称	主要化学成分	分布	备注
苍术 *Atractylodes lancea* 根茎含精油 北苍术： 1.03%～2.24% 南苍术： 3.30%～6.90% （菊科）	北苍术：苍术素、榄香醇、β-芹子烯、苍术酮、茅术醇、β-桉叶醇、香芹二烯醇、白术内酯A等	东北、华北以及山东、河南、陕西等。生长于山坡杂木林下或草地	精油可制成苍术硬脂，可配制皕香玉型香精，亦可入药，有健脾胃、祛湿等功效
	南苍术：β-芹子烯、苍术酮、茅术醇、β-桉叶醇、白术内酯A等	浙江、江苏、江西、山东、湖北、四川等。生长于山坡或灌丛、草丛	精油可配制化妆品香精，根茎入药与北苍术相同
东苍术 *A. japonica* 根茎含精油1.60% （菊科）	榄香醇、β-芹子烯、苍术酮、茅术醇、β-桉叶醇、香芹二烯酮等	东北。生长于山坡柞林下或灌丛	精油药用与北苍术相同
白术 *A. macrocephala* 根茎含精油 0.62%～1.42% （菊科）	β-芹子烯、苍术酮、茅术醇、β-桉叶醇、香芹二烯酮、白术内酯A等	华东、华中和西南有栽培	根茎入药，有补中益气、健脾燥湿之功效
中南蒿 *Artemisia simulans* 全草含精油0.31% （菊科）	1,8-桉叶油素、苯乙醇、β-侧柏酮、龙脑、4-松油醇、对伞花醇-8、辣薄荷醇、丙酸龙脑酯、丁酸龙脑酯、α-姜黄烯、2-甲基丁酸龙脑酯、愈创木醇等	我国中南部、西南部各省区。生长于低海拔山坡、荒地	

（续表）

植物名称	主要化学成分	分布	备注
准噶尔沙蒿 *A. songarica* 全草含精油0.56% （菊科）	α-蒎烯、桧烯、β-月桂烯、柠檬烯、cis-β-罗勒烯（Z）、β-罗勒烯（E）、γ-松油烯、3,4-二甲基-2,4,6-辛三烯、4-松油醇、丁酸-3-己烯酯、茵陈炔、橙花叔醇、α-甜没药醇氧化物、α-甜没药醇等	新疆北部准噶尔盆地。中亚地区亦有分布。生长于沙漠干旱砾质小丘	精油具抑杀仓库蛀虫的活性
西南大头蒿 *A. speciosa* 全草含精油0.40%～0.50% （菊科）	1,8-桉叶油素、蒿酮、丁酸-3-己烯酯、α-侧柏酮、α-侧柏酮异构体、cis-桧醇、樟脑、龙脑、4-松油醇、α-芹子醇、缬草酮、α-甜没药醇等	四川、青海、云南及西藏。生长于海拔3500～3800米的砾质荒坡和路旁	精油可开发提取侧柏酮
圆头蒿 *A. sphaerocephala* 全草含精油0.67% （菊科）	β-月桂烯、柠檬烯、β-罗勒烯、γ-松油烯、4-松油醇、3-壬烯醛、α-石竹烯、茵陈炔、橙花叔醇、α-甜没药醇氧化物B、α-甜没药醇等	华北和西北各省区。生长于海拔1000～2500米的荒漠和沙丘	瘦果入药作消炎和驱虫用
宽叶山蒿 *A. stolonifera* 全草含精油0.27% （菊科）	7-辛烯-4-醇、6-甲基-5-庚烯酮-2、对伞花烃、1,8-桉叶油素、芳樟醇、α-侧柏酮、樟脑、龙脑、4-松油醇、辣薄荷酮、香芹酚、β-石竹烯、β-荜澄茄烯、α-甜没药醇、橙花叔醇等	东北、华北、山东、江苏、安徽、浙江、湖北。日本、朝鲜、中亚亦有分布。多生于低海拔地区林缘路旁、湿润地区	精油可开发配制风油精

植物名称	主要化学成分	分布	备注
线叶蒿 A. subulata 全草含精油0.56% （菊科）	2-己烯醛、7-辛烯-4-醇、1,8-桉叶油素、α-松油烯、γ-松油烯、樟脑、龙脑、4-松油醇、α-松油醇、β-石竹烯、β-芹子烯、α-芹子醇等	东北、华北各省区。朝鲜、日本等亦有分布。多生于低海拔山坡、林缘、河岸、沼泽边缘	具有开发药用精油潜力
阴地蒿 A. sylvatica 全草含精油0.22% （菊科）	己醛、2-己烯醛、7-辛烯-4-醇、1,8-桉叶油素、芳樟醇、樟脑、龙脑、4-松油醇、α-松油醇、桃金娘烯醇、吲哚、β-波旁烯、β-石竹烯、别香树烯、β-荜澄茄烯、β-荜澄茄烯异构体、雅槛蓝烯、β-桉叶醇、缬草酮、α-甜没药醇等	除新疆、福建、广东、海南外全国均有分布。朝鲜、蒙古、俄罗斯东部亦有分布。生长于林缘灌丛荫蔽处	
南野艾 A. verlotorum 全草含精油0.57% （菊科）	7-辛烯-4-醇、3-辛醇、1,8-桉叶油素、蒿酮、樟脑、桃金娘烯醇、葛缕醇、β-石竹烯、金合欢烯、α-石竹烯、雅槛蓝烯等	除极干旱和旱寒地区外均有分布，全球热带、亚热带均有分布。生长于低中海拔地区。采摘地不同精油成分有差异	
毛莲蒿 A. vestita 全草含精油 0.30%～0.40% （菊科）	对伞花烃、1,8-桉叶油素、3,6,6-三甲基-2-降蒎醇、α-侧柏酮、α-侧柏酮异构体、侧柏醇、对异丙基苯酚、辣薄荷醇、橙花叔醇等	西北、西南、湖北西部、广西西北部。生长于中低海拔山坡、草地、灌木丛	全草入药有清热消炎、祛风利尿功效

（续表）

植物名称	主要化学成分	分布	备注
北艾 *A. vulgaris* 全草含精油1.20% （菊科）	7-辛烯-4-醇、1,8-桉叶油素、丁酸-3-己烯酯-α-侧柏酮、α-侧柏酮异构体、桧醇、樟脑、龙脑、4-松油醇、丁香酚甲醚、β-石竹烯、α-石竹烯、β-荜澄茄烯、橙花叔醇等	陕西、甘肃、青海、新疆、四川。蒙古、欧亚大陆也有分布。生长于海拔1500～3500米的路旁、草地、荒坡	
藏龙蒿 *A. waltonii* 全草含精油0.75% （菊科）	1,8-桉叶油素、芳樟醇、龙脑、枯茗醇、间-大茴香醛、丁香酚甲醚、茵陈炔、榄香脂素、茵陈炔酮、金合欢醇等	青海南部、四川西部及西藏。生长于海拔3000～4300米的灌丛、山坡、草原	精油具较强杀菌能力，有待开发成产品。西藏当地用于线香
内蒙古旱蒿 *A. xerophytica* 全草含精油0.54% （菊科）	6-甲基-5-庚烯-2-酮、月桂烯、1,8-桉叶油素、蒿酮、3,6,6-三甲基-2-降蒎醇、α-侧柏酮、α-侧柏酮异构体、cis-桧醇、樟脑、4-松油醇、橙花醇、香叶醇、cis-桧醇、橙花叔醇等	内蒙古及西北各省区。生于戈壁、半荒漠草原、半固定沙丘	为具开发潜力的药用精油资源
云南蒿 *A. yunnanensis* 全草含精油0.26% （菊科）	己醛、2-己烯醛、苯甲醛、7-辛烯-4-醇、1,8-桉叶油素、芳樟醇、α-侧柏酮、α-侧柏酮异构体、龙脑、樟脑、4-松油醇、桃金娘烯醇、苯并呋喃、香豆素等	青海、四川、云南。生长于3700米以下的干河谷或石灰岩地区	其精油具有开发成药用精油的潜力

（续表）

植物名称	主要化学成分	分布	备注
南岭紫菀 *Aster ageratoides* 全草含精油 0.22%～0.30% （菊科）	α-罗勒烯、芳樟醇、δ-榄香烯、β-榄香烯、α-古芸烯、β-石竹烯、α-石竹烯、β-荜澄茄烯、α-榄香烯、δ-杜松烯、β-马榄烯、γ-榄香烯、δ-杜松醇、γ-芹子醇、布黎醇、喇叭茶醇等	广东、贵州	精油富含倍半萜，较为清香，可开发成定香剂
艾纳香 *Blumea balsamifera* 全草含精油0.53% （菊科）	芳樟醇、L-龙脑、β-石竹烯、愈创木烯、β-桉叶醇等	云南、贵州、广西、广东、福建和台湾	叶、根入药有活血祛风、泻痢腹痛等功效，是"艾片"的来源
小蓬草 *Conyza canadensis* 全草含精油0.1%～0.3% （菊科）	α-蒎烯、β-蒎烯、月桂烯、柠檬烯、罗勒烯、石竹烯、金合欢烯、异金合欢烯、β-杜松烯、橙花叔醇、榄香烯等	原产于北美洲。现我国南北均有分布。生长于荒野、旷野、田边	全草入药，煎剂对痢疾、副伤寒有抑制作用
杯菊 *Cyathocline purpurea* 全草含精油0.14% （菊科）	1-辛烯-3-醇、乙酸橙花酯、麝香草酚、丁酸麝香草酯、异丁酸麝香草酯、戊酸麝香草酯等	云南、四川、贵州、广西。生长于山坡林下草地、田边水旁	全草入药，民间用于杀虫，并有消炎解毒、清热、止血等功效
飞机草 *Eupatorium odoratum* 全草含精油0.12% （菊科）	α-蒎烯、桧烯、β-蒎烯、α-罗勒烯、古巴烯、β-石竹烯、β-荜澄茄烯、γ-榄香烯、δ-杜松烯、愈创木醇等	原产于美洲。现云南、海南有分布	全草药用治外伤出血，无名种毒

（续表）

植物名称	主要化学成分	分布	备注
灰白银胶菊 *Parthenium argentatum* 叶含精油1.0% （菊科）	异丁醇、α-蒎烯、莰烯、桧烯、β-蒎烯、1,8-桉叶油素、薄荷脑、水杨酸甲酯、水杨酸丙酯等	原产于美洲。现我国南部有引种	为一种富含橡胶的资源植物
祁州漏芦 *Stemmacantiha uniflora* 干燥头状花序含精油0.02%~0.08% （菊科）	1-十三烯、古巴烯、cis-石竹烯、蛇麻烯、β-荜澄茄烯、十五烷、γ-榄香烯、α-依兰油烯、δ-杜松烯、氧化石竹烯、十六烷、β-桉叶醇、十五醛、六氢金合欢基丙酮等	东北、西北和西藏。生长于山坡、山脚、田埂	根入药清热解毒
紫苞风毛菊 *Saussurea purpurascens* 全草含精油 （菊科）	芳樟醇、β-芹子烯、雅榄蓝烯、γ-杜松烯、β-雪松烯、δ-杜松烯、4-甲基-2,6-叔丁基苯酚、十六烷、β-金合欢醇、β-金合欢醛、2,6-二叔丁基对苯醌、柏木烯醇、β-桉叶醇、δ-杜松醇、十七烷、1-十五烯、二氢去氢二木香内酯等	东北、西北、华东及华南，生长于海拔300~1800米的山坡草地、沟边路旁，日本、朝鲜也有分布	
西北绢蒿 *Seriphidium nitrosum* 全草含精油0.43% （菊科）	3-甲基丁酸乙酯、1,8-桉叶油素、α-侧柏酮、α-侧柏酮异构体、樟脑、龙脑、4-松油醇等	内蒙古西部、甘肃西北部、新疆。生长于海拔1500米以下荒漠草原	精油具抑杀仓库害虫，可开发天然α-侧柏酮资源

（续表）

植物名称	主要化学成分	分布	备注
伊犁绢蒿 S. transiliense 全草含精油 0.36%～0.48% （菊科）	1,8-桉叶油素、α-侧柏酮、α-侧柏酮异构体、cis-桧醇、樟脑、龙脑、4-松油醇、枯茗醇、cis-桧醇等	新疆北部。生长于海拔较低小丘下部砾质或黄土	精油具抑杀仓库害虫的活性
万寿菊 Tagetes erecta 全草含精油0.10% （菊科）	柠檬烯、β-罗勒烯X、β-罗勒烯、异松油烯、4-甲基-6-庚烯-3-酮、烯丙基环己烷、癸二烯醛、万寿菊烯酮、芳樟醇、t-β-石竹烯、乙酸-3-己烯酯、紫苏醛、柠檬醛、龙脑、α-松油醇、外异莰烷酮、乙酸-2-苯基乙酯、枯茗醇、马鞭草烯酮、广藿香烷、香叶醇、α-古巴烯-11-醇等	原产于南美洲墨西哥。全国均由国外引种，花色变化多样	为观赏花卉，花含精油，可配制化妆品、日化香精
细梗香草 Lysimachia capillipes 全草含精油 0.07%～0.10% （报春花科）	2-甲基-2-丁烯醛、3-甲基-2-戊酮、乙酸-1-乙氧基乙酯、2-己烯醛、芳樟醇、壬醛、10-十烯醇、十一醛、植醇等	贵州、云南、四川、湖北、湖南、河南、江西、浙江、广东、福建、台湾。生长于山旁、溪边旷野	全草药用可治流感、感冒、咳喘、月经不调等
灵香草 Lysimachia foenum-graecum 全草含精油0.21% （报春花科）	丁酸丁酯、3,5,5-三甲基己醇、β-蒎烯、丁酸戊酯、庚酸甲酯、紫苏醛、辛酸甲酯、异丁酸香叶酯、2-十一酮、癸烯酸甲酯、古巴烯、香树烯、β-芹子烯、t-β-金合欢烯、α-榄香烯等	广东、广西、云南。生长于林下及山谷阴晴地	精油和浸膏广泛用于烟香，精油配制入药有驱虫、清热、行气等功效

（续表）

植物名称	主要化学成分	分布	备注
宽萼岩风 Libanotis laticalycina 根含精油0.052% （伞形科）	己醛、己醇、庚醛、α-蒎烯、β-蒎烯、2-戊基呋喃、辛醛、t-2-辛烯、1-辛醇、2-壬醛、2-壬烯醛、t-2-癸烯醛、2-十烯醛、cis-石竹烯、花柏烯、去氢白菖烯、氧化石竹烯、蛇麻烯、α-姜黄烯等	山西。生于海拔1600米的山坡上	民间将根作防风之用
尖叶藁本 Ligusticum acuminatum 根含精油 （伞形科）	庚醛、α-蒎烯、桧烯、β-蒎烯、月桂烯、对伞花烃、柠檬烯、γ-松油烯、芳樟醇、薰衣草醇、4-松油醇、马鞭草烯酮、1-（4-甲氧基苯基）乙酮、乙酸龙脑酯、姜黄烯、γ-广藿香烯、3-丁基苯酞、瑟单烯内酯、新蛇床内酯等	云南、四川、湖北、河南、陕西等。生长于海拔1500～3000米的林下	为民间芳香草药
辽藁本 L. jeholense 根茎含精油1.31% （伞形科）	α-蒎烯、β-水芹烯、β-罗勒烯、乙酸松油酯、异松油烯、α-罗勒烯、蛇床酞内酯、4-松油醇、枯茗酸、胡椒烯酮、α-榄香烯、β-榄香烯、愈创木烯、β-愈创木烯、肉豆蔻醚、榄香脂素、亚丁基苯酞、藁本内酯等	我国东北、华北、华东地区。生长于阴坡草丛、山地林缘中，现栽培较多	根茎入药
膜苞藁本 L. oliverianum 根茎含精油 （伞形科）	α-蒎烯、β-蒎烯、月桂烯、对伞花烃、β-水芹烯、t-罗勒烯、薰衣草醇、柠檬醛、β-榄香烯、β-石竹烯、香树烯、肉豆蔻醚、榄香脂素、新蛇床内酯等	我国长江以南各省区均有分布。生长于山坡草丛河滩边，现已有栽种	根茎入药，治风寒头痛、泄泻、神经皮炎

（续表）

植物名称	主要化学成分	分布	备注
长茎藁木 *L. thomsonii* 全草含精油0.20% （伞形科）	3-己烯醇、α-蒎烯、苯甲醛、桧烯、柠檬烯、苯甲醇、苯乙醛、苯乙醇、4-松油醇、丁酸-3-己烯酯、乙酸龙脑酯、麝香草酚、丁香酚甲醚、榄香脂素、橙花叔醇、当归素、欧芹酚甲醚等	甘肃，青海，西藏。生于海拔2200～4200米的林缘、灌丛及草地	民间为芳香草药
藁本 *L. sinense* 根茎含精油 （伞形科）	α-蒎烯、β-蒎烯、月桂烯、对伞花烃、β-水芹烯、t-罗勒烯、薰衣草醇、马鞭草酮、柠檬醛、4-松油醇、β-榄香烯、β-石竹烯、β-金合欢烯、香树烯、α-蛇麻烯、α-姜黄烯、γ-依兰油烯、γ-广藿香烯、β-芹子烯、异肉豆蔻醚、榄香脂素、2-亚丁基酞、新蛇床内酯等	我国长江以南各省区。生长于山坡、草丛或河滩边，现已有栽种	根茎入药治风寒头痛、寒湿腹痛、泄泻、疥癣、神经性皮炎
大头姜活（姜活） *Notopterygium incisum* 全草含精油 0.25%～0.30% （伞形科）	3-己烯醇、β-月桂烯、对伞花烃、β-水芹烯、γ-松油烯、芳樟醇、乙酸龙脑酯、桂酸甲酯等	青海、甘肃、云南、陕西、四川等地。生长于海拔2900～3300米的山坡草丛中	为民间芳香草药
甘松 *Nardostachys chinensis* 根含精油2.5% （败酱科）	马兜铃烯、1(10)马兜铃烯、β-马榄烯、1,2,9,10-四脱氢马兜铃烯、甘松醇等	云南、四川。生长于山地草坡或河边	精油可配制香精

（续表）

植物名称	主要化学成分	分布	备注
桔梗 *Platycodon grandiflorus* 全草含精油0.05% （桔梗科）	己醛、2-戊基呋喃、2-呋喃醛、β-甜没药烯、1-十一炔、5-己烯酸、愈创木酚、2-苯基乙醇、2-己基噻吩、2-乙酸基吡咯、1-吡咯-2-甲醛、大茴香醛、9-癸烯酸丁香酚、丙氧基荷香醚、4-羟基-3-甲氧基苯甲醛等	东北、华北、华东、华中及广东、广西、贵州、云南、四川。生长于海拔2000米的草丛灌丛中	全草供药用，具排脓、祛痰、镇咳作用，花大美丽，为庭园常见花卉
毛麝香 *Adenosma glutinosum* 全草含精油 （车前科）	α-侧柏烯、α-蒎烯、间伞花烃、1,8-桉叶油素、γ-松油烯、芳樟醇、β-石竹烯、α-石竹烯、α-愈创木烯、α-芹子烯、β-甜没药烯、橙花叔醇、α-芹子醇等	云南、广西、广东、江西与福建。生长于海拔2000米以下的荒山或灌林中	全草药用可治风湿骨痛、跌打疮伤、湿疹、蚊虫伤
球花毛麝香 *A. indiana* 全草含精油0.40% （车前科）	α-蒎烯、β-蒎烯、柠檬烯、对伞花烃、1,8-桉叶油素、3-蒈烯、芳樟醇、小茴香酮、4-松油醇、α-松油醇、麝香草酚、t-葛缕醇、d-葛缕酮、α-古巴烯、β-榄香烯、斧柏烯、石竹烯、邻甲基茴香醛、别香树烯、乙酸二氢葛缕酯、α-蛇麻烯、广藿香醇等	广东、广西、云南等省区。生于海拔200～600米的背地干燥山坡	
大叶石龙尾 *Limnophila rugosa* 全草含精油 0.20%～0.43% （玄参科）	爱草脑、t-大茴香醚、cis-大茴香醚、大茴香醛、蛇麻烯、芳樟醇等	广东、云南、台湾。生长于水边	全草入药，性味辛平，用于清热解表，可治感冒，云南少数民族作腌清佐料

（续表）

植物名称	主要化学成分	分布	备注
接骨草 *Justicia gendarussa* 全草含精油0.42% （爵床科）	2-乙基呋喃、3-甲基丁醇、1,3,5-环庚三烯、乙醛、甲基吡嗪、3-己烯醛、3-己烯醇、3,5-二甲基吡嗪、苯甲醛、1-庚烯醇-3、3-辛酮、芳樟醇、3-（2-甲基丙基）-2-环己烯酮、2-羟基苯甲酸甲酯、α-紫罗兰酮、β-紫罗兰酮等	我国南部和西南部。亚洲热带地区也有分布。生长于林旁路边灌丛	
灰毛莸 *Caryopteris forrestii* 全草含精油0.24%～2.00% （马鞭草科）	α-蒎烯、β-蒎烯、β-月桂烯、α-松油烯、对伞花烃、柠檬烯、β-石竹烯、桧烯、莰烯、侧柏烯、δ-杜松烯、异松油烯等	四川、云南、贵州、西藏。生长于1700～3000米的山坡、路旁、荒地	全草民间用药有疏风解表、祛痰止咳、止痛功效，也是提取柠檬烯的原料
兰香草 *C. incana* 全草含精油0.24%～2.00% （马鞭草科）	α-蒎烯、桧烯、β-蒎烯、α-松油烯、β-月桂烯、对伞花烃、柠檬烯、t-β-罗勒烯、β-水芹烯、α-异松油烯、β-石竹烯、α-蛇麻烯、δ-杜松烯等	江苏、安徽、浙江、江西、湖南、湖北、福建、广东、广西。多生长于山坡、路旁	全草药用有疏风解表、祛痰止咳、散瘀止痛功效，可用于蛇咬伤等
毛球莸 *C. trichosphaera* 全草含精油0.24% （马鞭草科）	α-蒎烯、β-蒎烯、对伞花烃、柠檬烯、α-古巴烯、β-石竹烯、β-水芹烯、桧烯、莰烯等	四川西部，云南德钦、香格里拉，西藏昌都。生于海拔2700～3300米的山坡灌丛和河谷	全草民间用药，有疏风解表、祛痰止咳、散瘀止痛功效，也可发展为柠檬烯资源

（续表）

植物名称	主要化学成分	分布	备注
马樱丹 *Lantana camara* 枝叶含精油0.02%（马鞭草科）	1,8-桉叶油素、γ-松油烯、芳樟醇、樟脑、龙脑、古巴烯、α-石竹烯、β-荜澄茄烯、γ-榄香烯、δ-杜松烯、橙花叔醇、α-依兰油烯、雅槛蓝烯、β-古芸烯等	原产于美洲热带地区。我国台湾、广东、广西有栽种	根、叶、花药用，具清热解毒、散结止痛、祛风止痒等功效
藿香 *Agastache rugosa* 全草含精油0.28%（唇形科）	β-广藿香烯、β-石竹烯、α-愈创木烯、α-广藿香烯、β-愈创木烯、雅槛蓝烯、δ-愈创木烯、广藿香醇等	我国各地广泛分布，常见栽培	精油为一种名贵的天然香料，香气持久，多用于定香剂，茎叶为芳香健胃药，治感冒寒热头痛，水煎剂对钩端螺旋体具抑制作用
野香草 *Elsholtzia cyprianii* 全草（干）含精油0.81%（唇形科）	3-辛醇、1-辛烯-5-醇、苯甲醛、芳樟醇、苯乙酮、香薰酮萘、t-石竹烯、β-金合欢烯、β-去氧香薰酮、麝香草酚、氧化石竹烯等	陕西、河南、安徽、湖北、湖南、贵州、四川、广西、云南等省区。生于海拔400～2900米的田边路旁、河岸、林缘	全草入药，用于治疗感冒、疔疮等
益母草 *Leonurus japonicus* 全草含精油0.083%（唇形科）	1-辛烯-3-醇、3-辛醇、芳樟醇、古巴烯、cis-石竹烯、t-石竹烯、蛇麻烯、γ-榄香烯、δ-杜松烯、石竹烯氧化物、植物醇等	全国各地均有分布。生长于山野、旷野、河滩、草丛、溪边	全草入药有调经活血、祛瘀新生、利尿消肿功效

植物名称	主要化学成分	分布	备注
大花益母草 *L. macranthus* 全草含精油0.05% （唇形科）	1-辛烯-3-醇、古巴烯、β-波旁烯、t-石竹烯、β-荜澄茄烯、蛇麻烯、γ-榄香烯、金合欢基丙酮、植醇等	辽宁、吉林及河北。生长于海拔400米的草坡灌丛	精油有一定活性，具平喘作用，可治支气管炎
细叶益母草 *L. sibiricus* 全草含精油0.125% （唇形科）	1-辛烯-3-醇、芳樟醇、古巴烯、cis-石竹烯、t-石竹烯、蛇麻烯、γ-榄香烯、δ-杜松烯、石竹烯氧化物、植物醇等	内蒙古、山西、陕西、河北等省，蒙古中亚也有分布。生长于砂质草地、松林中	作用与大花益母草相同
辣薄荷 *Mentha piperita* （唇形科）	薄荷醇、薄荷酮、α-蒎烯、水芹烯、柠檬烯、1,8-桉叶油素、γ-松油烯、胡薄荷酮等	原产于欧洲。北京、南京有引种	可用于提取薄荷醇原料
留兰香 *M. spicata* 全草含精油0.60%～0.90% （唇形科）	柠檬烯、薄荷酮、薰衣草醇、薄荷醇、二氢葛缕酮、葛缕醇、葛缕酮、β-波旁烯、β-石竹烯、4-松油醇、t-二氢-葛缕酮、乙酸葛缕酯等	原产于南欧加那利群岛、马德拉群岛。河北、江苏、浙江、广东、广西、四川、云南、贵州有引种	全草入药，民间用于治疗感冒、中暑、头痛，外用治疗热痱皮炎、湿疹、疮疖、痔瘘下血等
姜味草 *Micromeria biflora* 全草含精油0.60% （唇形科）	β-蒎烯、柠檬烯、1,8-桉叶油素、薄荷酮、薄荷醇、异辣薄荷酮、异蒎莰酮、丁香酚甲醚、肉豆蔻醚、榄香脂素等	云南、贵州等省。生长于海拔3000～4000米的山坡、草丛、河流冲击处	全草药用，有祛风解表、温中除湿功效，用于治疗感冒咳嗽、肠炎、腹痛

（续表）

植物名称	主要化学成分	分布	备注
香薷 *Elsholtzia ciliata* 全草含精油0.08% （唇形科）	α-蒎烯、β-蒎烯、α-水芹烯、对伞花烃、γ-松油烯、4-松油醇、麝香草酚、香荆芥酚、t-α-香柠檬烯、蛇麻烯、β-甜没药烯等	除新疆和青海外遍布全国各地。生长于路旁、山坡、荒地、林内、河边	全草入药，民间用以治瘫痪、痨伤吐血、感冒、疮毒等
海州香薷 *E. splendens* 全草含精油0.70% （唇形科）	α-蒎烯、7-辛烯-4-醇、对伞花烃、t-罗勒烯、β-松油醇、4-松油醇、麝香草酚、β-石竹烯、β-金合欢烯等	辽宁、河北、河南、山东、江苏、浙江、江西、广东等。生长于草丛、沟边、山坡、路边	
木香薷 *E. stauntoni* 叶含精油1.30% （唇形科）	α-蒎烯、1,8-桉叶油素、苯乙酮、芳樟醇、4-松油醇、t-石竹烯、cis-石竹烯、δ-松油烯、α-松油醇	河北、河南、山西、陕西、甘肃。生长于海拔700~1600米的谷地、溪旁、河川沿岸	精油可治痢疾、肠胃炎、感冒
薰衣草 *Lavandula angustifolia* 全草含精油2.0%~2.30% （唇形科）	α-蒎烯、莰烯、β-蒎烯、β-月桂烯、柠檬烯、1,8-桉叶油素、芳樟醇、薰衣草醇、樟脑、4-松油醇、乙酸芳樟酯、β-石竹烯等	原产于地中海地区。现在我国多个地方有栽种	精油为配制化妆品、皂用、花露水重要原料，药用有消炎、防腐、镇痛、利尿作用
宽叶薰衣草 *L. latifolia* 全草含精油1.20%~1.70% （唇形科）	α-蒎烯、莰烯、β-蒎烯、2-辛酮、柠檬烯、1,8-桉叶油素、芳樟醇、樟脑、乙酸异丁酯、龙脑、α-松油醇、乙酸芳樟酯、苯甲酸、苯甲酯等	原产于法国、意大利、西班牙、保加利亚、巴尔干半岛。我国有引种，江苏、山东为宜	用途与薰衣草相同，为香料工业常用的植物精油

（续表）

植物名称	主要化学成分	分布	备注
石香薷 *Mosla chinensis* 全草含精油 0.40%～0.60% （唇形科）	α-蒎烯、莰烯、桧烯、β-蒎烯、β-月桂烯、间伞花烃、柠檬烯、乙酸龙脑酯、β-石竹烯、α-石竹烯、雅槛蓝烯、β-桉叶醇等	华东、中南及西南各省区。生长于海拔1400米以下的草坡、林下	茎叶入药治感冒、中暑、呕吐、腹痛、泄泻，外用治跌打瘀痛、湿疹
小鱼仙草 *M. dianthera* 全草含精油0.64% （唇形科）	月桂烯、柠檬烯、龙脑、薄荷酮、葛缕醇、t-石竹烯、2,6-二甲基-6-（4-异己烯基）二环[3,1,1]庚-2-烯、β-石竹烯、蛇麻烯、t-β-金合欢烯、cis-β-金合欢烯等	华东、中南、西南。生长于海拔176～2300米的山坡、路旁、水边	全草入药，民间用于感冒、发热、中暑、头痛、恶心等
台湾荠苧 *M. formosana* 全草含精油 0.18%～2.28% （唇形科）	莳萝脑、榄香脂素、葛缕醇、对伞花烃、葛缕酮、α-侧柏烯、柠檬烯、石竹烯等	台湾	
石荠苧 *M. scabra* 全草含精油 2.57%～3.50% （唇形科）	芳樟醇、2-甲基-5-（1-甲基乙基）-2，5-环己二烯1，4-二酮、香芹酚、麝香草酚、乙酸香芹酯、乙酸麝香草酯、古巴烯、β-石竹烯、α-佛手烯、α-石竹烯、β-荜澄茄烯异构体、α-金合欢烯（Z，E）、芹菜脑、芹菜脑异构体等	华东、中南以及四川、甘肃、陕西、辽宁等地。生长于山坡路旁	全草入药，具清暑热、祛风湿、消肿、解毒等功效
荆芥 *Nepeta cataria* 全草含精油 （唇形科）	7-辛烯-4-醇、柠檬烯、β-罗勒烯、α-罗勒烯、芳樟醇、苯乙酮、葛缕酮、α-柠檬醛、β-石竹烯、α-石竹烯、β-荜澄茄烯、α-金合欢烯、橙花叔醇、β-桉叶醇等	新疆、甘肃、陕西、河南、山西、山东、湖北、贵州、四川及云南。多生长于海拔2500米以下的宅旁灌丛中	全草入药，可治伤风感冒、头痛、咽喉肿痛、结膜炎等

（续表）

植物名称	主要化学成分	分布	备注
罗勒 *Ocimum basilicum* 全草含精油 0.90%~1.20% （唇形科）	α-罗勒烯、芳樟醇、小茴香醇、大茴香醚、丁香酚甲醚、丁香酚、α-石竹烯、α-甜没药烯等	新疆、吉林、河北、浙江、安徽、江西、湖北、湖南、广东、广西、台湾、云南等多为栽培	为重要栽培芳香植物，精油可调配皂用、牙膏用及化妆品用香精
丁香罗勒 *O. gratissimum* 全草含精油 0.80%~1.20% （唇形科）	丁香酚、古巴烯、β-石竹烯、β-波旁烯、β-榄香烯、β-荜澄茄烯、γ-依兰油烯、γ-依兰油烯异构体、β-荜澄茄烯异构体、γ-杜松烯、δ-杜松烯等	广东、广西、福建、浙江、江苏、上海等地区有栽种	精油是单离丁香酚重要的资源植物，也可用于调配日用香精原料
毛叶丁香罗勒 *O. gratissimum* var. *suave* 全草含精油 0.64%~1.27% （唇形科）	β-罗勒烯、α-罗勒烯、4-松油醇、丁香酚、β-石竹烯、β-荜澄茄烯等	江苏、浙江、福建、台湾、广东、广西、云南有栽种	精油可用于食品、医药及化工原料，也是单离丁香酚的资源植物
甘牛至 *O. marjorana* 全草含精油 0.20%~0.50% （唇形科）	α-蒎烯、莰烯、桧烯、β-蒎烯、间伞花烃、1,8-桉叶油素、cis-氧化芳樟醇（呋喃型）、t-氧化芳樟醇（呋喃型）、芳樟醇、樟脑、龙脑、4-松油醇、α-松油醇、乙酸芳樟酯等	原产于地中海沿岸及西亚。现广东、广西及上海有引种。法国、德国等地有产	常用于西式烹饪、酒类风味添加剂和肉食调味料

（续表）

植物名称	主要化学成分	分布	备注
紫苏 *Perilla frutescens* 全草含精油0.30% （唇形科）	6-甲基-5-庚烯酮-2、cis-氧化芳樟醇（呋喃型）、t-氧化芳樟醇（呋喃型）、芳樟醇、香茅醇、β-柠檬醛、香叶醇、α-柠檬醛、乙酸香叶酯、β-石竹烯、α-金合欢烯（Z、E）、橙花叔醇等	全国各地均有栽培	精油为提取柠檬醛原料，亦可直接用于调配皂用、食品用香精，全草入药治风寒、咳嗽、痰喘、便秘等，种子油可调节血液浓度
回回苏（鸡冠苏） *P. frutescens* var. *crispa* 全草含精油0.20%~0.26% （唇形科）	柠檬烯、芳樟醇、紫苏醛、紫苏醇等	全国各地均有栽种	茎叶入药治感冒、恶寒发热、胸腹胀满气郁、食滞等症
广藿香 *Pogostemon cablin* 全草含精油1.50% （唇形科）	广藿香醇、苯甲醛、丁香酚、桂醛、广藿香奠醇、广藿香吡啶、石竹烯、β-榄香烯、α-古芸烯、α-广藿香烯等	福建、台湾、广东、广西	精油具有强且浓的香气，是一种优良定香剂，茎叶入药，可健胃镇呕、解热等
夏枯草 *Prunella vulgaris* 全草含精油0.31% （唇形科）	α-蒎烯、β-蒎烯、月桂烯、α-水芹烯、1,8-桉叶油素、芳樟醇、1-壬烯醇-4、薄荷酮、1-对盖烯醇-8、胡椒酮、乙酸-1-对盖-8-基酯、乙酸芳樟酯、δ-榄香烯、乙酸香叶酯等	几乎产于全国各地林边、草地、湿地、田埂、路旁	全草入药，主治眼赤肿痛、甲状腺肿大、淋巴结核、乳腺增生、高血压等症

（续表）

植物名称	主要化学成分	分布	备注
迷迭香 *Rosmarinus officinalis* 全草含精油 0.48%～0.52% （唇形科）	α-蒎烯、莰烯、β-蒎烯、对伞花烃、1,8-桉叶油素、樟脑、龙脑、α-松油醇、乙酸龙脑酯、β-石竹烯等	原产于欧洲和北非，现我国有引种	全草入药，有健胃、发汗功效，精油对金黄色葡萄球菌、大肠杆菌有中度抑制
香紫苏 *Salvia sclarea* 全草含精油 0.10%～0.12% （唇形科）	莰烯、β-月桂烯、α-罗勒烯、芳樟醇、α-松油醇、乙酸芳樟酯、乙酸香叶酯、乙酸香叶酯异构体、乙酸龙脑酯、β-石竹烯、β-荜澄茄烯等	原产于欧洲。现陕西、河南、河北等地有栽培	花（干）广泛用于日用、食品中
山藿香（血见愁） *Teucrium viscidum* 全草含精油 （唇形科）	桧烯、β-蒎烯、1,8-桉叶油素、古巴烯、β-蒎烯、β-榄香烯、β-石竹烯、α-佛手烯、α-石竹烯、香树烯、β-荜澄茄烯、雅槛蓝烯、橙花叔醇等	长江以南各省区。日本、朝鲜和东南亚也有分布	全草入药，有凉血解毒、祛瘀生新等功效，治跌打疮毒、蛇咬伤
百里香 *Thymus mongolicus* 全草含精油 0.20%～0.50% （唇形科）	α-蒎烯、莰烯、7-辛烯-4-醇、β-月桂烯、2-蒈烯、间伞花烃、1,8-桉叶油素、γ-松油烯、芳樟醇、樟脑、龙脑、4-松油醇、对-叔丁基苯甲醇、麝香草酚、乙酸麝香草酯、香树烯异构体、β-甜没药烯、2-羟基-5-甲氧基苯乙酮、香树烯等	甘肃、青海、陕西、山西、山东、河北、内蒙古。生长于海拔1100～3600米的山地、沟谷、杂草丛	精油用于配制牙粉、爽身粉、香皂、洗涤等日用香精，具强杀菌能力

（续表）

植物名称	主要化学成分	分布	备注
地椒 *Thymus quinquecostatus* 全草含精油0.21%~0.58% （唇形科）	芳樟醇、α-蒎烯、莰烯、对伞花烃、1-辛烯-3-醇、麝香酚甲醚、龙脑、石竹烯、杜松烯、麝香草酚、香荆芥酚等	山东、辽宁、河北、河南及山西。生于海拔600~900米的山坡或海边低丘中	精油用于调配化妆品香精，全草入药用于止咳、消炎，治感冒、关节痛，亦用于提取麝香草酚和芳樟醇
云南草蔻 *Alpinia blepharocalyx* 种子含精油0.11% （姜科）	α-侧柏烯、β-蒎烯、2-羟基-3-甲基戊酸甲酯、间伞花烃、cis-氧化芳樟醇（呋喃型）、t-氧化芳樟醇（呋喃型）、芳樟醇、柠檬醛、α-松油醇、古巴烯、香叶醇、乙酸香叶酯、乙酸苯丙酯、金合欢醇等	云南南部和西部。生长于100~1000米的疏林中	果实民间入药有温胃健脾之功效，治腹冷痛、痞满吐酸、反胃寒湿、吐泻等
红豆蔻 *A. galanga* 根茎含精油 （姜科）	3-蒈烯、1,8-桉叶油素、龙脑、间大茴香醛、大茴香醚、桂酸乙酯、十五烷、cis-对甲氧基桂酸乙酯、t-对甲氧基桂酸乙酯等	广东、广西、云南等省区。广泛被栽种	根茎常作食品调味剂，入药用于行气止痛、止腹冷痛等
桂南山姜 *A. guinanensis* 种子含精油0.20% （姜科）	间伞花烃、1,8-桉叶油素、龙脑、t-葛缕醇、桃金娘烯醇、环癸醇、2-癸醇、乙酸龙脑酯、金合欢醇、喇叭茶醇等	广西隆安县。生长于石灰岩山坡灌丛中	种子精油具较强抑菌活性

（续表）

植物名称	主要化学成分	分布	备注
小草蔻 A. henryi 种子含精油0.09% （姜科）	月桂烯、间伞花烃、1,8-桉叶油素、芳樟醇、龙脑、4-苯基-2-丁酮、α-古巴烯、4-苯基-3-丁烯-2-酮、δ-杜松烯、喇叭茶醇、金合欢醇等	广东、云南、广西有分布。生长于山地密林中	广东、广西、云南等民间用种子代白豆蔻
山姜 A. japonica 种子含精油0.10% （姜科）	α-蒎烯、莰烯、β-蒎烯、1,8-桉叶油素、芳樟醇、樟脑、4-松油醇、α-松油醇、乙酸龙脑酯、石竹烯、邻烯丙基甲苯、蛇麻烯、愈创木烯、α-榄香烯、橙花叔醇等	我国东南、西南和南部各省区。生长于林下阴湿处	种子精油具抑菌活性，根茎民间入药用于治活脘冷霜、风湿骨痛、劳伤吐血
海南山姜（草蔻） A. hainanensis 种子含精油0.12% （姜科）	对伞花烃、1,8-桉叶油素、芳樟醇、樟脑、龙脑、4-松油醇、α-松油醇、3-苯基-3-丁烯酮-2、3-苯基-2-丁酮、β-金合欢烯、α-石竹烯、β-杜松烯、橙花叔醇、胡萝卜醇、杜松醇、金合欢醇等	广东、广西。生长于山地疏林或密林下	种子入药，民间用于治心腹冷痛、食滞寒湿、吐泻
假益智 A. maclurei 种子含精油0.09% （姜科）	β-蒎烯、间伞花烃、蒎葛缕醇、5-（1-甲基乙基）二环[3,1,0]己-2-酮、对异丙基苯甲醛、乙酸龙脑酯、2-丙烯基-3-亚甲基-4-环己烯醇、α-古巴烯、苯丙酸乙酯等	广东、广西、云南，越南亦有分布。生长于山地疏林或密林中	民间用于代益智
毛瓣山姜 A. malaccensis 种子含精油0.03% （姜科）	1,8-桉叶油素、香茅醇、4-苯基-3-丁烯酮-3、癸酸、乙酸香叶酯、橙花叔醇、十二酸、β-金合欢醇、α-金合欢醇、肉豆蔻酸、棕榈酸等	西藏、云南西双版纳、广东有栽培，生长于常绿阔叶林下，东南亚有分布	种子民间入药，具止痛、消食、催吐功效，常用于治疗胃肠疾病

（续表）

植物名称	主要化学成分	分布	备注
华山姜 *A. oblongifolia* 种子含精油 0.08%～0.11% （姜科）	α-蒎烯、β-蒎烯、柠檬烯、罗勒烯、芳樟醇、β-柠檬醛、香叶醇、α-柠檬醛、乙酸香叶酯、邻烯丙基甲苯、蛇麻烯、橙花叔醇、金合欢醇、乙酸金合欢酯等	东南部和西南部各省区。生长于海拔100～2500米的林阳下	种子精油具有强烈抑菌活性，民间入药治胃痛胀闷、嗳嗝、腹痛泄泻等
益智 *A. oxyphylla* 种子含精油2.00% （姜科）	α-侧柏烯、β-蒎烯、α-蒎烯、β-月桂烯、α-水芹烯、3-蒈烯、对伞花烃、4-松油醇、桃金娘烯醛、雅槛蓝烯等	广东、广西，近年云南、福建有栽种。生长于林下阴湿处	种子入药具温脾、止泻、摄唾、暖胃、固精、缩尿等功效
滑叶山姜 *A. tonkinensis* 种子含精油0.14% （姜科）	α-蒎烯、莰烯、β-蒎烯、柠檬烯、罗勒烯、2,6-龙脑二醇、t-蒎葛娄醇、桃金娘烯醛、桃金娘醇、乙酸龙脑酯、麝香草酚、乙酸香叶酯、广藿香烯等	广西	民间以种子入药作为山姜代用品
艳山姜 *A. zerumbet* 种子含精油 0.30%～0.50% （姜科）	α-蒎烯、β-蒎烯、对伞花烃、1,8-桉叶油素、α-侧柏醇、α-古巴烯、t-石竹烯、cis-t-金合欢醇等	我国东南至西南均有分布	民间以种子入药，有燥湿祛寒，健脾暖胃功效
红壳砂仁 *Amomum aurantiacum* 种子含精油 0.08%～1.20% （姜科）	α-蒎烯、β-蒎烯、芳樟醇、香叶醇、β-金合欢烯、橙花叔醇、广藿香醇等	云南。生长于海拔600米的山坡上、林下	民间以果实入药，具芳香健胃功效

（续表）

植物名称	主要化学成分	分布	备注
三叶豆蔻 *A. austrosinense* 种子含精油 0.60%～0.80% （姜科）	α-蒎烯、β-蒎烯、1,8-桉叶油素、柠檬烯、芳樟醇、α-松油醇、橙花叔醇等	广西、湖南。生长于海拔450～1000米的山地	民间以全草入药，治风湿骨痛、跌打肿痛、胃寒
海南砂仁 *A. longiligulare* 种子含精油1.47% （姜科）	柠檬烯、樟脑、异龙脑、龙脑、α-松油醇、乙酸龙脑酯、α-甜没药烯、橙花叔醇等	海南省澄迈、三亚、儋州，广东徐闻、遂溪有栽培。生长于山谷密林	果实可入药，有芳香健胃之效
九翅豆蔻 *A. maximum* 种子含精油0.50% （姜科）	1,8-桉叶油素、芳樟醇、蒎葛缕醇、樟脑、异龙脑、龙脑、t-氧化芳樟醇（呋喃型）、4-松油醇、枯茗醇、α-松油醇、桃金娘烯醇、乙酸龙脑酯、香兰素、橙花叔醇、植醇等	云南、西藏、广东、广西。生长于林中阴湿处	果肉可食，民间亦可代替砂仁、豆蔻
牛牯缩砂 *A. muricarpum* 叶含精油0.08% （姜科）	α-蒎烯、β-蒎烯、芳樟醇、α-龙脑烯醛、蒎葛缕醇、龙脑、桃金娘烯醇、马鞭草烯酮、t-葛缕醇、乙酸龙脑酯、喇叭茶醇等	广东、广西。菲律宾亦有分布。生长于海拔300～1000米的密林中	叶精油具中等抑菌活性
波翅豆蔻 *A. odontocarpum* 种子含精油 1.20%～1.40% （姜科）	α-蒎烯、β-蒎烯、β-月桂烯、1.8桉叶油素、γ-杜松烯、芳樟醇、月桂烯醇、4-松油醇、α-松油醇、γ-杜松烯等	广西。生长于海拔1500米的山坡疏林中	果实可供药用，代替豆蔻

（续表）

植物名称	主要化学成分	分布	备注
香豆蔻 *A. subulatum* 种子含精油 0.20%～0.30% （姜科）	α-蒎烯、β-蒎烯、β-月桂烯、1,8-桉叶油素、芳樟醇、4-松油醇、α-松油醇、桃金娘烯醇、甜没药烯、橙花叔醇等	西藏、云南、广西。生长于海拔300～1300米的林中阴湿地	果实为我国传统中药，有健胃、镇痛功效，亦可作调味品
草果 *A. tsaoko* 种子含精油 0.70%～0.90% （姜科）	α-蒎烯、α-水芹烯、芳樟醇、4-松油醇、α-松油醇、柠檬醛、香叶醇、2-苯基-2-丁烯醛、乙酸桧酯、α-柠檬醛、2-癸烯醛、1-乙基丙基苯、2-甲基桂醛、乙酸香叶酯、2-十二烯醛、橙花叔醇等	云南、广西、贵州等省区。野生于疏林下，多为栽种	果实为常用调味料并对食品有保鲜作用，也入药用于祛寒湿，内阻脘腹痛
砂仁（阳春砂仁，长泰砂仁） *A. villosum* 种子含精油 2.50%～3.00% （姜科）	α-蒎烯、莰烯、β-月桂烯、柠檬烯、樟脑、异龙脑、龙脑、乙酸龙脑酯、α-古巴烯、橙花叔醇等	福建、广东、广西和云南。野生在山地林中，多为人工栽种	果实为我国传统芳香健胃药，用于行气调中、和胃、食滞，叶精油入药功效与果实相似
缩砂密（绿壳砂） *A. villosum* var. *xanthioides* 种子含精油3.00% （姜科）	柠檬烯、樟脑、异龙脑、乙酸龙脑酯、古巴烯、α-佛手烯、α-蒎烯、β-蒎烯、莰烯、芳樟醇、4-松油醇等	云南南部，广东也栽培。生长于海拔600～800米的林下潮湿处	果实入药功效与砂仁相同
郁金（姜黄） *Curcuma aromatica* 块根含精油6.10% （姜科）	1-α-姜黄烯、1-β-姜黄烯、莰烯、樟脑、脱甲氧基姜黄素、莪术醇、芳姜黄酮、葛缕酮、水芹烯等	我国东南部至西南各省区。野生长于林下，多为栽培	块基入药，用于行气、解郁、凉血、破瘀

（续表）

植物名称	主要化学成分	分布	备注
温郁金 *C. wenyujin* 根茎含精油 （姜科）	1,8-桉叶油素、异龙脑、龙脑、樟脑、石竹烯、丁香酚、芳姜黄烯、姜黄烯、姜烯、莪术醇、莪术酮、姜黄酮、大香叶酮、莪术二酮等	浙江瑞安。栽培于上层深土排水良好的砂质土壤	精油中所含莪术醇对宫颈癌有一定疗效
广西莪术 *C. kwangsiensis* 根茎含精油 （姜科）	α-蒎烯、莰烯、β-蒎烯、柠檬烯、1,8-桉叶油素、异龙脑、龙脑、松油醇、丁香酚、芳姜黄烯、姜黄烯、莪术醇、莪术酮、姜黄酮、芳樟油酮、牻牛儿酮、莪术二酮等	广西、云南。野生于山坡草地或灌木丛，也有栽培	精油中所含莪术醇对宫颈癌有一定疗效
姜黄 *C. longa* 根茎含精油 （姜科）	α-蒎烯、β-蒎烯、1,8-桉叶油素、松油烯、石竹烯、芳姜黄烯、姜黄烯、姜油烯、莪术醇、莪术酮、姜黄酮、芳姜油酮、牻牛儿酮、莪术二酮等	台湾、福建、广东、广西、云南、西藏。多生长于向阳处，或多有栽培	精油中莪术醇对宫颈癌有一定疗效，根茎能破血行气，也有用于治疗高血压以及一些妇科病
莪术 *C. phaeocaulis* 根茎含精油 （姜科）	α-蒎烯、莰烯、β-蒎烯、1,8-桉叶油素、龙脑、樟脑、乙酸龙脑酯、石竹烯、丁香酚、芳姜黄烯、姜烯、莪术酮、芳姜酚、姜黄酮、脱水莪术酮等	台湾、福建、江西、广东、广西、四川、云南等省区。野生长于林下荫处，有栽培	精油具抗菌活性，根茎入药用于行气破血、消积、止血，并治宿食不消、跌打损伤等

（续表）

植物名称	主要化学成分	分布	备注
山黄姜 C. zedoaria 根茎含精油，种子含精油1.50% （姜科）	1,8-桉叶油素、3,6-二甲基-1,6-辛二烯醇-3、樟脑、异龙脑、龙脑、4-松油醇、α-松油醇、（＋）-蒈烯-cis-醇-4、α-石竹烯、β-石竹烯、姜黄烯酮、莪术酮、吉玛酮、姜黄烯醇等	福建、广东、广西、台湾、四川、云南等省区。野生长于山谷林边，有栽培	根茎入药，有行气破血、消积止痛效果
红茴砂 Etlingera littoralis 叶含精油0.25%～0.30% （姜科）	α-蒎烯、莰烯、β-蒎烯、1,8-桉叶油素、芳樟醇、α-松油醇、α-龙脑烯醛、蒎葛缕醇、乙酸龙脑酯、橙花叔醇等	广东。马来西亚亦有分布。生长于海拔200～300米的林下	叶精油具中药型香气，有待开发
姜花 Hedychium coronarium 花含精油 （姜科）	α-蒎烯、桧烯、β-蒎烯、β-月桂烯、1,8-桉叶油素、d-罗勒烯、β-罗勒烯、苯甲酸甲酯、芳樟醇、cis-氧化芳樟醇（呋喃型）、t-氧化芳樟醇（呋喃型）、α-松油醇、α-金合欢烯、3-（4,8-二甲基-3,7-壬二烯基）呋喃等	台湾、广东、湖南、广西、云南、四川等省区。生长于林区，较多栽培	花香宜人，为华南重要芳香切花花卉
姜 Zingiber officinale 块茎含精油0.10%～0.2% （姜科）	三环烯、α-蒎烯、莰烯、β-蒎烯、桧烯、4-蒈烯、月桂烯、柠檬烯、β-水芹烯、1,8-桉叶油素等	我国中部东南部至西南各省区均有栽种	姜油可作食品和化妆品香精原料，根茎为常用调味香料，供药用有抑菌、祛风、发汗功效

（续表）

植物名称	主要化学成分	分布	备注
大蒜 *Allium sativum* 鳞茎含精油0.50% （百合科）	二烯丙基二硫醚、甲基烯丙基三硫醚、3-乙烯基-1,4-甲基-1,2-二硫杂-3-环戊烯、二烯丙基、3-乙烯基-1,2-二硫杂-5-环己烯、二烯丙基三硫醚、甲基烯丙基四硫醚、3-乙烯基-1,2-二硫杂-4-环己烯等	原产于亚洲西部及欧洲。现我国各地均有栽种	可作蔬菜、调味品，其精油多用于制作罐头食品，可预防动脉硬化，精油有杀菌功效，药用具利尿、祛痰、镇静、消炎等功效
菖蒲 *Acorus calamus* 全草含精油1.50%～3.20% （天南星科）	β-细辛醚、白菖烯、白菖醇、细辛酮、莰烯、β-蒎烯、细辛醛等	我国各省区均有分布。生长于海拔260米以下的水边、沼泽、湖泊、浮岛	根茎有化痰、开窍、健脾、利湿等功效，精油可调配酒类香料
金钱蒲 *A. gramineus* 全草含精油0.30% （天南星科）	β-荜澄茄烯、β-石竹烯、异丁香酚甲醚、α-荜澄茄烯、丁香醚甲醚、愈创木醇、橙花叔醇、2,10,11-三甲基-2,4,11-十二碳三烯酮-6等	浙江、江西、湖北、河南、广东、广西、陕西、甘肃、四川等省区。生长于1800米以下的水边湿地，多有栽种	精油具杀菌防霉腐作用，是提取胡椒酚甲醚的原料
茴香菖蒲 *A. macrospadiceus* 全草含精油1.50%～2.0% （天南星科）	α-蒎烯、β-小茴香烯、β-水芹烯、伪柠檬烯、对伞花烃、3-蒈烯、异龙脑、胡椒酚甲醚、大茴香脑、侧柏醇、α-榄香烯、β-石竹烯等	广西融水苗族自治县江洞	精油具杀菌防霉腐作用，是提取胡椒酚甲醚的原料

植物名称	主要化学成分	分布	备注
黑三棱 *Sparganium stoloniferum* 块茎含精油 0.04%~0.06% （黑三棱科）	糠醛、糠醇、5-甲基呋喃醛、己酸、2-乙酰基吡咯、苯乙醇、3-乙基苯酚、2-羟基-5-甲苯乙酮、2,3-二氢苯并呋喃、5-己基二-2（2H）-呋喃酮、肉豆蔻醚、对苯二酚、8-羟基-3-甲基-3,4-二氢化-1H-2-苯并吡喃-1-酮、β-榄香烯、二氢去氢木香内酯、去氢木香内酯等	东北、黄河流域、长江中下游各省区及西藏。生长于沼泽、湖池及水沟	块茎入药治疗气血凝滞、心腹疼痛、肋下胀痛、闭经、产后腹痛、跌打损伤
香根鸢尾 *Iris pallida* 根茎含精油 0.50%~0.80% （鸢尾科）	鸢尾酮、苯甲酸、癸醛、苯乙酮、乙醛、丁二酮、丁香酚、肉豆蔻酸甲酸、苯甲醇、苯甲醛、苯乙醛、香叶醇等	原产于欧洲。我国庭园常有栽种	精油可配制化妆品香精，是庭园观赏芳香花卉
柠檬草 *Cymbopogon citratus* 全草含精油 （禾本科）	甲基庚烯酮、月桂烯、芳樟醇、香茅醛、柠檬烯氧化物、α-柠檬醛、β-柠檬醛、乙酸香叶酯等	福建、台湾、云南有栽培	精油可直接用作皂用香精，也是很多日化用品的原料（柠檬醛）
芸香草 *C. distans* 全草含精油0.20% （禾本科）	辣薄荷酮、4-莔烯、香叶醇、乙酸香叶酯、柠檬烯等	西南及甘肃、陕西。多生长于山坡草地	精油可直接用作皂用香精，具杀菌消毒功效，可提取辣薄荷，用于合成薄荷脑
香茅 *C. citratus* 全草含精油 0.37%~0.40% （禾本科）	α-蒎烯、莰烯、β-蒎烯、柠檬烯、芳樟醇、异胡薄荷酮、香茅醛、香茅醇、香叶醇、乙酸香茅酯、乙酸香叶酯等	原产于东南亚。华南、台湾有栽种	精油提取香叶醇和香茅醇用于调配各种化妆品香精

（续表）

植物名称	主要化学成分	分布	备注
爪哇香茅 *C. winterianus* 全草含精油 1.20%～1.40% （禾本科）	香茅醛、香叶醇、柠檬醛、荜澄茄烯、丁香酚甲醚、异丁醇、丁二酮、黑胡椒酚、倍半萜香茅烯等	原产于斯里兰卡。我国南部和台湾有栽种	精油在香料工业中占有重要地位，除直接用于皂用香精外，还可提取香茅醛、香叶醇

五、香花精油资源

我国常用的香花精油资源除木犀科桂花、茉莉花等外，不少是引种归化的种，如木兰科木兰属荷花玉兰（*Magnolia grandiflora*）、含笑属的白兰（*Michelia alba*）、黄兰（*M. champaca*），蔷薇科玫瑰（*Rosa rugosa*）等，但其他有待开发的香花精油资源潜力很大，详细参看表1-5。

表1-5　我国香花精油资源一览表

植物名称	精油主要化学成分	分布	备注
玉兰 *Magnolia denudata* 花蕾含精油 0.29%～0.67% （木兰科）	桧烯、β-月桂烯、α-柠檬烯、1,8-桉叶油素、对伞花烃、β-石竹烯、α-蒎烯、榄香醇、乙酸香茅酯、乙酸香叶酯、t-橙花叔醇、β-桉叶醇等	浙江、安徽、江西、湖南南部、广东北部，现黄河以南均有栽种	花蕾（苞）入药用于治疗头痛、骨痛、牙痛等
荷花玉兰 *M. grandiflora* 鲜花含精油 （木兰科）	乙酸甲酯、α-蒎烯、坎烯、桧烯、β-蒎烯、6-甲基-5-庚烯-2-酮、1,8-桉叶油素、苯甲酸甲酯、芳樟醇、苯乙醇、辛酸甲酯、松樟酮、桃金娘烯醛、马鞭草烯酮、香叶醇、α,β-柠檬醛、癸酸甲酯、茉莉酮、月桂酸甲酯等	原产于美洲东南部。我国长江以南各省区均有栽种	是常见庭园芳香绿化树种，叶入药治高血压，也是材用树种

（续表）

植物名称	精油主要化学成分	分布	备注
紫基玉兰 *M. purpurella* 鲜花含精油 （木兰科）	桧烯、β-蒎烯、柠檬烯、α-罗勒烯、cis-氧化芳樟醇（呋喃型）、t-氧化芳樟醇（呋喃型）、芳樟醇、α-金合欢烯、β-石竹烯、苯甲酸甲酯、苯乙醇、2-甲基-6-亚甲基-1,7-辛二烯-3-酮等	湖南望城。生长于海拔150～300米的山地	为我国特有树种，可开发庭园绿化芳香树种
凹叶木兰 *M. sargentiana* 花蕾含精油0.30% （木兰科）	芳樟醇、t-石竹烯、石竹烯氧化物、丁香酚甲醚、榄香醇、香榧醇、γ,β,α-桉叶醇等	四川、云南。生长于高海拔林地中	
武当木兰 *M. sprengeri* 花蕾含精油0.20%～0.30% （木兰科）	芳樟醇、t-石竹烯、香茅醇、石竹烯氧化物、香榧醇、α,β,γ-桉叶醇、榄香醇、丁香酚甲醚等	湖北。生长于山地杂木林	花蕾含精油，枝叶亦含精油
睦南木莲 *Manglietia chevalieri* 鲜花含精油 （木兰科）	3-蒈烯、桧烯、β-蒎烯、柠檬烯、1,8-桉叶油素、芳樟醇、萘、奠、β-石竹烯等	原产于越南和老挝。广东、广西有引种	为优良庭园绿化树种和建筑用材
木莲 *M. fordiana* 鲜花含精油 （木兰科）	柠檬烯、β-蒎烯、桃金娘烯醇、橙花醇等	华南和西南各省区	为材用树种
海南木莲 *M. hainanensis* 鲜花含精油 （木兰科）	β-蒎烯、柠檬烯、α-蒎烯、蒎葛缕醇、橙花醇、桃金娘烯醛、马鞭草烯酮、橙花醇、香叶酸甲酯等	海南特有种	为优良庭园绿化树种，船用木材

（续表）

植物名称	精油主要化学成分	分布	备注
大叶木莲 *M. megaphylla* （木兰科）	桧烯、β-蒎烯、1,8-桉叶油素、α-蒎烯	广西和云南南部	为材用树种
毛桃木莲 *M. moto* 鲜花含精油 （木兰科）	1,8-桉叶油素、4-松油醇、α-松油醇、古巴烯、β-杜松烯、橙花叔醇	湖南、广东、福建	木材纹理细，细木工用材
白兰 *Michelia alba* 鲜花含精油0.30% （木兰科）	乙酸甲酯、cis-氧化芳樟醇、t-氧化芳樟醇、芳樟醇、丁香酚甲醚、异戊酸-β-苯乙酯、苯乙醇等	原产于印度尼西亚。现广东、广西、云南、四川、福建大量栽种	叶含精油主含芳樟醇，花为重要的香花资源，枝叶亦含精油0.20%～0.28%
苦梓含笑 *M. balansae* 鲜花含精油 （木兰科）	乙酸丁酯、丁酸乙酯、3-甲基丁酸乙酯、己酸乙酯、己酸-2-甲基丙酯、乙酸-1-乙氧基乙酯等	广东、广西、海南、福建、云南南部	庭园绿化树种，可开发花香资源
黄兰（缅桂） *M. champaca* 鲜花含精油 （木兰科）	3-戊醇、3-己醇、氧化芳醇（呋喃型）、氧化芳樟醇（吡喃型）、苯甲酸甲酯、苯乙醇、苯甲酸丙烯酯、金合欢醇、橙花叔醇、α-紫罗兰酮等	原产于东南亚。现广东、云南和长江以南有栽种	可提取精油和浸膏
含笑 *M. figo* 鲜花含精油 （木兰科）	2-甲基丙酸乙酯、乙酸丁酯、1,3-丁二醇、2-甲基丙酸丁酯、己酸乙酯、己酸-2-甲基丙酯等	华南各地庭园及民间广泛栽种	庭园家庭绿化树种

（续表）

植物名称	精油主要化学成分	分布	备注
金叶含笑 *M. foveolata* 鲜花含精油 （木兰科）	2-戊醇、2-甲基-2-丁烯醇、2-甲基丁酸甲酯、乙酸-1-乙氧基乙酯、cis-3-己烯醇、α-蒎烯、β-蒎烯、辛酸甲酯苯乙醇、2-辛烯酸甲酯、1-甲氧基-3,7-二甲基-2,6-辛二烯等	贵州东部、云南南部、江西、广东、湖南南部和越南	为材用树种
醉香含笑 *M. macclurei* 鲜花含精油 （木兰科）	2-甲基丁酸甲酯、2-亚甲基丁酸甲酯、α-蒎烯、桧烯、β-蒎烯、苯甲酸甲酯、2-甲基-6-亚甲基-1,7-辛二烯酮-3、1,2-二甲氧基苯、柠檬烯等	广东、广西。生长于海拔1000米以下的林地中	花芳香美丽，为庭园绿化树种
展毛含笑（火力楠） *M. macclurei* var. *sublanea* 鲜花含精油 （木兰科）	2-亚甲基丁酸甲酯、庚醛、桧烯、β-蒎烯、柠檬烯、苯甲酸甲酯、芳樟醇、1,2-二甲氧基苯、乙酸龙脑酯、γ-榄香烯、β-荜澄茄烯、β-石竹烯等	广东西南、广西南部	花香美丽，为庭园绿化观赏树种
深山含笑 *M. maudiae* 鲜花含精油0.18% （木兰科）	α-蒎烯、莰烯、β-蒎烯、1,8-桉叶油素、芳樟醇、4-松油醇、α-松油醇、β-石竹烯、γ-榄香烯、橙花叔醇等	广东、海南。生长于海拔600～1000米的密林中	有待开发庭园观赏树木
白花含笑 *M. mediocris* 鲜花含精油 （木兰科）	5,5-二甲基-2-呋喃酮、对伞花烃、1,8-桉叶油素、t-氧化芳樟醇、cis-氧化芳樟醇、1,2-二甲氧基苯、马鞭草烯酮、苯甲酸甲酯等	广东、广西。越南、柬埔寨也有分布。生长于海拔400～1000米的杂木林中	庭园观赏树种
广西含笑 *M. guangxiensis* 鲜花含精油 （木兰科）	乙酸丁酯、丁酸乙酯、2-甲基丁酸乙酯、2-甲基丙酸-2-甲基丙酯、己酸乙酯、己酸-2-甲基丙酯	广西苗儿山	庭园观赏树种

（续表）

植物名称	精油主要化学成分	分布	备注
云南含笑（皮袋香） *M. yunnanensis* 鲜花含精油 （木兰科）	柠檬烯、樟脑、乙酸龙脑酯、茉莉酮、柏木烯、龙脑、小茴香烯、4-莰烯	分布于云南松杉林下或红土的灌丛中	花可提取精油，叶捣碎可用于调味料
观光木 *M. odora* 鲜花含精油 （木兰科）	2-甲基丁酸甲酯、乙酸-1-乙氧基乙酯、2-庚酮、5-甲基-2-己醇、4-甲基-2-庚酮、乙酸-4-己烯酯、cis-氧化芳樟醇（呋喃型）、t-氧化芳樟醇（呋喃型）、2-壬酮、壬醇-2、十五烷等	广东、广西、江西、海南，生于海拔100～1000米的山地林缘或疏林中	树干直，花美丽芳香，宜作庭园观赏及行道树
单性木兰 *Woonyoungia* *septentrionalis* 鲜花含精油 （木兰科）	乙酸-1-乙氧基乙酯、cis-氧化芳樟醇（呋喃型）、t-氧化芳樟醇（呋喃型）、苯甲酸-3-己烯酯、2-甲酸氨基苯甲酯等	广西东北、贵州东南。生长于石灰岩林地	高大乔木
鹰爪花 *Artabotrys* *bexapetalus* 鲜花含精油0.75% （番荔枝科）	2-甲基丙酸乙酯、丁酸丁酯、丁酸乙酯、2-甲基丁酸乙酯、2-甲基丙酸丙酯、乙酸-2-甲基丁酯、2-甲基-2-丙烯酸、2-甲基丙酯、异丁酸丁酯、3,3-二甲基丙烯酸叔丁酯等	浙江、台湾、福建、江西、广东、广西、云南多见栽培	常见华南地区庭园芳香观赏树种，野生不多见
依兰 *Cananga odorata* 花含精油 0.5%～1.0% （番荔枝科）	对甲氧基苯甲醇、苯甲酸甲酯、芳樟醇、乙酸苄酯、乙酸香叶酯、乙酸桂酯、异丁香酚、γ-杜松烯、苯甲酸苯甲酯等	原产于印度尼西亚、菲律宾和马来西亚。台湾、福建、广东、广西、云南有栽种，品种较多	其精油称卡南卡油，是一种高价值、配制香精原料

(续表)

植物名称	精油主要化学成分	分布	备注
假鹰爪 *Desmos chinensis* 鲜花含精油0.24% （番荔枝科）	苯甲醛、间伞花烃、1,8-桉叶油素、苯甲醇、γ-松油烯、苯乙酮、芳樟醇、桂醛、古巴烯、β-石竹烯、δ-杜松烯、苯甲酸苯甲酯等	广东、广西、云南、贵州。生长于丘陵山坡、林缘灌丛中或低海拔荒野	为庭园芳香观赏树种，其浸膏可开发香精香料，可造酒曲
上思瓜馥木 *Fissistigma shangtzeense* 鲜花含精油0.04%～0.05% （番荔枝科）	1,8-桉叶油素、氧化芳樟醇（呋喃型）、2,4-蓋二烯、芳樟醇、氧化芳樟醇（吡喃型）、α-松油醇、香茅醇、香叶醇、2,5-十八碳二烯酸甲酯、石竹烯、橙花叔醇、肉豆蔻酸甲酯等	广西、云南山地林中	鲜花香，成分为一具开发潜力的天然香料资源
莲 *Nelumbo nucifera* 鲜花含精油 （睡莲科）	1,4-二甲基苯、β-石竹烯、α-石竹烯、十七烷、十七炔、苯并噻唑、α-松油醇、芳樟醇、苯甲酸甲酯等	我国南北各省区均有种养	观赏水生植物，栽种广泛，栽培品种较多
睡莲 *Nymphaea* 'Fragrant Hybrid' 鲜花含精油0.11% （睡莲科）	苄醇，6.9-十七碳二烯、2-十七烷酮、8-十七碳烯、正十五烷、正十六烷酸、二十一烷、叶绿醇、环十六烷、9,12,15-十八碳三烯酸、1,9-十四碳二烯、9,12-十八碳二烯酸、2-十五烷酮、4-(2,6,6-三甲基-1-环己烯基)-3-丁烯-2-酮等。	我国南北各省区均有种养。	备注：为盆栽、园林、庭院、水生观赏或插花观赏。鲜花可供生食，亦可泡茶、浸酒或随其他食物炖煮。可开发香精香料。
珠兰 *Chloranthus spicatus* 花含精油 （金粟兰科）	α-蒎烯、莰烯、桧烯、β-蒎烯、月桂烯、cis-β-罗勒烯、t-β-罗勒烯、2-甲基-6-亚甲基-1,7-辛二烯酮-3、异蒎樟脑酮、β-橙花叔醇、茉莉酮酸甲酯等	云南、四川、贵州、广东、福建。生长于山坡沟谷密林下，多为栽培	南方常见芳香花卉，花可熏茶

（续表）

植物名称	精油主要化学成分	分布	备注
麝香石竹 *Dianthus caryophyllus* 花含精油0.5% （石竹科）	丁香酚、芳樟醇、α-松油醇、3-己烯醇、苯甲醇、苯甲醛、α-松油醇、茉莉酮、苯甲酸cis-3-己烯酯、苯甲酸苯甲酯、水杨酸苯甲酯、茉莉酮酸甲酯等	我国常见庭园花卉	芳香庭园花卉
散沫花 *Lawsonia inermis* 鲜花含精油 （千屈菜科）	己醛、2-己烯醛、3-己烯醇、2-己烯醇、乙酸-3-己烯酯、cis-氧化芳樟醇（呋喃型）、t-氧化芳樟醇（吡喃型）、芳樟醇、α-松油醇、紫苏醛、香叶醇、橙花叔醇、苯甲酸苯甲酯等	广东、广西、云南、福建、江苏、浙江等热带、亚热带地区	供庭园观赏花卉，叶可用作红色天然染料
待霄草 *Oenothera stricta* 花含精油 （柳叶菜科）	cis-氧化芳樟醇、t-氧化芳樟醇、芳樟醇、6-十一酮、苯并噻唑、3,4-二氢-2,5-二甲基2H吡喃甲醛-2、R-二氢猕猴桃内酯、壬酸等	原产于南美洲，我国东北有栽种	香花可提浸膏，用于配制香精
陕甘瑞香 *Daphne tangutica* 鲜花含精油 （瑞香科）	2-呋喃甲醛、2-呋喃甲醇、苯甲醛、苯甲醇、cis-氧化芳樟醇（呋喃型）、t-氧化芳樟醇（呋喃型）、芳樟醇、壬醛、苯乙醇、cis-氧化芳樟醇（吡喃型）、t-氧化芳樟醇（吡喃型）、苯乙醇、壬醛、苯并噻唑、3-苯基-2-丙烯酸乙酯、苯甲酸苯甲酯、丁香酚甲醚等	青海、甘肃、西藏、湖北、四川、陕西。生长于海拔1400～3900米的山坡林下和岩石缝中	鲜花浸膏可用于配制香精
海桐 *Pittosporum tobira* 花含精油 （海桐花科）	2-甲基-2-丁烯醇、苯甲醛、苯甲醇、cis-氧化芳樟醇（呋喃型）、t-氧化芳樟醇（呋喃型）、芳樟醇、乙酸苯甲酯、吲哚、2,6-二叔丁基-对甲酚、橙花叔醇等	长江以南各省区有栽培	为常见庭园观赏芳香植物

（续表）

植物名称	精油主要化学成分	分布	备注
水翁 *Cleistocalyx operculatus* 花蕾含精油0.18% （桃金娘科）	α-蒎烯、β-蒎烯、月桂烯、β-罗勒烯（Z）、β-罗勒烯（E）、蛇麻烯、γ-依兰油烯、α-愈创木烯、δ-愈创木烯、别-香树烯、δ-杜松烯、金合欢醇、cis-石竹烯、3,6,8,8-四甲基八氢-7-亚甲基薁等	广东、广西、云南。多生于水边	花蕾清香宜人，可泡茶消滞食，枝叶亦含精油
丁子香 *Syzygium aromaticum* 花蕾含精油15.30% （桃金娘科）	丁香酚、苯甲醛、石竹烯、α-蛇麻烯、水杨酸甲酯、香荆芥酚等	原产于坦桑尼亚桑给巴尔，我国有引种栽培	为重要芳香药用植物，可消痰止痛，叶亦含精油
毛叶蔷薇 *Rosa mairei* 花含精油 （蔷薇科）	1,1-二乙氧基乙烷、cis-氧化芳樟醇、t-氧化芳樟醇、芳樟醇、苯乙醇、水杨酸甲酯、α-松油醇、乙基苯酚、香茅醇、橙花醇、香叶醇、丁香酚甲醚、二氢-β-紫罗兰酮、丁香酚、十四醛等	云南、四川、西藏、贵州。生长于海拔2300～4180米的山坡向阳处	鲜花有待开发成天然香原料
玫瑰 *R. rugosa* 鲜花含精油 （蔷薇科）	芳樟醇、玫瑰醚、香茅醇、香叶醇、乙酸香茅酯、丁香酚、乙酸香叶酯、2-十二酮、香树烯、β-荜澄茄烯、2-十三酮、2-十三醇、橙花叔醇、金合欢醇（Z.E）、金合欢醇、十五酮-2、苯甲酸苯甲酯等	原产华北和日本、朝鲜半岛，现我国各地均有栽种，栽培种十分多	花浸膏精油可供食品和高级香水用，花瓣亦供食用和泡茶
苦水玫瑰 *R. sertata* 花含精油0.03% （蔷薇科）	cis-玫瑰醚、t-玫瑰醚、香茅醇、香叶醇、乙酸香茅酯、丁香酚、丁香酚甲醚、别香树烯、γ-依兰油烯、2-十三酮、2-十三醇、cis-金合欢醇、t-金合欢醇、石竹烯醇、竹石烯氧化物等	西北地区大范围种植	可提取苦水玫瑰精油或浸膏，供调配香精

（续表）

植物名称	精油主要化学成分	分布	备注
贵州刺槐 *Robinia pseudoacacia* 鲜花含精油 （蝶形花科）	7-辛烯-4-醇、cis-氧化芳樟醇（呋喃型）、愈创木烯、芳樟醇、苯乙醇、cis-氧化芳樟醇（吡喃型）、香叶醇、檀香醇、橙花叔醇、肉豆蔻酸乙酯、柏木烯醇、壬内酯	原产于北美，世界各地均有栽种	花浸膏可调制各类花香型香精，花有止血功效（内出血）
啤酒花 *Humulus lupulus* 干谢花含精油 0.90%～1.0% （大麻科）	β-月桂烯、2-甲基丙酸、2-甲基丁酯、6-甲基庚酸甲酯、辛酸甲酯、壬酸甲酯、2-十一酮、4-癸烯酸甲酯、β-石竹烯、9-十二烯甲酯、α-石竹烯、十二烯甲酯、2-甲基丙酸香叶酯	新疆北部有野生，东北、华北、山东有栽种	干花用于酿造啤酒，雌花序入药助消化
尖果沙枣 *Elaeagnus oxycarpa* 花含精油 0.2%～0.4% （胡颓子科）	苯甲醇、苯乙醇、桂酸甲酯、t-桂酸乙酯、苯乙酸乙酯、己酸乙酯、cis-桂酸乙酯、苯丙醇、异丁酸苯乙酯、异戊酸-β-苯乙酯等	华北、西北及内蒙古等。生长于海拔400～600米的戈壁沙地、低洼潮湿地	香气独特持久，可用于配制化妆品、皂用香精，叶、根、果均入药
代代花 *Citrus aurantium* 鲜花含精油0.28% （芸香科）	β-蒎烯、柠檬烯、cis-氧化芳樟醇（呋喃型）、t-氧化芳樟醇（呋喃型）、芳樟醇、t-氧化芳樟醇（吡喃型）、cis-氧化芳樟醇（吡喃型）、2-氨基苯甲酸芳樟酯、香叶醇、乙酸香叶酯异构体、橙花叔醇、金合欢醇等	长江流域及以南均有栽种	花及枝叶均含精油，可调制食用和化妆香精，果可入药
柚 *C. grandis* 花含精油 （芸香科）	水芹烯、β-蒎烯、柠檬烯、α-罗勒烯、芳樟醇、3-甲基-4-庚酮、玫瑰呋喃、邻氨基苯甲酸甲酯等	长江以南均有栽种	主要水果，品种多，果、木、叶均含精油，用于配制香精，柚可提果胶

植物名称	精油主要化学成分	分布	备注
柠檬 *C. limon* 果皮含精油1.56% （芸香科）	β-蒎烯、甲基庚烯酮、柠檬烯、罗勒烯、芳樟醇、香茅醛、橙花醇、α-柠檬醛、香茅醇、β-柠檬醛、乙酸香叶酯、乙酸香茅酯、乙酸香叶酯、丙酸香茅酯等	云南、广西、广东、福建、四川、贵州，有野生和栽种	果实可制凉果，果汁可配饮料，叶含精油，可配制皂用香精
四季米仔兰 *Aglaia duperreana* 花含精油 0.2%～0.4% （楝科）	β-罗勒烯、t-氧化芳樟醇、cis-氧化芳樟醇、芳樟醇、吲哚、金合欢烯、石竹烯、香树烯、β-甜没药烯、γ-依兰油烯、β-愈创木烯、t-二氢氧化金合欢醇、cis-二氢氧化金合欢醇等	广东、广西、福建有栽种	为南方地区常见庭园芳香花卉，浸膏为我国特有天然香原料
米仔兰（树兰） *A. odorata* 鲜花含精油 （楝科）	α-蒎烯、β-罗勒烯、t-氧化芳樟醇、cis-氧化芳樟醇、乙酸芳樟酯、吲哚、α-古巴烯、β-榄香烯、t-β-金合欢烯、蛇麻烯、γ-杜松烯、l-芳樟醇等	福建、广东、广西、云南、四川。生长于林中，多见栽培	花是我国特有香原料，可调配化妆品、皂用，可用作定香剂，也可熏花茶
素馨花（大花茉莉） *Jasminum grandiflorum* 鲜花含精油 （木犀科）	苯甲醇、对甲酚、芳樟醇、乙酸苯甲酯、1H吲哚、丁香酚、茉莉酮、异丁香酚、α-金合欢烯、橙花叔醇、苯甲酸苯甲酯、3-己烯醇等	原产地中海沿岸、法国摩洛哥，基本已人工栽种，我国有引种	花精油和浸膏为名贵天然香原料，可配制高级化妆品香精
素芳花 *J. officinale* 花含精油 （木犀科）	乙酸苯酯、乙酸芳樟酯、芳樟醇、橙花醇、橙花叔醇、松油醇、苯甲醇、苯甲醛、丁香酚、茉莉酮等	云南、四川和西藏，现有栽种	素芳花香气宜人，可配制高级化妆品香精

（续表）

植物名称	精油主要化学成分	分布	备注
厚叶素馨 *J. pentaneurum* 鲜花含精油（木犀科）	苯甲醇、cis-氧化芳樟醇、t-氧化芳樟醇、芳樟醇、2-羟基苯甲酸甲酯、茉莉酮、橙花叔醇、3-（4,8-二甲基-3,7-壬二烯基）呋喃、十八碳二烯酸甲酯、丁酸-3-己烯酯等	广东、广西。生长于疏林和灌木丛中	鲜花为有待开发的天然香原料
茉莉花（小花茉莉）*J. sambac* 鲜花含精油（木犀科）	乙酸-3-己烯酯、cis-3-己烯醇、芳樟醇、苯甲酸甲酯、乙酸苯甲酯、石竹烯、苯甲醇、苯甲酸-cis-3-己烯、杜松烯、邻氨基苯甲酸甲酯、金合欢烯等	广东、广西、云南、福建、四川、贵州，大多已人工栽种	净油和浸膏为配制日化皂用香精原料，亦为常见庭园芳香花卉
山指甲 *Ligustrum sinense* 鲜花含精油（木犀科）	苯甲醛、柠檬烯、1,8-桉叶油素、苯乙醇、苯乙醛、罗勒烯、苯甲酸甲酯、芳樟醇、壬醇、樟脑、N-苯基甲酰胺等	长江流域以南各省区。生长于山地疏林，路旁沟边常见栽种	浸膏可配制日化用香精，叶药用可消热、消肿痛
桂花（银桂）*Osmanthus fragrans* 鲜花含精油（木犀科）	t-氧化芳樟醇（呋喃型）、cis-氧化芳樟醇（呋喃型）、芳樟醇、壬醇、α-松油醇、6-乙基四氢-2,2,6-三甲基吡喃-3-醇、橙花醇、二氢-β-紫罗兰酮、香叶醇、α-紫罗兰酮、苯乙醇、橙花叔醇、4,2-辛二烯醛、水杨酸甲酯等	长江下游、广西、广东、云南等广泛有栽种	为庭园芳香绿化树种，浸膏可配制化妆品、食用香精，花亦可直接食用。

（续表）

植物名称	精油主要化学成分	分布	备注
金桂 *O. fragrans* var. *thunbergii* 鲜花含精油 （木犀科）	甲基庚二烯酮、6-二乙酸氧基己烷、5-苯甲氧基戊醇、环己烯-3-甲醇、薄荷酮、乙氧基甲醛、t-氧化芳樟醇（呋喃型）、cis-氧化芳樟醇（呋喃型）、丁酸己酯、葛缕酮、α-紫罗兰酮、二氢-β-紫罗兰酮、γ-癸内酯、β-紫罗兰酮等	我国南方各地园林竞相引种栽培，花金黄色	为庭园观赏树种
丹桂 *O. fragrans* var. *aurantiacns* 鲜花含精油 （木犀科）	β-罗勒烯（E）、cis-氧化芳樟醇（呋喃型）、t-氧化芳樟醇（呋喃型）、芳樟醇、cis-氧化芳樟醇（吡喃型）、t-氧化芳樟醇（吡喃型）、α-紫罗兰酮、二氢-β-紫罗兰酮、5-己基二氢-2(3H)-呋喃酮、β-紫罗兰酮等	栽培种，我国各地均有栽培。花橘红色	香气较银桂、金桂淡
暴马丁香 *Syringa reticulate* subsp. *amurensis* 花净油收得率4.0% （木犀科）	cis-氧化芳樟醇、t-氧化芳樟醇、芳樟醇、苯乙醇、2,6,6-三甲基-2-乙烯基-5-羟基吡喃、癸酸乙酯、δ-杜松烯、橙花叔醇、肉豆蔻酸乙酯等	我国辽宁、黑龙江、吉林、河北、宁夏等省区。生长于河岸、林缘混交林内	树皮树枝入药治咳嗽、哮喘，净油可配制化妆品香精
鸡蛋花 *Plumeria rubra* 'Acutifolia' 花含精油 （夹竹桃科）	苯甲醛、三环[3,2,1,0,1,5]辛烷、苯甲酸甲酯、芳樟醇、香叶醇、柠檬醛、橙花叔醇、苯甲酸苯甲酯、水杨酸甲酯等	原产于美洲热带地区，我国有栽种	为热带、亚热带芳香观赏植物，可用作凉茶原料

（续表）

植物名称	精油主要化学成分	分布	备注
红鸡蛋花 *P. rubra* 花含精油 （夹竹桃科）	苯甲醛、苯甲醇、苯乙醇、香叶醇等	原产于美洲，我国有栽种	
栀子 *Gardenia jasminoides* 鲜花含精油 （茜草科）	cis-3-己烯醇、惕各酸甲酯、乙酸-3-己烯酯、苯甲酸甲酯、芳樟醇、异丁酸-cis-3-己烯酯、丁酸-cis-3-己烯酯、葛缕醇、惕各酸-cis-3-己烯酯、惕各酸苯甲酯、苯甲酸-cis-3-己烯酯等	我国南部、西南和中部。生长于山野间，现多栽种	为庭园芳香观赏植物，精油、浸膏可配制各种香型香精
白蟾 *G. jasminoide* var. *fortuniana* 花含精油 （茜草科）	乙基丙基醚、甲基丁基醚、乙基丁基醚、3-甲基-2-丁烯酸甲酯、乙酸-2-甲基环戊酯、6-甲基-5-庚烯酮-2、氧化芳樟醇（呋喃型）、芳樟醇、丁酸-3-己烯酯、氧化芳樟醇（吡喃型）、2-甲基-2-戊烯醇、1-甲基-2-（2-丙基）环戊烷、己基环己烷等	我国中部以南各省区有栽培，多见于大中城市	常见庭园芳香花卉
滇丁香 *Luculia pinceana* 鲜花含精油 （茜草科）	2,4-二甲基-3-戊酮、α-蒎烯、莰烯、β-月桂烯、β-蒎烯、柠檬烯、t-氧化芳樟醇（呋喃型）、cis-氧化芳樟醇（呋喃型）、芳樟醇、樟脑、龙脑、水杨酸甲酯、雅兰槛酮、β-杜松烯、金合欢烯、2,6-二叔丁对甲酚、β-荜澄茄烯等	云南、广西。生于海拔1300米以上的山坡林中或灌木丛中	云南、广西山区有名的香花，有待开发

（续表）

植物名称	精油主要化学成分	分布	备注
华南忍冬（山银花） *Lonicera confusa* 花含精油 （忍冬科）	3-己烯醇、2-亚甲基丁酸甲酯、6-甲基-5-庚烯酮-2、苯甲醇、α-罗勒烯、cis-氧化芳樟醇（呋喃型）、t-氧化芳樟醇（呋喃型）、苯甲酸甲酯、芳樟醇、苯乙醇、cis-氧化芳樟醇（吡喃型）、邻氨基苯甲酸甲酯、α-金合欢烯等	四川、广东、广西、湖南、贵州、云南等省区。生长于疏林和灌木林中	花入药治感冒、痢疾、腹泻，叶可治风湿
忍冬（金银花） *L. japonica* 花含精油0.4% （忍冬科）	莰烯、1-己烯、cis-己烯-3-醇、cis-氧化芳樟醇、t-氧化芳樟醇、芳樟醇、α-松油醇、异橙花醇、橙花醇、香叶醇、苯甲醇、β-苯乙醇、香荆芥酚、丁香酚等	北起辽宁，西至陕西，南达湖南、云南、贵州。生长于山坡灌丛和疏林中，有大量栽培	为常用清热解毒中药，也是庭园观赏花卉
三裂蟛蜞菊 *Wedelia trilobata* 鲜花含精油 （菊科）	α-蒎烯、莰烯、桧烯、β-蒎烯、β-月桂烯、对伞花烃、柠檬烯、苯甲酸甲酯、樟脑、龙脑、乙酸桧酯、异丙基苯酚、麝香草酚、t-4-羟基-3-甲基-6-异丙基-2-环己酮、t-4-羟基-3-甲基-6-异丙基-2-环己烯酮异构体、α-石竹烯、β-荜澄茄烯等	为归化植物，华南地区较常见	植株可供地表绿化和观赏
夜香树 *Cestrum nocturnum* 花含精油 0.3%~0.6% （茄科）	乙酸乙酯、1-乙氧基-2-甲基丙烷、乙氧基丁烷、乙酸戊酯、苯甲醛、苯甲醇、苯甲酸甲酯、苯乙醇、2-甲氧基-4-（2-丙烯基）苯酚、乙酸苯乙酯、水杨酸甲酯、邻氨基苯甲酸甲酯、4-甲基-6-庚烯酮-3、金合欢烯等	原产于南美洲，现被引种世界各地。广东、广西、福建、云南有栽种	为庭园观赏芳香花卉，晚上开放，香气具消毒、杀菌功能

植物名称	精油主要化学成分	分布	备注
臭牡丹（臭茉莉）*Clerodendrum philippinum* var. *simplex* 鲜花含精油（马鞭草科）	1-乙氧基戊烷、乙酸、1-乙氧基乙酯、苯甲醛、7-辛烯醇-4、6-甲基-5-庚烯-2-酮、1,8-桉叶油素、苯乙醛、cis-氧化芳樟醇（呋喃型）、t-氧化芳樟醇（呋喃型）、芳樟醇、苯乙醇、2-羟基苯甲酸甲酯、橙花醇、香叶醇、苯甲酸苯甲酯等	浙江、江西、湖南、福建、广东、广西、贵州、云南以及日本等地有栽种	叶、根药用能祛风活血、强筋壮骨
姜花 *Hedychium coronarium* 花含精油（姜科）	月桂烯、桧烯、1,8-桉叶油素、α-罗勒烯、β-罗勒烯、cis-氧化芳樟醇（呋喃型）、t-氧化芳樟醇（呋喃型）、苯甲酸甲酯、芳樟醇、苯乙醇、α-松油醇、吲哚、茉莉酮、α-金合欢烯、苯甲酸戊酯、3-（4,8-二甲基-3,7-壬二烯基）呋喃等	台湾、广东、湖南、广西、云南、四川等省区。生长于林中或栽种，喜水边潮湿，现多见人工栽种	为华南和台湾普遍栽种芳香花卉
水仙 *Narcissus tazetta* var. *chinensis* 花含精油（石蒜科）	1,3-二甲氧戊环-2-甲醇、乙酸-3-甲基-2-丁烯酯、α-蒎烯、β-蒎烯、柠檬烯、1,8-桉叶油素、β-罗勒烯、cis-氧化芳樟醇（呋喃型）、t-氧化芳樟醇（呋喃型）、十一烷、芳樟醇、乙酸苯甲酯、α-松油醇、乙酸苯乙酯、乙酸苯丙酯、乙酸葛缕酯等	原产于亚洲沿海温暖地区，现基本人工栽培，福建最多	花香气优雅，为春节过年重要芳香花卉，但不宜放于室内过久
晚香玉 *Polianthes tuberosa* 鲜花含精油（龙舌兰科）	α-蒎烯、苯甲醛、桧烯、β-蒎烯、柠檬烯、1,8-桉叶油素、苯甲酸甲酯、2-羟基苯甲酸甲酯、吲哚、邻氨基苯甲酸甲酯、β-月桂烯、6-甲基-5-庚烯酮-2等	原产于墨西哥，现我国各地均有栽种	浸膏可配制高级香精原料，也是一种常见芳香插花

植物名称	精油主要化学成分	分布	备注
墨兰 *Cymbidium sinense* 花含精油 （兰科）	庚烷、3-甲基-3-戊醇、3,4-二甲基-3-戊酮、2,4-二甲基-3-戊醇、3,4-二甲基己醇、癸烷、1,8-桉叶油素、α-罗勒烯、苯甲酸甲酯、3-苯基丙醛、古巴烯、二氢-β-紫罗兰酮、β-紫罗兰酮、2,6-二叔丁基对甲酚、cis-t-金合欢醇等	我国各地均有栽种，品种、变种甚多，为蕙兰的一种	为庭园厅堂常见芳香花卉，香气幽香宜人
鼓槌石斛 *Dendrobium chrysotoxum* 花含精油 （兰科）	3-蒈烯、苯甲醛、1,7,7-三甲基二环[2,2,1]庚二烯、3,7-二甲基-1,3,7-辛三烯、苯乙醛、戊基环丙烷、苯乙醇、侧柏烯醇-2、龙脑、蒽、乙酸辛酯、香茅醇、马鞭草酮等	云南、西南部地区。附生于树上，各地植物园有引种	花色美丽，香气宜人，为庭园芳香花卉

第二节　我国开发的植物精油资源

从上节我国植物精油概况列举的逾500种芳香植物中，在不同历史年代，陆续引入我国的外来芳香植物超过50种，如桃金娘科的桉属植物，木兰科的白兰、黄兰、荷花玉兰，菊科的飞机草、万寿菊、马缨丹，伞形科的芫荽、茴香，禾本科的香茅、枫茅，还有胡椒、香叶天竺葵、大蒜等它们已经落地生根，再经人工栽培繁殖育出我国重要植物精油资源，如柠檬桉（*Eucalyptus citriodora*），其枝叶精油和香茅（*Cymbopogon citratus*）全草精油成为香茅醛（citronellal）主要来源；蓝桉（*Eucalyptus globulus*）、窿缘桉（*E. exserta*）成为1,8-桉叶油素（1,8-cinule）的主要来源；柠檬草（*Cymbopogon citratus*）则是柠檬醛（citral）主要来源之一，还有唇形科薄荷（*Mentha canadensis*）、留兰香（*M. spicata*）等也成为我国重要的植物精油资源，特别是薄荷在江苏崇明一带大量栽种，目前我国薄荷脑的产量占世界产量的60%以上，尚有很多香花资源如木本的白兰、黄兰、荷花玉兰，草本的大花茉莉、玫瑰类等不但是我国重要天然香料原料，也是各地公园、庭园的绿化芳香花卉。

一、右旋龙脑（天然冰片）的开发

随着我国经济发展，科技进步，研究者们开发出不少植物精油资源。如中科院研究人员在樟科樟属种内发现不同化学类型的两种植物：梅片树和龙脑樟。它们的枝叶精油富含右旋龙脑（d-borneol）。其中梅片树为阴香的一个化学类型，枝叶精油含右旋龙脑51%；龙脑樟枝叶精油含右旋龙脑81%。为此这一发现改变了我国右旋龙脑以往全依赖从印度尼西亚进口的局面，而且也改变了砍伐龙脑香树（*Dryobalanops aromatica*）木材从而提取获得天然冰片的方式。现由于长期开发加上天灾、资源缺乏、再生周期漫长，梅片树、龙脑樟只用树叶提取，每年都可以采收，再生能力强，也不会影响当地的自然环境，经有关药政部门审核已批准作为中药材，载入《中华人民共和国药典》。

二、柠檬醛资源的开发

樟科木姜子属山鸡椒（*Litsea cubeba*），别名山苍子，多年前，已开发利用其果实皮精油，精油富含α，β-柠檬醛（α，β-citra）。初期砍树取果蒸油，现已改变方式，只砍带果枝条，保护植株，把它们围起来管理。另外从研究黄樟种内多样性及不同化学类型中，发现其枝叶精油富含柠檬醛70%以上的姜樟，利用枝叶蒸油是今后开发植物精油的基本策略，其再生周期不到一年，而伐树挖根取油取脑起码需要20年以上才能恢复。柠檬醛是合成紫罗兰酮、维生素E的主原料，柠檬醛还具有抑菌效果（图1-1）。

（a）果　　　　　　　　　　　　　（b）植株

图1-1　山鸡椒

三、花香型右旋芳樟醇的开发

在研究樟属种内多样性及不同化学类型时，发现黄樟中的大叶芳樟枝叶精油含有右旋芳樟醇化学类型，枝叶精油含右旋芳樟醇高达90%，其精油可直接调配各类香精。这一发现颇有学术研究价值，因为樟科特别是樟属中含芳樟醇的较为普遍，但都是木香型的左旋芳樟醇，以往尚未见有报道发现花香型的右旋芳樟醇，而且含量高达90%实为罕见，目前已作为植物精油种质资源保存起来（图1-2）。

（a）花　　　　　　　　　　　　　　　　（b）枝叶

图1-2　黄樟

四、木本芳香树种的开发

我国从国外引进不少乔木型芳香植物，其中有不少是植物精油资源种类中的佼佼者，如从美洲引进来的松属树种湿地松（*Pinus elliottii*）、加勒比松（*P. caribaea*）、火炬松（*P. taeda*），三者作为建筑用材树种引进，但同样也能生产松脂，例如湿地松在我国南方生长迅速，并成功在佛山红岭林场大面积造林，与此同时在南方多地区推广，湿地松生长快，树干笔直，是建材好原料，同时产松脂较多，而松脂产松节油高达19%～28%，从各方面来看湿地松经济价值都大大优于马尾松。

五、互叶白千层的引进和开发

从澳大利亚引进的白千层属木本芳香小乔木，互叶白千层（*Melaleuca alternifolia*），其枝叶精油（茶树油）富含具杀菌能力很强的4-松油醇（terpinen-4-ol），目前纯化后广泛应用于护肤品、香波类、洗涤类等。4-松油醇本身气味十分清淡，适合添加于香精、化妆品和洗涤品中。但茶树油还有其他成分，如1,8-桉叶油素，其气味浓郁，需除去或降低其含量，才能被广泛应用。由于互叶白千层是种子引进，变异较大，因此生产的茶树油质量差异较大，这要引种者特别注意。目前广东西部和南部均有种植，今后应加强优质品种的选育才能更好推广这一植物精油资源（图1-3）。

（a）枝叶　　　　　　　　　　（b）花

图1-3　互叶白千层

六、薰衣草品种的引进和开发

在20世纪60年代，我国从南欧和地中海引进了唇形科薰衣草品种。在气候方面，地中海型气候是冬季寒冷湿润，夏季炎热干燥，这与我国大多数地区正好相反，冬季寒冷干燥，夏季则炎热潮湿，所以在选择宜栽地时必须考虑到这方面。经科研人员的努力，选择在我国西北新疆和河西走廊种植，目前已大面积栽种，生产出优质薰衣草精油，改变了我国调配薰衣草类型香精的原料全靠进口的局面，但这类精油生产受国际需求影响甚大，反而制约了这类精油的发展（图1-4）。

（a）收割作业　　　　　　　　　　　（b）植株

图1-4　云南薰衣草花田（王耀奇摄）

七、苦水玫瑰提取新工艺和综合利用

近年来在西北甘肃一带生长的苦水玫瑰（*Rosa sertata*）是钝齿蔷薇和我国传统玫瑰的自然杂交种，在甘肃永登县苦水镇种植已有200多年历史，故得此名。由于当地红粘土质、气候等自然因素非常适宜苦水玫瑰栽种，这里所产的玫瑰不论香型、产花量、含油量、抗逆性等均具有优良玫瑰品种的特性，但苦水玫瑰油的生产加工历史也只是50年左右，在此期间不论种植面积还是苦水玫瑰油加工技术都在不断发展，仅2023年甘肃永登县苦水玫瑰种植面积就达10.16万亩，鲜花年产量2800万千克，生产玫瑰精油1200千克、干花蕾260万千克、纯露1600吨。花精油已由水汽馏发展到CO_2（二氧化碳）超临界流体萃取，使苦水玫瑰精油质量不断提高，并发展综合利用，利用萃取后的花渣提取玫瑰黄酮（图1-5、图1-6）。

图1-5　苦水玫瑰花田　　　　　图1-6　苦水玫瑰提取设备

八、檀香的引进与发展

檀香（*Santalum album*）原产于南太平洋帝汶岛，后来遍布东南亚诸群岛，以至中南半岛、印度半岛南部、斯里兰卡岛、澳大利亚北部，其心材富含精油，称檀香油或白檀油。檀香木主要用于熏香、木雕、入药以及提取檀香精油，是一种非常名贵的芳香植物。檀香与佛教有着千丝万缕的关系，佛教称它为佛树、旃檀，从树本身到檀香木各种类产品如佛像婆萨雕像、佛珠都是礼佛上等供品，浴佛、参禅、诵经等佛事都少不了檀香。檀香木由于木材坚硬、色泽光亮、纹理细密，是细工雕琢的上等材料，檀香扇在以前是贵妇、淑女的一种身份象征。檀香木为古老名贵中药材，《中华人民共和国药典》记载其性味为辛、湿，归经为归脾、胃、心、肺，功能主治为行气温中、开胃止痛，用于寒凝气滞、胸痛、腹痛、胃痛、食少、冠心病、心绞痛等。檀香也是多种中成药原料，檀香精油是美容、护肤、消除疲劳、芳香镇痛的名贵天然香料产品，檀香精油也是配制名贵香精不可多得的香原料，而且留香时间长久，随着人工合成檀香醇的大量出现，天然檀香油受到影响，但天然檀香精油还是不失其价值（图1-7、图1-8）。

图1-7　檀香花　　　　　　　　　图1-8　檀香木横截面

檀香这一精油资源在20世纪五六十年代经中科院华南植物园的研究人员努力，已经在广东等地引种成功。由于檀香是半寄生类型树种，需要搭配其他植物作为寄主，在国内特别在广东地区选择与檀香树相匹配的寄主至关重要，而且它从幼苗开始到中苗至成年树需草本、灌木、乔木分阶段选择寄主，研究人员成功选出草本的假蒿、金钱草等，灌木为山毛豆、九里香、米仔兰等，乔木为黄皮、苏木、木麻黄、台湾相思等。这使得檀香在我国广东地区能生长迅速。在原产地，檀香树要经30～40年成材，才能进行砍伐，现在看来我国种植的檀香树早已达到或将达到成材砍伐年限。我国生产的檀香木作为一种植物精油资源，单从精油化学成分组成表分析，香气等方面还有待进一步提高，根据华南植物园提供的檀香精油来看，α，β-檀香醇含量只有不到50%，而进口檀香精油一级品达90%。

近年来，在广东、云南、广西、福建、海南等地区已大量种植檀香树，据不完全估计，我国成为世界上仅次于澳大利亚的种植檀香面积最大的国家。但檀香树应该在我国种植多少年才能成材砍伐的问题另当别论，据报道2001年种的檀香树在其树干中心已经有檀香心材出现。在1994年曾用生长2年的檀香树注射激素加快心材形成，取出的檀香油和进口的大致相同。

在世界上能产生檀香心材的树种很多，目前我国主要引进两种，即檀香（*Santalum album*）和巴布亚檀香（*S. papuanum*）（图1-9、图1-10）。

图1-9　种植20年的檀香木

图1-10　檀香原木材

九、沉香人工加速结香

我国分布有土沉香（*Aquilaria sinensis*）和云南沉香（*Aquilaria yunnanensis*），产自广东、海南、广西、福建和云南。产于马来西亚、印度尼西亚等地的为马来沉香（*Aquilaria malaccensis*）。产于柬埔寨、缅甸、越南等地的为 *Aquilaria agallocha* 沉香。现在公认沉香的形成主要是沉香树由于外界昆虫侵扰、火烧、雷电打击、真菌侵蚀和外物击入，受伤后自我愈合产生的分泌物（图1-11、图1-22）。

图1-11　传统的沉香树开香门取香法

图1-12　传统的钻孔打火钎得到的香（褐色部分）

沉香自然结香时间是漫长的，从十多年至几十年，甚至上百年，由于长年的开采，沉香资源越来越稀少，在二十世纪四五十年代天然沉香基本绝迹，在八十年代，中科院华南植物研究所研究人员将从沉香中分离出的真菌菌丝体接种到白木香树干中，经3～5年观察，白木香茎材中出现棕褐色斑纹，这种斑纹被证明为沉香树脂，可惜研究成果未能被广泛推广。经过多年研究实践，陆续出现多种注射异物加速沉香结香的技术，这些技术发展非常迅速，5年以上的白木香树在接种菌种10个月到1年多的时间里能结出可开采的沉香（图1-13），这一结香技术能使内皮层成环状结香，现期已在茎木中心开始结香试验，这对提升白木香树结香产量和加快加工提取沉香速度更有利。

目前沉香精油资源获取的相关技术正不断发展，包括人工加速结香技术、白木香种植技术、精油提取技术（如加压水汽蒸馏、水渗透法蒸馏、超临界CO_2流体提取和分子蒸馏等）。由于目前人工结香方式五花八门，产品质量参差不齐，除药用沉香标准已经由国家林业和草原局制定外，作为国际使用最多的香料——沉香精油应尽快制定符合我国国情、能达到国际认同的标准。这有助于今后这一宝贵精油资源更快发展和壮大。

图1-13　人工菌种通体结香

第三节　樟科樟属植物枝叶精油主要成分的多样性研究

一、樟科樟属种内枝叶精油成分多样性

早在20世纪50年代，日本藤田安二和我国焦启源教授在它们的著作中已经提到，樟科樟属中种内枝叶精油成分的多样性表现在同一种枝叶精油的主要成分的差异很大，当时焦启源教授称之为种内不同生理类型。往后这类研究报道不断出现在有关刊物中，而且这一发现多出现在同一地域，但在异地也有被发现，因此，根据这一研究在樟属中发现了不少有开发价值的精油成分，如在阴香和樟的枝叶精油中发现了宝贵的中药材右旋龙脑（天然冰片）；在樟的枝叶精油中发现了天然右旋樟脑；在黄樟枝叶精油中发现高纯度的右旋芳樟醇（花香型）；在产自云南的樟的枝叶精油中发现了高纯度黄樟油素等等。上述这些精油成分为现在和将来开发樟属精油提供了有力依据。樟科樟属种内化学类型（或生理类型）详见表1-2。

据资料介绍，樟属植物枝叶精油已发现30多种，现介绍其中20种。其枝叶精油的化学成分可以分为：脑樟型，主成分为樟脑；油樟型，主成分为1,8-桉叶油素；黄樟型，主成分为黄樟油素；姜樟型，主成分为柠檬醛；芳樟型，主成分为芳樟醇；丁香酚型，主成分为丁香酚和丁香酚甲醚；桂醛型，主成分为t-桂醛；龙脑型，主成分为右旋龙脑等。这些主成分与大多数樟属植物枝叶精油成分是相吻合的，但个别种的主成分含苯甲酸苄酯，则较为少见。

二、樟科樟属种内不同化学类型的种质资源库和种苗基地

从近期研究的樟属种内化学类型中，发现不少具有开发价值的化学类型，有必要深入研究。基于此，首先应建立这一类化学类型的种质资源库，以便为深入研究提供材料，把优良型通过无性繁殖，使其子代保留母体枝叶精油成分特征，进一步扩大种群，建成资源库，并逐渐将资源库扩大为无性繁殖的扦插圃，使其成为不断提供优质化学类型种苗的造林基地。整

个流程如下：

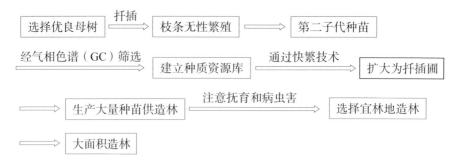

图1-14　建立樟属种内化学型种质资源库和种苗基地流程图

气相色谱（GC）筛选：通过GC检测枝叶精油主成分，选出母树剪一年生枝条进行扦插。

快繁技术：选好的枝条用生根药处理后在保湿保温塑料大棚扦插圃进行扦插繁殖，从种质资源库中移栽三年树龄的种苗以提供大量扦插枝条。

由于所营造的基本是单一纯林，要特别注意病虫害，且应在一年内抽检植株精油，查看有无变化，发现存在变异则须马上清除问题树。

第一节　参考文献

[1] 库斯托娃.精油手册[M].刘树文，胡宗蕃，译.北京：中国轻工业出版社，1982.

[2] 何兰，姜志宏.天然产物资源化学[M].北京：科学出版社，2008.

[3] 金英花，刘蕊，Lee Sung-je，等.长白山区芳香植物资源及精油成分调查[J].延边大学农学学报，2013，35（2）：107-115.

[4] 李宝灵，朱亮锋，林有润，等.中国蒿属植物化学分类的初步研究——精油化学成分与系统分类的相关性[J].华南植物学报，1992（1）：87-100.

[5] 李毓敬，李宝灵，曾幻添，等.湖南油樟的化学类型[J].植物资源与环境，1993，2（3）：7-11.

[6] 李毓敬，朱亮锋，陆碧瑶，等.天然右旋龙脑新资源——梅片树的研究[J].植物学报（英文版），1987，29（3）：527-531.

[7] 梁忠云，王国聪.草蒿脑资源及生物活性应用研究进展[J].广西林业科学，2010，39（1）：49-51.

[8] 刘布鸣，赖茂祥.广西药用精油植物资源与应用[J].广西中医学院学报，2005，8（2）：83-87.

[9] 刘布鸣，魏绣枝，赖茂祥.广西精油资源与药用[J].广西林业科学，2004（6）：28-29.

[10] 刘志秋，陈进，许勇.黄樟素植物资源的开发利用现状及前景[J].香料香精化妆品，2001（4）：14-19.

[11] 陆碧瑶，李毓敬，麦浪天，等.黄樟油素新资源——香楠的精油成分研究[J].林产化学与工业，1986，6（4）：7-12.

[12] 陆碧瑶，李毓敬，朱亮锋.樟属植物不同生理类型精油主要化学成分的研究[C]//中国科学院华南植物研究所.中国树木提取物化学与利用学术讨论会论文集.广州：中国科学院华南植物研究所，1986（11）：22-27.

[13] 吕洪飞.药用芳香植物资源的开发和研究[J].中草药，2000，3（9）：

[14] 711-715.

[15] 田玉红，张祥民，黄泰松，等.植物精油资源及其在烟草工业中的应用[J].广东化工，2007，34（10）：73-75.

[16] 吴航，王建军，刘驰，等.黄樟化学型的研究[J].植物资源与环境，1992，1（4）：45-49.

[17] 吴航，朱亮锋，李毓敬.阴香种内化学型的研究[J].植物学报，1992，34（4）：302-306.

[18] 吴卓珈，徐哲民，李春涛.芳香植物的研究进展[J].安徽农业科学，2005，1（2）：6-8.

[19] 张文莲，朱亮锋，陆碧瑶，等.银木种内化学类型研究[J].热带亚热带植物学报，1995，4（1）：61-64.

[20] 赵海燕，哈登龙，杨庆华，等.鸡公山自然保护区芳香植物资源调查研究[J].现代农业科技，2011（19）：237-238.

[21] 朱亮锋，曾幻添，李毓敬，等.异大茴香脑新资源——齿叶黄皮的研究

[J]. 植物学报（英文版），1987，29（4）：413-421.

[22] 朱亮锋，李宝灵，李毓敬，等. 精油资源研究概况[C]//中国科学院华南植物研究所. 中国科学院华南植物研究所集刊（第八集）. 北京：科学出版社，1992.

[23] 朱亮锋. 我国樟属精油资源的研究近况[J]. 植物资源与环境，1994，3（2）：51-55.

[24] 陆碧瑶，李毓敬，朱亮锋. 樟属植物不同生理类型精油主要化学成分的研究[C]//中国科学院华南植物研究所. 中国树木提取物化学与利用学术讨论会论文集. 广州：中国科学院华南植物研究所，1986（11）：22-27.

第二节　参考文献

[1] 曾幻添，董文忠，吴质朴. 沉香的人工结香[J]. 中草药通讯，1978，12：38-42.

[2] 弓宝，杨云，黄立标. 沉香精油化学成分和药理学研究关系的探讨[J]. 化学香精化妆品，2012，10（5）：43-48.

[3] 郭晓玲，田佳佳，高晓霞，等. 不同产区沉香药材挥发油成分GC-MS分析[J]. 中药材，2009，32（9）：1354-1358.

[4] 郝兰芳，李慧芬，周烨. 沉香挥发油化学成分分析[J]. 中草药，2010，14（2）：210-211.

[5] 李应兰. 檀香引种研究[M]. 北京：科学出版社. 2003.

[6] 李毓敬，朱亮锋，陆碧瑶，等. 天然右旋龙脑新资源——梅片树的研究[J]. 植物学报（英文版），1978，29（5）：527-531.

[7] 梁永枢，刘军民，魏刚，等. 沉香药材挥发油成分的气相色谱-质谱联用分析[J]. 时珍国医国药，2006，17（12）：2518.

[8] 林峰，梅文莉，吴娇，等. 人工结香法所产沉香挥发性成分的GC-MS分析[J]. 中药材，2010，35（2）：222-225.

[9] 林励，魏敏，肖省娥，等. 外界刺激对檀香挥发油含量及质量的影响[J].

中药材，2000，23（3）：152-154.

[10] 林奇艺，神永强，袁亮，等.激素对檀香生长的影响[J].中药材，2000，23（8）：457-459.

[11] 林奇艺，蔡岳文，袁亮，等.外界刺激檀香"结香"试验研究[J].中药材，2000，23（7）：376-379.

[12] 陆碧瑶，李毓敬，朱亮锋.樟属植物不同生理类型精油主要化学成分的研究[C]//中国科学院华南植物研究所.中国树木提取物化学与利用学术讨论会论文集.广州：中国科学院华南植物研究所，1986（11）：22-27.

[13] 丘雁玉，李飞飞，邓超宏，等.广东省3种野生香茅属植物精油的化学成分及含量分析[J].植物资源与环境学报，2009，18（1）：48-51.

[14] 汪科元.中药瑰宝——沉香[M].广州：南方日报出版社，2005.

[15] 吴航，王建军，刘驰，等.黄樟化学型的研究[J].植物资源与环境，1992，1（4）：45-49.

[16] 吴航，朱亮锋，李毓敬.阴香种内化学型的研究[J].植物学报，1992，34（4）：302-308.

[17] 徐金富，朱亮锋，陆碧瑶，等.中国沉香精油化学成分研究[J].植物学报，1988，30（6）：635-638.

[18] 许明英，李跃林，任海，等.檀香在华南植物园的引种栽培[J].经济林研究，2006，24（3）：39-42.

[19] 颜仁梁，林励.檀香的研究进展[J].中药新药与临床药理，2003，14（3）：218-220.

[20] 杨峻山，陈玉武.国产沉香化学成分研究[J].药学学报，1983，18（3）：191-198.

[21] 杨峻山，沉香化学成分的研究概况[J].天然产物研究与开发，1998，10（1）：99-103.

[22] 国家林业和草原局.沉香（Aqarwood）：LY/T 2904—2017[S].北京：中国标准出版社，2018.

[23] 朱亮锋，陆碧瑶，李宝灵，等.芳香植物及其化学成分（增订版)[M].海口：海南出版社，1993.

[24] ZHU LF，DING DS. New resources of essential oil in China[J]. Perfumer &

Flavorist，1991，16（7-8）.

[25] Maheshwari ML，Jain TC，Bhattacharyya SC. Structure and absolute configuration of α-agarofuran，β-agarofuran and dihydroagarofuran[J]. Tetrahedron，1963，19（6）：1079-1090.

[26] Nagashima T，Yoshida IT，Nakaniski T，etc. New Sesquiter penoids from agarwood[R]. sixth International oil Congress.Singapore.12-16.

第三节　参考文献

[1] 程必强，喻学俭，丁靖铠，等.中国樟属植物资源及其芳香成分[M].昆明：云南科技出版社，1997：8-9.

[2] 焦启源.芳香植物及其利用[M].上海：上海科学技术出版社，1953.

[3] 李毓敬，李宝灵，曾幻添，等.湖南油樟的化学类型[J].植物资源与环境，1993，2（3）：7-11.

[4] 陆碧瑶，李毓敬，朱亮锋.樟属植物不同生理类型精油主要化学成分的研究[C]//中国科学院华南植物研究所.中国树木提取物化学与利用学术讨论会论文集.广州：中国科学院华南植物研究所,1986(11)：22-27.

[5] 藤田安二.植物精油的基础研究[M].东京：小川香精株式会社.1951.

[6] 王羽梅.中国芳香植物精油成分手册（上、中、下三册）[M].武汉：华中科技大学出版社，2015.

[7] 吴航，王建军，刘驰，等.黄樟化学型的研究[J].植物资源与环境，1992，1（4）：45-49.

[8] 吴航，朱亮锋，李毓敬.阴香种内化学型的研究[J].植物学报，1992，34（4）：302-308.

[9] 张文莲，朱亮锋，陆碧瑶，等.银木种内化学型研究[J].热带亚热带植物学报，1995，4（1）：61-64.

[10] 朱亮锋，李泽贤，郑永利.芳香植物（自然珍藏图鉴丛书)[M].广州：南方日报出版社，2009.

第二章　植物精油的化学成分

第一节　植物精油的化学组成

　　植物精油由一类非常复杂的混合油状物组成，它是在植物生长发育过程中的一种代谢产物。植物精油不同于植物生长发育过程中的营养物质，诸如蛋白质、脂肪和糖类，其成分是植物细胞原生体（protoplasma）所分泌和积累的产物，并贮存在特定的器官中，诸如腺毛、油室、油管、分泌腺或树脂道中。因此植物体内的精油成分，即芳香挥发性物质，被认定为是植物体内的"排泄物"，是植物生长发育过程中的生理排泄废渣，亦可称为代谢产物或次生物质。

　　然而事实上，精油在植物生长发育过程中有其存在的作用，如植物本身受到外界入侵，如火烧、电击、撞击、昆虫和微生物侵害等，精油会起一种抵抗和保护作用，以及驱除功能，释放出芳香性物质——精油中的某些成分。另外精油的某些成分能引诱昆虫传播花粉，或在果实成熟时散发出芬芳诱人的香味等。这类物质是在植物生长发育过程中不断变化和积累的，不同种属以及同一种属不同器官的积累方式和途径也不一样。例如，当柑橘果实未成熟时，青绿色果皮的精油含大量柠檬醛（citral），柠檬醛缺少成熟果实芳香气味，而且还具有极强的抑杀微生物活性，但到柑橘果实成熟、果皮变金黄色时，柠檬醛基本不存在，而是具有成熟果香的柠檬烯（limonene）和其他酯类大量存在果皮精油中。

　　植物精油与另一些植物次生物如生物碱、黄酮、甾体等不同，从化合物角度来归类，它们是由多种类别的化合物组成，主要是萜类，其次是芳香族、脂肪族以及含氮、硫的杂环化合物，但它们的相对分子量都在280以下，结构比较简单，各类别的形成和积累不尽相同。

一、萜类化合物

萜类（terpene）从其化学结构来说，凡由甲戊二羟酸衍生，而其分子式符合$(C_5H_8)_nO$者，此通式的衍生物都称为萜类化合物，也称异戊二烯类化合物（isoprenoid），萜类化合物从其生源来说：它是由若干个异戊二烯单位聚合构成的，即所谓"经验异戊二烯法则"，而到1938年，瑞士人路奇斯卡（Ruzicka）将这一法则拓展为"生源异戊二烯法则"，而当时只是假设所有存在于自然界的全部萜类的前提是"活性的异戊二烯"。实际上后来人们认为异戊二烯的生物合成是由细胞质中的甲羟戊酸（mevalonic acid，MVA）途径而来的，但在20世纪90年代，大量植物次级代谢研究证明在质体中存在着磷酸甲基赤藓糖（methyl erythritol phosphate，MEP）途径。目前的研究表明，几乎所有的单萜、二萜和类胡萝卜素等异戊二烯类化合物均在质体中由MEP途径合成，而MVA途径则承担倍半萜和三萜等化合物合成。

萜类化合物的结构多种多样，从最简单直链无环碳氢化合物到单环、多环结构萜类化合物，低级萜类如单萜、倍半萜类及其衍生物主要存在于高等植物、藻类、苔藓和地衣等的各种器官中，昆虫和微生物中也有萜类。按照组成异戊二烯单元数目来划分$(C_5H_8)_n$，n为1称为半萜；n为2称为单萜；n为3称为倍半萜，如此类推二萜、二倍半萜、三萜等，详见表2-1。

表2-1　萜类化合物分类及其存在情况

名称	n	碳原子数目	主要分布
半萜	1	5	异戊二烯
单萜	2	10	精油
倍半萜	3	15	苦味素、精油、树脂
二萜	4	20	苦味素、树脂、乳汁、植物醇
二倍半萜	5	25	海绵、菌类、地衣
三萜	6	30	皂苷、树脂、乳汁
四萜	8	40	植物色素

单萜和倍半萜主要存在于高等植物中，具体来说主要存在于芳香植物的精油（芳香挥发成分）中，据不完整统计，在这些植物精油中单萜、倍半萜占2/3到3/4。其中单萜化合物占比最高，从其结构来划分可分为无环、单环和双环单萜以及三环单萜。

1. 其中无环单萜除萜烯化合物及其含氧衍生物（醇、醛、酮、醚）等，它们主要存在于木兰科、樟科、桃金娘科、芸香科、禾本科、龙脑香科、夹竹桃科、伞形科、败酱科、唇形科、毛茛科、松科、柏科、姜科、菊科等精油的低沸点部分。

无环单萜（acyclic monoterpenoid）中代表性化合物主要是含氧衍生物单萜醇、单萜醛和单萜酮等，它们主要存在于柑橘类精油、芳樟油、香茅油、香叶天竺葵油以及一些香花类精油中，如芳樟醇linalool（1）、香茅醇citronellol（2）、香叶醇geraniol（3）和橙花醇nerol（4）等。

而单萜醛主要是存在于山苍子油、姜樟叶油的柠檬醛citral（5），它是混合异构体，香叶醛geranial（6）、橙花醛neral（7）和柠檬烯（8），另外香茅醛（citronellal）主要存在于柠檬桉叶油、枫茅、香茅油中。

单环单萜（monocyclic monoterpenoid）中代表性化合物有单萜烯、单萜醇、单萜醛、单萜酮、单萜醚等，其中伞花烃（聚散花素）cymene广泛存在于精油的低沸点部分，如松节油、桂（皮）油、柠檬桉和蓝桉油、芫荽子油、樟油、土荆芥油等。由于伞花烃大量存在于孜然芹的种子中，故又称孜然芹烯（P-cymene）（9），另一单环单萜烯松油烯醇-4（terpinen-4-ol）（10）是主要的杀虫成分，澳洲产的互叶千层叶精油（茶树油）含70%～80%。另外沙地柏（*Sabina vulgaris*）叶精油同样以4-松油醇为主要成分。薄荷醇（脑）（11）也是单环单萜的重要成分，在天然薄荷油中，只含有右旋体薄荷醇（12）和左旋体薄荷醇。薄荷酮（13）在薄荷油中含10%～25%，为左旋体。香芹酮（carvone）（14）或称藏茴香酮，存在于藏茴香（葛缕子）（*Carum carvi*），在莳萝（*Anethum graveolens*）果实精油中高达60%，为右旋体，而在绿薄荷（*Mentha viridis*）精油中则含左旋体高达60%，它是泡泡糖、口香糖的重要香原料。胡椒酮（piperitone）（15）又称辣薄荷酮，存在于日本产的薄荷油中，为右旋体，也存在于胡椒桉叶油（eucalyptus piperita）中。1,8-桉叶油素（16）也是单环单萜的重要精油成分，它主要存在于蓝桉叶油（Eucalyptus globulus），高达95%，是生产桉叶糖的主要原料。此外，白树、白千层（*Melaleuca leucadendron*）等桃金娘科桉属和千层属叶精油均广泛分布；在土荆芥油中的主要成分驱蛔素（ascaridole）（17）含量达70%，是有效的驱蛔虫药物。

双环单萜（bicyclic monoterpenoid）较为多样，它们从平面结构可分为侧柏烷、蒈烷、蒎烷、莰烷四种。

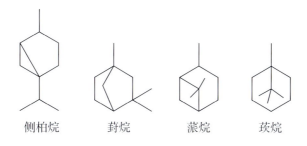

侧柏烷　　　莐烷　　　蒎烷　　　莰烷

在我国产量最大的松节油中，α-蒎烯 α-pinene（18）占 70% 以上，β-蒎烯占 20%（19）。此外，在很多植物精油中，单萜类精油中低沸点馏分很多都含有蒎烯，如八角茴香、百里香、蓝桉、柠檬、茴香、芫荽、薄荷等是合成龙脑、樟脑以及一些香料的重要原料。马鞭草油中存在的马鞭草烯酮（verbenone）（20）是合成一种抗癌药紫杉烷的骨架。而小茴香（一种食用辛辣香原料）和侧柏精油中则含有大量的具旋光性的侧柏酮（thujone）（21）。龙脑（Borneol）是一种中药材，其右旋体（22）则存在于龙脑香树（*Dryobalanops aromatica*）根茎中以及阴香（梅片树）和龙脑樟的枝叶精油中，为中药梅片，它的左旋体称为艾片（23），同样是一种中药材，但药效与右旋梅片不同，主要存在于菊科草本艾纳香中。樟脑存在于菊蒿（*Tanacetum vulgare*）精油中，为左旋体，而大量存在于樟脑油中的为右旋体，在酊水剂、膏霜等制剂中应用甚广，但现今多为合成的消旋体代替。

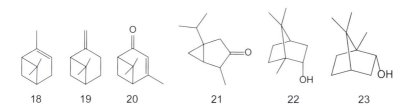

18　　　19　　　20　　　　21　　　　22　　　　23

除了上述三类单萜成分外，尚有䓬酚酮类，这是植物精油不常见的一类单萜成分，它们不符合异戊二烯法则，它们存在于柏科植物心材中，如存在于北美乔柏（*Thuja plicata*）、北美崖柏（*T. occidentalis*）、罗汉柏（*Thujopsis dolabrata*）的茎木心材中的 α-崖柏素（α-thujaplicin）（24）、γ-崖柏素（γ-thujaplicin）（25）、台湾扁柏（*Chamaecyparis taiwnensis*）的心材含有扁柏醇（Hinokitiol）(26)和其氧化产物β-多位布林（β-dolabrin）(27)，这些化合物均有抗微生物活性。

环烯醚萜类化合物虽然在茜草科、龙胆科、唇形科都有存在，并发现这类化合物多达800余种，但在我们常见到的植物精油中的确很少，本文不再赘述。

2. 重要的倍半萜类化合物。倍半萜类（sesquiterpenoids）从平面结构分，目前所发现的有41种，可分为无环、单环、双环、三环、四环5个类型；按环上碳原子数目则可分为五元环、六元环、七元环直到十二元大环。现按结构类型列出主要的倍半萜类。

无环倍半萜化合物（acyclic sesquiterpene）金合欢烯（farnesene）和它的含氧衍生物金合欢醇，广泛存在于多种香花精油中，如金合欢花油（*Acacia farnesiana*），还有橙花叔醇（nerolidol）同样存在于橙花油等多种香花精油中。

α-金合欢烯　　β-金合欢烯　　金合欢醇　　橙花叔醇

单环倍半萜（monocyclic sesquiterpene）类型较复杂，其中生物活性较显著的有没药烷型、蛇麻烷型、牻牛儿烷型，详见表2-2。

表2-2 重要单环倍半萜

结构类型	主要化合物	植物来源
	姜烯	姜、黄姜、百里香精油
没药烷型	γ-没药烯	红没药树脂精油、樟油、杨树嫩芽精油
	姜油酮	姜黄根茎精油
蛇麻烷型	α-蛇麻烯 β-蛇麻烯 蛇麻二烯醇 蛇麻二烯酮	蛇麻（啤酒花精油）
牻牛儿烷型	牻牛儿酮	牻牛儿苗科大根老鹳草、杜鹃花科兴安杜鹃叶精油

（续表）

结构类型	主要化合物	植物来源
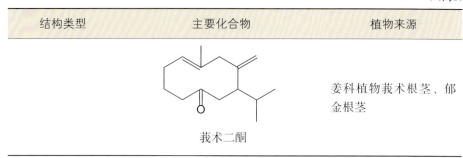 莪术二酮		姜科植物莪术根茎、郁金根茎

双环倍半萜类化合物结构类型很多，按生源途径和基本碳骨架分为萘基型衍生物类如艾里莫芬烷型、杜松烷型、拉松烷型、桉烷型等，以及薁型（azulene）衍生物类型。每类又以其骨架含氧衍生物如醇、醛、酮、酯等来划分化合物类型。

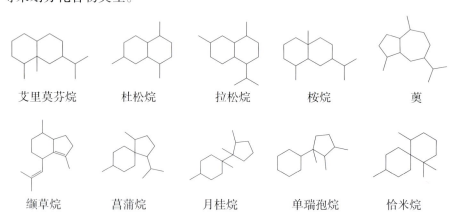

上述10个双环碳化合物及其含氧衍生物广泛存在于菊科、桃金娘科、马桑科、兰科、卫矛科等植物。

而薁、菖蒲烷、月桂烷、单瑞孢烷、恰米烷等结构式属于双环倍半萜类。其中薁型化合物为非苯芳烃化合物，它们存在于一些植物精油的高沸点部分，出现在蓝色或绿色的馏分中，常称为蓝油烃。

愈创木醇（guaiol）为重要薁型化合物，存在于愈创木（Guajacum officinale）的木材精油中。愈创木醇用硫或硒脱氢可产生1,4-二甲基-7-异丙基薁，亦称愈创木薁（guaiazulene）和2,4-二甲基-7-异丙基薁，可用于合成抗炎、抗溃疡药物（见下页附图）。

愈创木薁　　　　　愈创木醇　　　　2,4-二甲基-7-异丙基薁

此外薁类还包括广藿香酮、洋甘菊薁等，其结构如下：

广藿香酮　　　　　　　　洋甘菊薁

倍半萜内酯类是较为重要的内酯类化合物，倍半萜内酯源于牻牛儿内酯（germacranolide），由此能演化多种结构类型化合物。内酯类化合物广泛存在于菊科、伞形科、木兰科、卫矛科、芸香科、葫芦科、苏木科、蝶形花科以及含羞草科等植物中，它们多具生物活性，如山道年草（Artemisia cina）、蛔蒿含的左旋-α-山道年（1-α-santonin）是很好的天然驱蛔药。木香根（Aucklandia costus）含有的木香内酯和黄花蒿（artemisia annua）含有抗恶性疟疾的有效成分青蒿素（artemisinin），它们的多种衍生物已成为临床药物。还有变色马兜铃（Aristolochia versicolor）、绵毛马兜铃（A. mollissima）根茎精油和关木通（A. manshuriensis）茎中的一种十二元大环内酯，对人体肝癌细胞Qgy7703有抑制作用。详见如下附图：

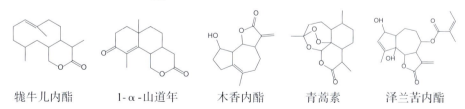

牻牛儿内酯　　　1-α-山道年　　　木香内酯　　　青蒿素　　　泽兰苦内酯

二、芳香族化合物

芳香族化合物在精油化学成分中仅次于萜类，然而它们在挥发性物质（香气）中占重要地位，不少是一些小分子萜源衍生化合物，如麝香草酚（thymol），它的芳香骨架是孟烷（薄荷烷）脱氢产物，同样α,β-姜黄烯

（curcumene）等是萜类演变而来的。另外苯丙烷具有C_6H_3碳骨架，从苯酚或其酯类结构演变而来，如桂醛（cinnamaldehyde）、茴香脑（anethole）、丁香酚（eugenol）、α和β-细辛脑（α和β-asarone）、爱草脑（estragole）；另一些还有C_6H_1和C_6H_2的碳骨架，如茴香醛（anisaldehyde）、花椒油素（xanthoxylin）等（见如下附图）。

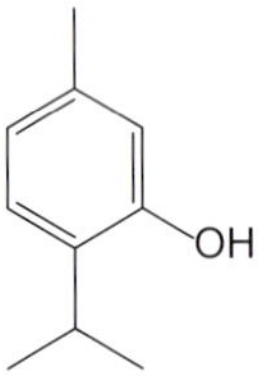

麝香草酚 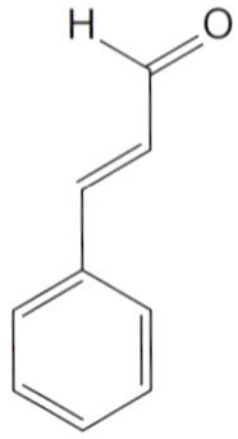　　　　　桂醛 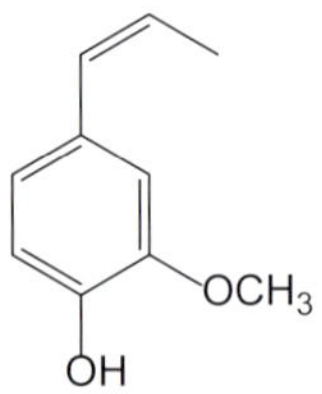　　　　　丁香酚 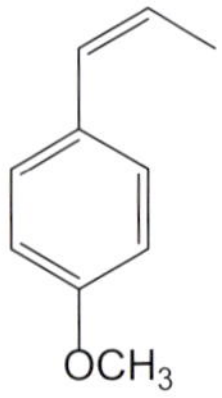　　　　　胡椒酚

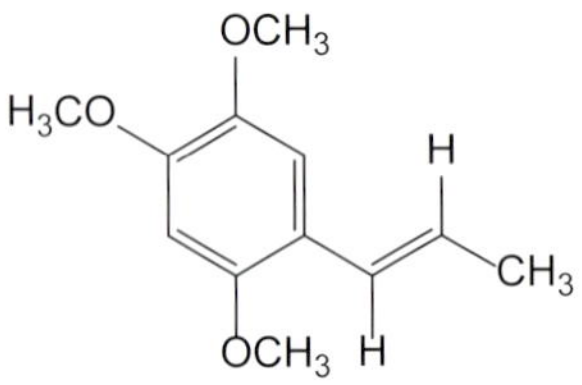

α-细辛脑 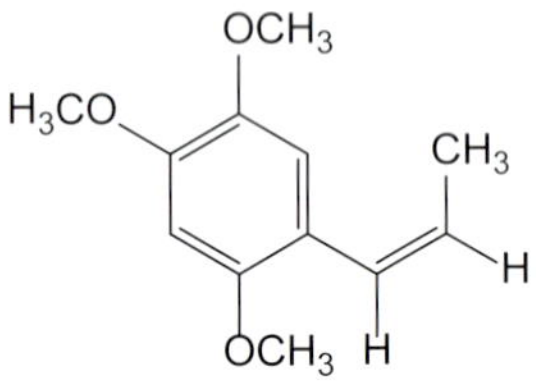　　　　　　β-细辛脑 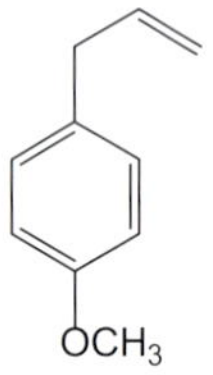　　　　　爱草脑

香豆素（coumarin）是植物精油中另一类芳香族化合物，含苯骈α-吡喃酮母核的天然成分，为邻羟基桂皮酸的内酯，它在植物体内起着植物激素作用，在低浓度时能刺激发芽和促进生长，高浓度时则可抑制发芽和抑制生长。香豆素广泛存在于双子叶植物中，主要为伞形科、芸香科、菊科、茄科、木樨科、苏木科、蝶形花科、含羞草科等。香豆素除具有植物激素活性外，其α-吡喃酮类化合物还具有多种生物活性。

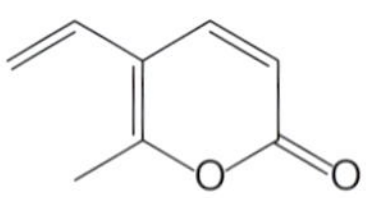

苯骈α-吡响酮 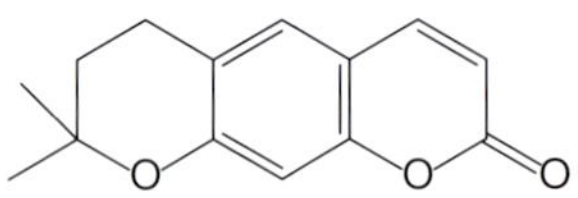　　　　　花椒内酯 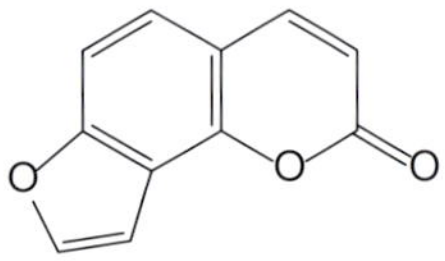　　　　　白芷素

三、脂肪族化合物

在精油中存在一些小分子的脂肪族化合物，如正癸烷（n-decane）存在

于桂花中，以及它们的含氧衍生物，如甲基正壬酮（methyl nonylketone），正壬醇（*n*-nonylalcohol）。有研究表明，它们主要是脂肪酸的代谢产物。

四、杂环类化合物

这类含硫和含氮化合物并不普遍存在于植物精油中，较为常见的是存在大花茉莉和小花茉莉中的邻氨基苯甲酸甲酯（methyl anthranilate）、吲哚（indole），存在于大蒜中的含硫化合物大蒜辣素（allicin）等。这类化合物一般很少被划入植物精油成分主要研究范围，只有对某一种含有上述含硫、含氮化合物单独研究时才会提及。

第二节　植物精油成分分离和鉴定

不同芳香植物的植物精油成分获取是用不同的方法，在下一章节中将较详细介绍这些方法，使用水汽蒸馏是最为普遍的，但获取的精油样品中会有少量易溶于水的组分，如含有羟基的组分需要再用有机溶剂萃取、干燥、浓缩回收溶剂才算获得较为完整的精油，对鲜花用水汽蒸馏法则"头香"会损失，只能用大孔树脂吸附法才能真正收集到它们的全部芳香成分。另一种获得精油成分的方法是用低沸点有机溶剂，如石油醚、己烷等，一般在常温下萃取，蒸去溶剂后获得浸膏，再用冷冻乙醇除去花蜡和色素后获得净油。这一方法主要用于鲜花，该方法所用设备及工艺要求较为烦琐，并且设备投资较高，涉及防火、防爆的措施和设备。

一、分离方法

1. 蒸馏法（分馏、精馏）

用于植物精油成分分离，由于精油存在于各种芳香植物的不同器官中，来源不同，它们的成分也不相同。它们多为油状液体，而且沸点不算高，C_{10}主沸点为140～180℃（40～100℃/10 mmHg）（1 mmHg=133.3 pa），C_{15}主沸点为250～280℃（140～180℃/8 mmHg）。根据每种精油实际情况可先进行减压分馏或先用化学方法把精油中酸性部分分离出，也可以用吸附方法把精油中萜烯类和含氧部分分开，然后再细分。

2. 色层分析法

1）薄层分离法：对某些含单萜类为主的精油如松科、杉科枝叶精油、柑橘类果皮精油，可直接用薄层层析中极性小的展开剂检查。可先使用实验室精密分馏（塔板数在30～40），收集不同沸点馏分，再用薄层层析或气液色谱检测各馏分情况，根据结果再进一步分离纯化。若纯度达到鉴定要求，可制备衍生物或光谱测定。

2）柱层层析分离：层析分离精油样品可直接进行氧化铝或硅胶柱层析，也可在分馏基础上选择某些馏分进行柱上层析，吸附剂即固定相可根据实际精油样品来决定，如分离单萜烯类则选择吸附能力较强的吸附剂，如果分离含氧的或芳香族化合物则不一定需要吸附剂有太强的吸附能力，一般多用活性硅胶，填装柱多用湿法。展开剂多是先用极性低的溶剂逐步改用极性较高的溶剂，分段收集，用薄层层析检查每一馏分。展开剂选用也可先用薄层层析结果来参考。如果获得的纯度不高，不宜做衍生物鉴定，只能用光谱或波谱测试。如需要分离的样品太少，只能用薄层层析进行分离，获得少量纯样品也只能用光谱或波谱进行鉴别（红外光谱等）。

另一种柱层层析法是反向柱层层析法，反向柱层与常用柱层层析分离不同，常用柱层样品是在柱顶投入，利用展开剂通过重力使样品自上而下慢慢移动，直到出口。反向柱层层析法（如图2-1），样品是从柱底部由展开剂慢慢向上推进，被分离样品则从顶部流出。这一方法分离效果较常用方法分离效果好。但操作比较麻烦，应用较不方便。但还有一种柱层层析方法，先把样品从柱顶投入后，用展开剂展开至底部时则停止加入展开剂，用紫外灯或荧光灯照，样品将会有荧光反应，可把固定相推出分段切割收集所需的一段，也可不用玻璃柱，用透明塑料膜软管，则不用把固定相推出分段切割收集所需的一段，可直接用刀分段切割，但所需样品成分必须有紫外或荧光反应。

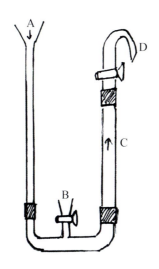

图2-1　反向柱层层析设备示意图

A.溶剂入口　B.样品入口　C.装吸附剂柱　D.层析液出口

3.吸附法

早期吸附方法收集植物精油，吸附剂多用活性炭、硅胶、脂肪等，除脂肪吸附鲜花生产香脂外，已改用大孔树脂为吸附剂，用于收集鲜花头香在水中的挥发成分非常成功，结合CO_2超临界萃取方法，从大孔树脂萃取头香成分效果非常好，而大孔树脂可再生重复使用，方法将在第三章节中详细讲述。

4.CO_2超临界流体萃取

超临界流体萃取这一方法不断发展和改进，已经广泛伸展到各个领域中，已被用于提取贵重的植物精油，如沉香、檀香、鸢尾、啤酒花等，结合分子蒸馏这类研究和生产已被越来越多研究报道。当然一种方法是高压，另一方法则是高真空，这些设备比较昂贵，目前广泛应用于精油方法是不可行的。（在第三章节中将较详细讲述）

5.化学方法

这一方法主要是针对精油中含氧的衍生物，首先是要把不含氧如萜烯分出，用精密分馏方法把低沸点成分蒸出。如果精油样太少可改用硅胶柱层析，柱口径较粗的为100～120毫米，柱长约300毫米，精油用1：10至1：15低沸点石油醚稀释后，直接倒入填装好的硅胶柱，使混合溶液迅速通过硅胶柱，不含氧精油随石油醚一起被冲出，含氧组分留在硅胶柱中，再用乙酸乙酯把留在硅胶柱中的含氧组分全部冲洗出来，把两部分溶剂回收，则获得含氧精油和萜烯两部分。

1）含氧组分如含有酚类，则用3%～5%氢氧化钾或氢氧化钠溶液萃取用精制乙醚稀释的含氧组分，碱萃取液用盐酸中和至酸性，则析出酚性物，也会有含羧基的化合物，可用弱碱碳酸钠萃取，可分离出含羧基的化合物，酚类化合物可通过柱层层析进一步纯化。含氧组分亦可在用碳酸钠直接把酚性物提取纯化后，用紫外可见光谱（UV）和红外光谱（IR）来对照鉴定，可直接查阅标准图谱和已知数据（对应已知化合物）。

2）组分含有一定醇类化合物，可用丙二酸单酰氮反应，产生相应的单酯，这类单酯溶于碳酸钠溶液中，用乙醚把不溶于碳酸钠溶液的非醇类物提取出来，再用盐酸酸化后用乙醚把单酯提取出来。干燥后把乙醚蒸去得到单酯，再用20%氢氧化钠溶液回流水分解，用乙醚萃取，蒸去乙醚则获得醇类化合物。

另一方法，从精油含氧组分中分离提取含醇组分，基于醇类化合物能与氯化钙生成复盐。复盐溶于水，除去油层杂质后，或用稀乙醇萃取醇类，留下的为萜烯；有一部分醇类能与洋地黄皂苷（digitonin）反应，生成复合物析出。上述方法最大的缺点是损耗大，操作麻烦，应用这些方法的不多。

用上述方法获得醇类组分，如果有一定数量，则可以进行精密分馏，做进一步分离纯化。获得较高纯度者则可以制备衍生物或进行光谱测试，纯度未达到制备衍生物者可再通过柱层层析分离纯化或制备薄层加以再进一步纯化。衍生物的制备，可根据组分中醇类种类选择衍生物制备方法，一般常用3,5-二硝基苯甲酸酯、3,5-二硝基苯甲酸酯的α-萘胺、α-萘胺甲酸酯，它们的产物均为结晶体，较易重结晶纯化，并有固定的熔点，有文献可查阅对照，但这类衍生物不易还原成原来的醇类。

组分含有羰基化合物（醛、酮），这类含羰基组分除了萜醛和萜酮外，还有直链和有支链脂肪族醛、酮类和芳香族醛、酮，它们能与亚硫酸氢钠发生加成反应，生成结晶性产物，并且大多数可溶于水，这样可用乙醚等有机溶剂把未加成的非羰基组分分离出来，再加入酸或碱则可以把羰基组分还原成醛酮。这一方法较为简单常用，但必须控制还原时间和温度，否则反应会出现不可逆，浪费全部样品。另外有些酮类不一定起这一反应。

另一方法是使用含羰基化合物与吉拉德（Girard）T或P试剂加成反应生成亲水性加成产物[Girard试剂为分子内带有酰肼及季铵基团的总称，已配制成吉拉德T（P）试剂成品出售]，它可以与任何羟基化合物起反应，一般是把加成组分除去酚、酸类后干燥，加入亚硫酸氢钠饱和溶液搅拌，或根据一般情况下有加成物结晶析出，用酸或碱处理结晶，再用乙醚萃取，乙醚萃取液水洗至中性后再用无水硫酸钠干燥，回收乙醚后则获得醛酮的化合物；如果此操作不能析出羰基化合物，则此时可在油样中加入吉拉德试剂T或P的乙醇溶液以及10%乙酸，这可使反应迅速顺利进行，加热回流1小时，则生成缩氨脲衍生物，用乙醚萃取反应液以除去未反应非羰基化合物，水溶部分用酸处理，则可用乙醚萃取回收羰基组分。所得混合羰基组分通过柱层层析或制备薄层进一步分离纯化，获得较纯组分醛类则可制备衍生物、缩氨脲、2,4-二硝基苯肼、对硝基苯肼等，详见附图如下：

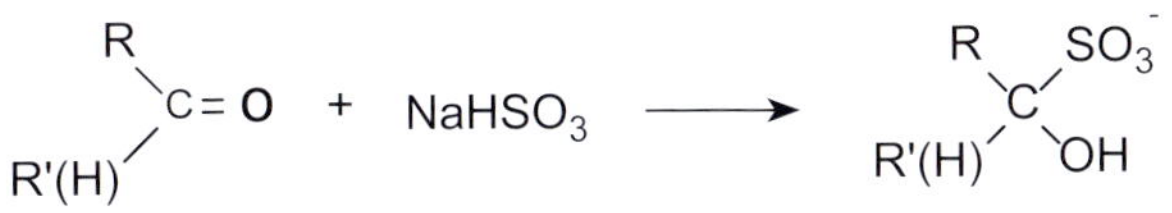

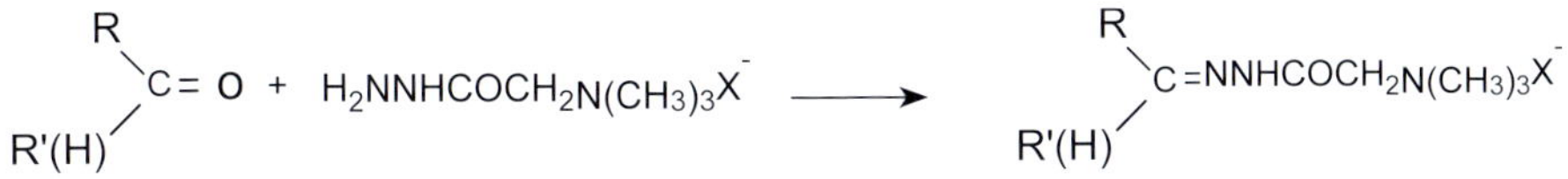

羰基化合物　　　　　吉拉德试剂T　　　　　　　　　吉拉德腙

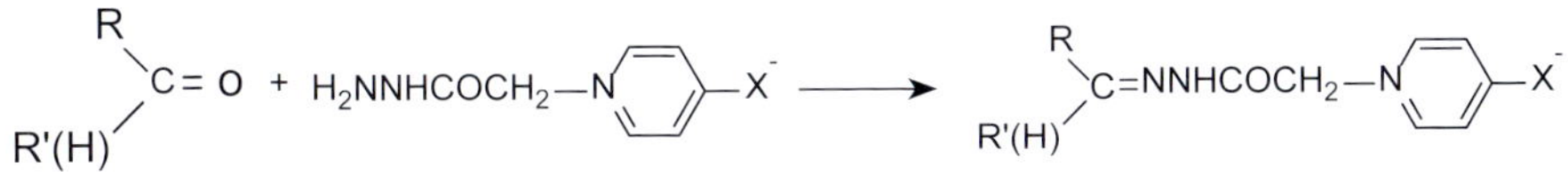

羰基化合物　　　　　吉拉德试剂P　　　　　　　　　吉拉德腙

6. 酯类和内酯类：酯类除水杨酸甲酯外绝大多数是以游离状态存在于植物精油中，脂肪族醇类与脂肪族酸结合成酯，如丁酸乙酯，辛酸、戊酸等的甲酯、乙酯、丙酯等；如脂肪族醇与芳香族酸结合成苯甲酸酯类；又如芳香族醇类与脂肪族酸结合成如乙酸苯甲酯、甲酸苯甲酯等，也有芳香族醇与芳香族酸结合成苯甲酸苯甲酯、苯乙酸苯甲酯等，它们经水解则还原成相对应的醇和酸。两者分离后醇类则可制备衍生物，光谱鉴定，酸可制备异羟肟酸衍生物，进行纸上层析，用三氯化铁显色。内酯在植物精油中并不非常普遍，主要是以倍半萜内酯存在于精油中，最著名的如菊科山道年（santonin）、青蒿素（artemisinin）、马兜铃内酯（caristololactone），这一类内酯与碱作用生成溶于水的开环盐类，而再酸化后又还原析出内酯，根据红外光谱，在1700～1800 cm⁻¹区间有内酯类的特征吸收峰。

用化学方法鉴定内酯是用羟胺与内酯反应生成羟肟酸，它能与氯化铁生成复盐，呈紫色：

$$RCOOR_1 + NH_2OH \longrightarrow RCONHOH + R_1OH$$

7. 苷类：在植物精油中萜类、脂肪族等与糖类结合成苷，这类化合物多溶于水以及低分子醇类如甲醇、乙醇等。如果精油中含苷类如水杨苷、苦杏仁苷等，可用甲醇或乙醇萃取，加水后再以乙醚或石油醚萃取以除去脂溶部分，再用正丁醇萃取，回收正丁醇获得混合苷，经水解后获得苷元，再用层析分离纯化做进一步鉴定。

二、精油成分分析和波谱技术应用

波谱方法应用在精油研究方面主要包括紫外可见光谱（UV）、红外光谱（IR）、核磁共振波谱（NMR）、质谱（MS）以及气液层析与质谱联用（GC/MS）。其中紫外光谱和红外光谱早在20世纪三四十年代就被应用在植物精油成分研究方面。

1. 紫外光谱。紫外光谱是检测精油中组成的分子被紫外光（200～300纳米）照射产生振动的一种吸收光谱图，它是通过紫外光谱检测仪被记录下来。检测的样品是用乙醇为溶剂，样品约为1毫克，在透光长度为1厘米的石英吸收池中测量。紫外光的波长为200～400纳米，并自动记录出这一波段数据中吸收位置和强度（峰值），吸收峰最大值（λ_{max}），通过这两组数字可以了解到羰基前的双键位置，是否共轭双键。测量精油样品中某些成分含量也可通过吸收池中样品浓度、吸收池透光距离（一般为1厘米），在紫外光谱图中读出光密度等数据则可以利用公式$\sum=D/C1$，测量到样品中某一含量。$\sum$为分光密度，D为测量出光密度，C为被量溶液浓度，测量时透光长度一般为1厘米。而上一公式可改为E1%/1cm=D/C1。这时C则为百分浓度，即100毫升溶液中溶解的样品量（克），一般有双键的有机化合物才能在紫外光区域中显示出吸收峰，因而常被用来确定样品中的共轭系统。因此存在双键的同类化合物中它们的紫外吸收是一致或类似的，甚至除共轭外基本骨架也可能相同（实例应用见图2-2，是一种精油成分的混合物谱图）。

图2-2　紫外光谱示意图

2. 红外光谱。红外光谱是精油中有机化合物分子被波长2～15微米红外光照射后其分子内键振动而产生的吸收光谱，测量时可把样品与KBr结晶研磨混合后压片，也可采用涂膜方式，亦可用氯仿等有机溶剂溶解后测量。在波长2～15微米的红外光照射后，样品分子内振动被红外光谱仪自动记录出的红外光谱图，见图2-3。

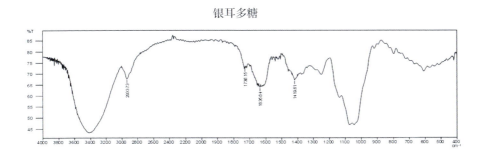

银耳多糖

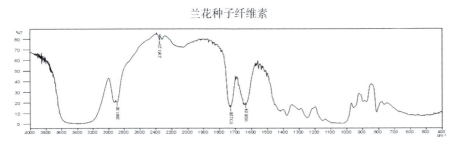

兰花种子纤维素

图2-3　红外光谱示意图

红外光谱图中波长范围为 $4000\sim400\,cm^{-1}$，其中 $600\sim1600\,cm^{-1}$ 代表单键区，即 N—H、C—C、C—H、C—N、O—H 的振动吸收，$1500\sim1900\,cm^{-1}$ 代表双键区，即 C=C、C=O、C=N 的振动吸收，$2000\sim2400\,cm^{-1}$，为三键区，$C\equiv C$、$C\equiv N$ 的振动吸收，$2500\sim3700\,cm^{-1}$ 为氢键区，表示 N—H、O—H、S—H、C—H 的伸缩振动吸收。$1400\sim4000\,cm^{-1}$ 也称功能团区；$400\sim600\,cm^{-1}$ 也称指纹区。每种化合物在这一区都有其特征吸收峰。这就是说两个同一种化合物可叠加重复。目前标准红外光谱图集可以查阅到已知化合物，加上红外光谱图库中已经用识别软件功能，则更容易解决查阅问题。

3. 核磁共振波谱法。核磁共振波谱仪是研究处于磁场中的原子核对射频辐射（Radio-frequency Radiation）的吸收，它是对各种有机和无机物的成分、结构进行定性分析的最强有力的工具之一，有时亦可进行定量分析。在强磁场中，自旋量子数（I）不为零的原子核发生自旋能级分裂（能级极小：在 1.41T 磁场中，磁能级差约为 $25'10-3J$），当吸收外来电磁辐射（$109\sim1010\,nm$，$4\sim900\,MHz$）时，原子核将在不同能级间发生核自旋能级的跃迁，从而产生所谓核磁共振现象。其中，氢谱是记录有机化合物中各个质子在外加磁场中受照射频率作用所产生不同的共振频率，用化学位移来表示，一般以四甲基硅烷（TMS）或六甲基二硅醚（HMDS）为内标测定各质子共振频率与它的相对距离，这个相对距离值称化学位移，常用 δ 表示，有时也用 ζ 来表示，四甲基烷的 $\delta=0$ 或 $\zeta=10$，通常可用 $\zeta=10-\delta$ 表示，质子周围的环境不同，产生不同的化学位移，质子因自旋偶合而产生裂分，根据不同的化学位移与偶合常数（J）就可以解析化合物的结构。由于有机化合物中碳原子数目比氢少得多，而精油成分更是如此，通过碳谱能了解分子中碳的分布，特别是化合物的骨架情况，因此碳谱对解释结构有更大帮助。具体操作上用 20 毫克样品溶解于 0.5 毫升不含氢的溶剂中，因为氢（质子）会干扰核磁共振，常用四氯化碳、三氯氟碳（$CDCl_3$）、氘代苯等作溶剂，这通过核磁共振仪可获得显示各种信号即峰形的氢图谱，各个峰的面积积分总和应该相当于样品氢的总和，图谱将同时显示出一条积分线，核磁共振图谱中，在不同化学位移位置可出现单峰、双裂峰和三裂峰甚至多裂峰等。这些峰代表某个质子相邻几个形成 n+1 个峰，这样可以推算出核磁共振图谱中化合物的各个质子即氢的分布情况，有助于了解化合物的结构，从中获得非常重要的信息。精油基本都是单萜、倍半萜有机小分子，结构多种多样。在 ¹H-NMR 中，峰的数目标志分子中磁不等性质子的种类；峰的强度表明每类质子的数目（相对）；峰的位移值（δ）表示每类质子所处的化学环境、化合物中位置；峰的裂分数表明相邻碳原子上质子数；偶合常数（J）表明化合物构型。结合 ¹³C-NMR 和 DEPT135 提供的"碳骨架的个数以及 CH_3、CH_2、CH 和季碳的个数和分布"的信息，基本可以确定精油小分子的结构组成。核磁共振波谱仪主要分析比较纯的化合物，精油成分也不例外，因此分析精油成分，需要将精油从植物中提取出来并通过纯化分离才可以进行核磁共振波谱分析。以下图2-4～图2-15为精油核磁共振波谱：

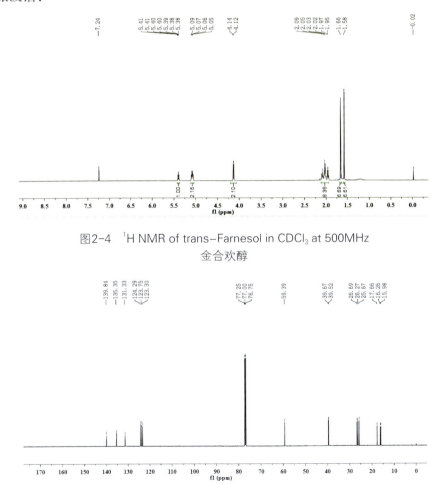

图2-4 ¹H NMR of trans-Farnesol in CDCl₃ at 500MHz
金合欢醇

图2-5 ¹³C NMR of trans-Farnesol in CDCl₃ at 125MHz

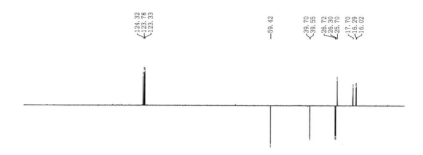

图2-6　Dept135 of trans-Farnesol in CDCl₃ at 125MHz

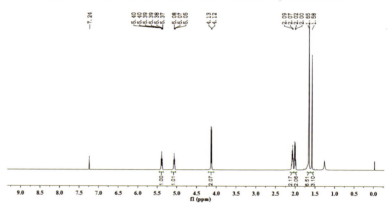

图2-7　¹H NMR of Geraniol in CDCl₃ at 500MHz

香叶醇

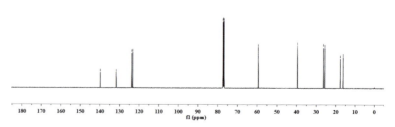

图2-8　¹³C NMR of Geraniol in CDCl₃ at 125MHz

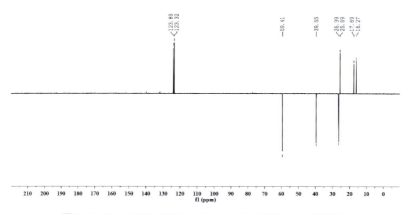

图2-9　Dept135 NMR of Geraniol in CDCl$_3$ at 125MHz

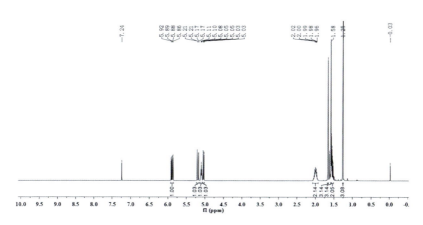

图2-10　^{1}H NMR of Linalool in CDCl$_3$ at 500MHz

芳樟醇

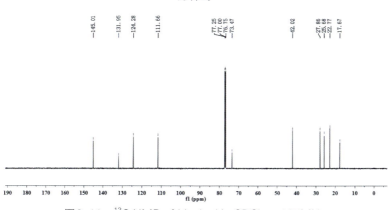

图2-11　^{13}C NMR of Linalool in CDCl$_3$ at 125MHz

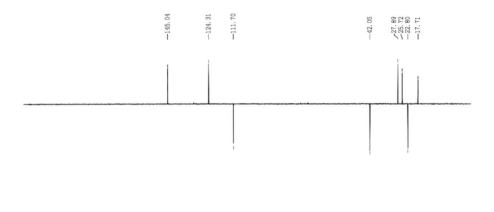

图2-12 Dept135 NMR of Linalool in CDCl₃ at 125MHz

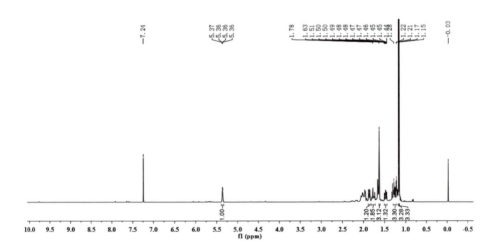

图2-13 ¹H NMR of α-Terpineol in CDCl₃ at 500MHz

α-松油醇

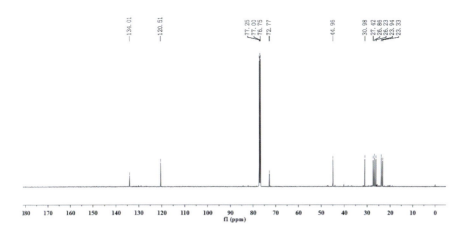

图2-14　^{13}C NMR of α-Terpineol in CDCl$_3$ at 125MHz

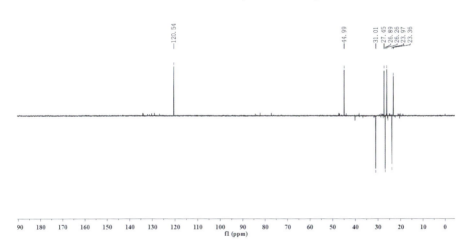

图2-15　Dept135 NMR of α-Terpineol in CDCl$_3$ at 125MHz

4. 气相色谱-质谱联用法。该法总的来说是气相色谱进行分离，质谱进行检测。色谱法是一种分离分析技术，气相色谱法是流动相为气相的一种色谱分析方法。它的本质在于色谱柱高选择性的高效分离作用与高灵敏度检测技术的结合。所有在加热条件下易气化并且稳定的有机化合物都可以用气相色谱来分析检测。其工作原理为：所分析的混合组分在仪器的进样口气化，在载气的携带下，同一时刻进入色谱柱，由于流动相和固定相之间溶解、吸附、渗透或离子交换等作用的不同，随流动相在色谱柱中运行时，在流动相和固定相中反复多次进行分配，使得分配系数本来只有微小差别的组分，产生了保留能力明显差异的效果，进而各组分在色谱柱中的移动速度就不同，经过一定长度的色谱柱之后，混合物彼此很好地分离开来。在组分的分离作用中色谱柱的固定相起着决定性的作用。经色谱柱分离的物质按顺序离开色谱通过检测器，先后流出的各组分将按其理化特征进行检测，即按各组分含量的多少转化成易于测量的电信号，产生的离子流信号经放大后，在记录器上绘出各组分的色谱峰，从而达到定性或定量分析的目的。其定性依靠保留时间，定量则依靠峰面积或峰高。其中记录器可以是质谱和各类检测器。质谱（MS）仪是利用带电粒子在磁场或电场中的运动规律，按其质荷比实现分离分析，测定离子质量及其强度分布。每个峰代表一个质量数，一般可根据质谱提供的分子离子峰和碎片峰质荷比确定分子量并推断被测量的分子结构。对于精油成分中的单萜、倍半萜在质谱图中会显示出强碎片峰有152、154、204、202、220 等，质谱仪除在精油研究中常用电子轰击电离源外还有化学源、场致电离源、同位素"氙"气源等，它们这些离子源是针对某一类特定化合物所需要的。其中电子轰击电离源得到的谱图，提供丰富的化合物结构信息，是化合物的"指纹谱"，有庞大的标准谱库供检索，谱库中的谱图是在 70 eV 条件下获得的，谱图重复性好，被称为经典的EI谱。质谱测量有机化合物所需样品数量是非常少的，一般只需微克甚至到纳克，如此少的量有利于与气液层析联用。气相色谱-质谱联用使植物精油的研究有着飞跃性的发展。因为它把分离技术和鉴定方法联合起来，使植物精油化学成分研究显得得心应手，由于植物精油绝大部分是油状液体，分离和纯化目前来说比较困难，气相色谱-质谱联用仪的出现大大地解决了这个问题。特别是发展成气液分离，使样品用量少，分离效率高。而气相色谱-质谱联用仪基本解决分离每组分再用联用的质谱进行鉴定的问题，整个植物精油样品十多个成分乃至几十个成分可以通过气相色谱-质谱联用（GC/MS）进行定性与定量分析，而且这类联用仪器基本上都与各类数据库相连并进行自动检索。近几十年来发表的有关植物精油成分研究的文章基本离不开GC/MS，目前常用数据库为NIST/EPA/MSDC 系统磁盘（美国国家标准局数据库）。

　　目前看来，虽然有关植物精油成分研究的文章基本离不开GC/MS，但也不乏使用上述经典方法制备衍生物测定它们熔点或使用其他方法进行对照做一步鉴定，并用四大光谱测定精油某一成分的实例。有关GC/MS谱图和检索请见图2-17。

表2-3至表2-7列出了植物精油的主要成分。

表2-3　植物精油主要成分一览表（萜类化合物）（Ⅰ）

名称	化学名称	特性
1　月桂烯myrcene	α-7-甲基-3-亚甲基1,7-二烯 β-7-甲基-3-亚甲基1,4-辛二烯	BP（沸点）：136~138℃ 由于双键位置不同，有α-和β-异构体，多数以混合体存在于松科、杉科、柏科以及多种双子叶植物，植物精油中普遍存在月桂烯，但含量不高
2　罗勒烯ocimene	α-3,7-二甲基-1,3,7-辛三烯 β-3,7-二甲基-1,3,6-辛三烯	BP：176~178℃ 无色油状液体，具有令人愉快的香气，广泛分布于双子叶植物精油中，特别是草本植物，单子叶植物则在精油中很少出现，罗勒烯因共3个双键而聚合
3　香叶醇geraniol	3,7-二甲基-2,6-辛二烯-1-醇	BP：229℃ 无色，带玫瑰香气油状液体
4　橙花醇nerol	3,7-二甲基-2,6-辛二烯-1-醇	BP：229~230℃ 无色，带新鲜玫瑰香气，两者在自然界中是混合存在，主要存在于鲜花、水果、香叶等精油中

（续表）

名称	化学名称	特性
5 香茅醇（香草醇）citronellol 	α-3,7-二甲基-6-辛烯-1-醇 β-3,7-二甲基-6-辛烯-1-醇	BP：224～225℃ 无色，具有比香叶醇更柔和的玫瑰香气，有不对称碳原子，故是旋光异构d、l、dl，由于羟基位置不同，具有α-和β-两种异构体，但在天然中多为混合存在，主要存在于香茅油、柠檬桉叶中
6 芳樟醇linalool	α-3,7-二甲基-1,6-辛二烯-3-醇 β-3,7-二甲基-1,6-辛二烯-3-醇	BP：197～199℃ 芳樟醇具不对称碳原子，故有d、l、dl异构体，同时由于双键位置不同，存在α-、β-两种异构体，它多以混合体存在，主要分布于芳樟油及其他樟属植物精油中，同样存在于玫瑰油、橘类油、芙蓉油、香柠檬油、橙花油、薰衣草油、依兰油等花和果精油中。在旋光异构体中，l-型为木香型，d-型侧为鲜花型
7 柠檬醛citral 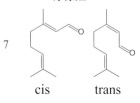 cis trans	α-3,7-二甲基-2,6-辛烯-1-醛（香叶醛)-反式 β-3,7-二甲基-1-辛烯-2,7-醛（橙花醛)-顺式	BP：229℃ 为无色或淡黄色具较浓柑橘类香气的油状液体，主要存在于芸香科果皮精油、叶精油、柠檬桉油、柠檬草油中
8 柠檬烯limonene（dipentene） 	1-甲基-4-异丙烯基-环己烯	BP：175～176℃ 为无色、具清淡柠檬香气的油状液体，具不对称碳原子，故存在旋光异构l、d、dl三个光学异构体，主要存在于柑橘类精油中，以及大多数精油的低沸点部分

（续表）

名称	化学名称	特性
9 水芹烯phellamdrene α β	α-（1,5-对蓝二烯） β-（1(7)-5-对蓝二烯）	α-异构体BP：173～175℃ β-异构体BP：175～177℃ 存在不对称碳原子，有3个旋光异构体l、d、dl，为无色，具愉快香气，存在于小茴香油、水芹油、某些桉叶油和松节油以及一些植物精油的低沸点部分
10 松油醇terpineol α β γ	α-（1-甲基-4-异丙基-8-环己烯-1-醇） β-（1-甲基-4-异丙烯基-1-环己醇） γ-（1-甲基-4-异亚丙基-1-环己醇）	α-异构体BP：219℃　熔点：35℃ 具不对称碳原子，有3个旋光异构体l、d、dl，具丁香香气，无色油状液体 β-异构体BP：206℃　熔点：32～33℃ 为无色、具风信子香气油状液体 γ-异构体BP：218℃　熔点：68～78℃ 常温结晶体具紫丁香味
11 4-松油醇	1-甲基-4-异丙基-8-环己烯-4-醇	无色，具清淡轻微的愉快香气，具有较强的杀菌活性，主要存在于澳洲纯叶千层（茶树油）中（含量在60%以上），另还广泛存在于各种植物精油中，但一般含量不高
12 薄荷脑menthol l-薄荷醇　d-薄荷醇 l-异薄荷醇　l-新薄荷醇	1-甲基-4-异丙基环己醇	薄荷脑有三个不对称碳原子，已知有四个反式和顺式异构体，分别为薄荷脑、异薄荷脑、新薄荷脑、新异薄荷脑，具有强烈薄荷香气，后三者较薄荷脑气味差些，有些略带微苦。存在于薄荷中，也有少量存在于胡薄荷油和辣薄荷油中

（续表）

名称	化学名称	特性
13 薄荷酮menthone	1-甲基-4-异丙基-3-环己酮	BP：207℃ 存在顺式和反式立体异构两个，每个异构均有旋光异构体dl、l、d，它们均为无色，具薄荷脑特有香气，而dl-异构体具水果香味
14 樟脑camphor	1,7,7-三甲基-2-二环[2,2,1]庚酮	BP：204℃　熔点：178℃ 樟脑为白色半透明晶体，具特殊气味，它现存在两个不对称碳原子，应有4个旋光异构体，但目前只发现d、l和dl异构体，天然樟脑多为d-异构体，合成者则为dl-异构体，主要存在于樟属植物精油中，其他植物精油也有但含量不大
15 龙脑borneol  d-boneol	1,7,7三甲基-2-二环[2,2,1]庚醇	BP：212℃ 龙脑为白色结晶体，熔点为208℃，具不对称碳原子，故存在旋光异构体，d-异构体为梅片，是名贵中药材，l-异构体为艾片，药性与d-异构体不同，合成龙脑为dl，药性较d-异构体差些，它们的几何异构体为异龙脑，d-龙脑产于龙脑香科、樟属龙脑樟和梅片树中，l-异构体产于菊科艾纳香
16 α-蒎烯α-pinene	4,7,7-三甲基-3-二环[3,1,1]庚烯	BP：155.9℃ 无色、有淡淡松节油气味的油状液体，它分布极广，主要分布于松节油中，同样分布于400多种精油中，松科植物中含量最多

（续表）

名称	化学名称	特性
β-蒎烯β-pinene 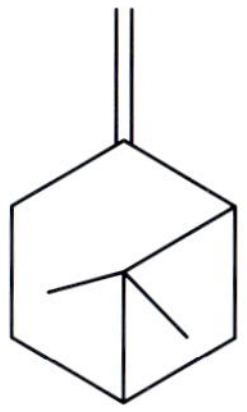	4,7-二甲基-4-亚甲基[3,1,1]庚烷	BP：166℃ β-蒎烯分布比α-蒎烯少，思茅松针油中含量较丰富，高达26%
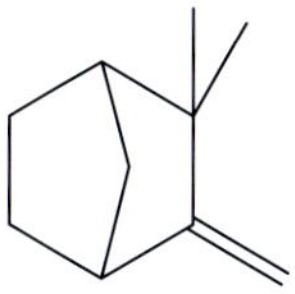 蒈烯camphene	3,3-二甲基-2-亚甲基二环[1,2,2]庚烷	BP：160～161℃　熔点：51～52℃ 蒈烯存在不对称碳原子，具旋光异构体，常温为白色结晶，具樟脑气味，主要分布于樟属和多种精油的低沸点部分
葑酮（小茴香酮）fenchone 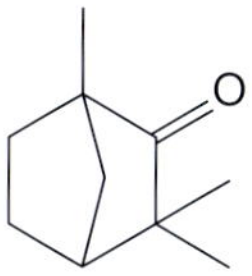	1,3,3-三甲基-2-双环[2,2,1]庚酮	BP：192～194℃ 葑酮存在不对称碳原子，具旋光异构体d、l、dl，d-异构体熔点6.1℃，l-异构体熔点5.2℃，都具有樟脑的气味，存在于小茴香油中，在别的精油中也有少量存在
香芹酮carvone 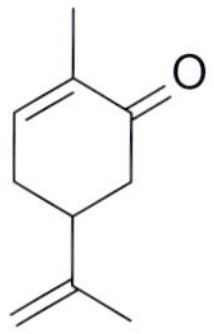	1-甲基-4-异丙烯-1-酮	BP：230～231℃ 存在不对称碳原子，具旋光异构体l、d、dl，为无色、具葛缕子特有香气的油状液体，主要存在于葛缕子油中，唇形科留兰香油中含量45%～65%
驱蛔素ascaridole 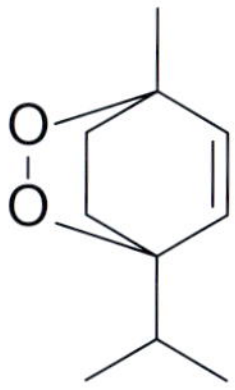	1-甲基-4-异丙基-1,4-环己烯-5-环氧	BP：96～98℃　纯熔点：25℃ 存在于香黎油，通过真空分离得纯驱蛔素，常温为无色油状液体

序号 17, 18, 19, 20, 21

（续表）

名称	化学名称	特性
22 菌烯-33-carene	3,7,7-三甲基二环[4,1,0]-3-庚烯	BP：169.5～170℃ 为无色、具清淡香气油状液体，在菌烷素中，菌烯在植物精油中分布最广
23 1,8-桉叶油素 1,8-cineole	1-甲基-4-异丙基-1,8-环己烷内醚	BP：174～177℃ 为无色，具较浓清凉龙脑气味，较为广泛存在于植物精油中，其同类物1,4-桉叶油素（1-甲基-4-异丙基1，4-环己内醚）则较少分布于精油中
24 金合欢醇farnesol	3,7,10-三甲基十二碳（2,6,10）三烯-1-醇	BP：283.4℃ 有四个几何异构体，为无色、具优雅香气黏稠性油状液体，含量不高但对香韵起较重要作用，具温和细腻带有铃兰特征的花香，有良好的定香作用，存在于依兰、柠檬草、木樨、金合欢花、枇杷叶、苍术等植物精油中
25 橙花叔醇nerolidol	3,7,10-三甲基十二碳（1,6,10）三烯-3-醇	BP： 橙花叔醇为金合欢醇的异构体，较容易脱成金合欢烯，橙花叔醇较金合欢醇更具优雅香韵，精油中含量不高，对精油香韵起较重要作用。存在于各种花果精油中，如卡鲁瓦油（cabreuva oil）中含80%，还有秘鲁香等

（续表）

名称	化学名称	特性
26　蛇麻烯humulene α-蛇麻烯　β-蛇麻烯	四甲基十一碳环三烯 1,1,4,8四甲基环十一碳三烯	BP：260.9℃ 蛇麻烯为倍半萜烯，有两个α-和β-几何异构体，存在于啤酒花中
27　石竹烯（丁香烯）caryophyllene	β-丁香烯，反式石竹烯，IR.95-8-亚甲基-4,11,11-三甲基双[7,2,0]十一碳-4-烯	BP：254～257℃ 石竹烯是一个双环倍半萜，它主要存在于丁香油、乌尾松的重质油中，其他植物精油中亦有发现，但含量不高
28　长叶烯longifolene	[Is-Ia,3ab,4a,8ab]-十氢-4,8,8-三甲基-9-亚甲基-1,4-甲桥	BP：254～256℃ 长叶烯是一种三环倍半萜，为重要倍半萜之一，存在于松节油的高沸点馏分中，为很多合成香料的原料，最早被发现于长叶松油中
29　柏木烯（雪松烯）cedrene	2,6,6,8-四甲基-三环[3,3,1,0]十一烷	BP：263.1℃ 柏木烯为三环倍半萜，有α-、β-两个几何异构体，主要存在于柏科植物的木材和根部中，是合成香料的原料
30　柏木脑（雪松脑）cedrol	2,6,6,8-四甲基三环[5,3,1,0]十一烷-8-醇	BP：290～292℃ 柏木脑存在于柏木油和几种刺柏属植物精油中，分馏后，再经冷冻则可获得白色结晶柏木脑，熔点86℃，用65%冷乙醇亦可提出柏木脑，柏木脑可杀虫，也是调制木香型香精的原料

（续表）

名称	化学名称	特性
31 桉叶醇eudesmol 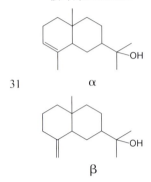 α β	α-1甲基-4-亚甲基十氢萘-7-异丙基-2-醇 β-1,5-二甲基-4-烯-八氢萘-异丙基-2-醇	BP: 301.1℃ 桉叶醇为白色结晶，熔点82～83℃，有旋光异构体和几何异构体，α-、β-异构体，存在于各种桉树油重油部分，可重结晶纯化，气味清淡，具较强驱除蚊虫活性
32 檀香醇santalol	2,3-二甲基三环[2,2,1,2,6]-3-庚基-2-甲基-2-戊烯-1-醇	檀香醇为无色、具檀香木特有香气的黏稠油状液体，产自檀香木，在沉香木树脂中也发现檀香醇，但含量有限，目前大多数使用的檀香醇多为合成，其香气比天然檀香醇略差

表2-4　植物精油主要成分一览表（芳香族化合物）（Ⅱ）

名称	化学名称	特性
1 苄醇benzyl alcohol 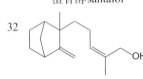	苯甲醇	BP: 205℃ 为无色、具弱愉快香气的油状液体，它是植物精油中分布最广的芳香族醇类之一，主要存在于茉莉油、晚香玉油、依兰油、丁香油等植物精油中，除游离存在外也有以酯形式出现，主要存在于风信子油、秘鲁香、安息香树脂油中
2 β-苯乙醇phenethyl alcohol	2-苯乙醇	BP: 222℃ 无色黏稠液体。在苹果、杏仁、香蕉、桃子、梨子、草莓、可可、蜂蜜等天然植物中发现。它具有清甜的玫瑰样花香，遇明火、高热可燃。

（续表）

名称	化学名称	特性
3　桂醇（肉桂醇） cinnamic alcohol 	β-苯丙烯醇 3-苯基-2-丙烯-1-醇	BP：256.5℃ 桂醇为白色结晶体，熔点34℃，具温和、持久而舒适的香气，类似风信子香
4　麝香草酚（百里香酚） thymol 	3-甲基-6-异丙基苯酚	BP：233℃ 麝香草酚为白色或淡黄色结晶性粉末，唇形科地椒含量达40%，现多用于牙膏杀菌
5　爱草脑（异大茴香脑） anethole 	对甲氧基异丙烯基苯	BP：216℃ 无色油状液体，具轻微茴香气味，具有强烈抑杀微生物活性，主要存在于齿叶黄皮精油中，在齿叶黄皮中含量高达93%
6　大茴香脑anethole 	1-丙烯基-4-甲氧基苯	BP：234～237℃ 大茴香脑具反、顺两异构体，常温下为白色糊状物，在22.5℃以上时则具甜味和强烈的茴香味，主要存在于八角茴香油中
7　苯甲醛（安息香醛） benzaldehyde 	苯基甲醛	BP：179℃ 苯甲醛无色，具强烈苦杏仁味，苯甲醛以苷的形式存在于苦杏仁油中，它占该油80%左右，亦存在于蔷薇的核仁中，具有一定的抑杀细菌活性

（续表）

名称	化学名称	特性
8　桂醛 cinnamic aldehyde	3-苯基-2-丙烯醛	BP：252℃（有部分分解） 桂醛为无色，具强烈桂皮香气和辛辣味，主要存在于桂皮油、桂叶油中，是可口可乐、百事可乐主要的香精原料，亦可作调味料，桂油是我国主要的出口植物精油之一
9　苯丙醛（氢化桂醛） 3-henylpropion-aldehyde	3-苯基-2-丙醛	BP：244℃ 苯丙醛为无色，具有较温和的桂醛辛辣味
10　香兰素（香草醛） vanillin	4-羟基-3-甲氧基苯甲醛	BP：235℃ 香兰素为无色，具香爽兰豆香气，其针状熔点为77～79℃，四方晶体熔点为81～83℃，香兰素存在于兰科植物的豆荚内，一般经发酵才能获得，现大多使用合成香兰素，但香气逊于天然香兰素
11　黄樟油素 safrole	3，4-二氧亚甲基苯丙烯	BP：235.9℃ 黄樟油素为无色或淡黄色、有着轻微水杨酸甲酯香气的油状液体，存在致癌因素，现已不能用于食品
12　水杨酸甲酯	2-甲酯苯酚	BP：222℃ 为无色、具强烈水杨醛气味的油状液体，大量用于配制沙示型饮料，具一定的杀菌活性

（续表）

名称	化学名称	特性
13　丁香酚eugenol OH OCH₃	2-甲氧基-4-烯丙基苯酚	BP：252.7℃ 丁香酚为无色至淡黄色，具强烈丁香气，主要存在于丁香干花蕾、叶油以及罗勒丁香油中，其他植物精油亦有少量，罗勒丁香油中丁香酚是合成香兰素的原料
14　丁香酚甲醚（甲基丁香酚） eugenyl methyl ether （methyl eugenol） OCH₃ OCH₃	1,2-二甲氧基-4-烯丙基苯	BP：254.7℃ 丁香酚甲醚在植物精油中分布极广，如木兰科花、樟科枝叶、菊科等，还有一些草本芳香植物精油 丁香酚甲醚具一定的抑菌活性，并有长效功能
15　紫罗兰酮ionone （三个异构体）		紫罗兰酮以三种异构体形式存在于植物精油中，为无色、经稀释后有类似紫罗兰香气的黏稠油状液体
16　α-紫罗兰酮α-ionone O	1-（2，2，6-三甲基-3-环己烯-5基-1）丁烯-1-酮	BP：211℃ 无色黏稠油状液体，稀释后有类似紫罗兰香气
17　β-紫罗兰酮β-ionone O	1-（2，2，6-三甲基-3-环己烯-6基-1）丁烯-酮	BP：127～128℃ 无色黏稠油状液体，比α-紫罗兰酮具有更为鲜明的柏木香韵
18　γ-紫罗兰酮γ-ionone O	1-（2，2-二甲基-6-次甲基-3-环己基-1-）丁烯-1-酮	γ-紫罗兰酮为无色黏稠油状液体，其香气较类似α-紫罗兰酮，但更为浓郁，更为刺鼻

（续表）

名称	化学名称	特性
19 花椒素xanthoxylin 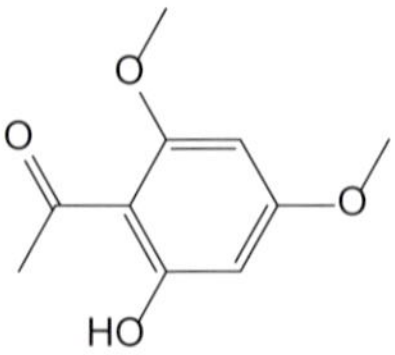 	1-羟基-3,5-二甲氧基-8-甲基-7-甲基苯甲酮	熔点：80～84℃ BP：355℃ 溶于热甲醇

表2-5　植物精油主要成分一览表（脂肪族化合物）（Ⅲ）

	名称	化学名称	特性
1	壬醇nonanol	九碳烷基-1-醇	BP：212℃ 壬醇为无色、具果味油状液体，主要存在于玫瑰油、柠檬油、甜橙油等植物精油中，以游离或酯的形式存在
2	癸醇decanol	十碳烷基-1-醇	BP：228～232℃ 癸醇为无色、具略带橙花和玫瑰香味的油状液体，主要存在于黄葵种子油中
3	叶醇3-hexen-1-ol	3-己烯-1-醇	BP：156～157℃ 为无色油状液体，广泛存在于各种植物精油中，以游离或酯的形式存在
4	叶醛3-hexen-1-al	3-己烯-1-醛	BP：47～48℃ 为无色或淡黄色易挥发的轻油状液体，主要存在于茶叶、桑叶、萝卜叶中
5	癸醛decanol	癸烷基-1-醛	BP：208～209℃ 癸醛为无色，具类似甜橙花的辛香气，主要存在于芫荽子油、橘子油、柠檬油、酸橙油和柠檬草油中

（续表）

名称	化学名称	特性
6　甲基庚烯酮 methyl heptenone	2-甲基庚烯-2-酮-6	BP: 173～174℃ 为无色、具水果新鲜清香气的油状液体，广泛存在于花果类精油中，甲基庚烯酮可直接合成柠檬醛和芳樟醇
7　鸢尾酮irone α β γ	α1-（2,2,3,6-四甲基-3-环己烯-5-基-1）丁烯酮 β-1（2,2,3,6-四甲基-3-环己烯-6-基-1）丁烯酮	BP: 278.7℃ 鸢尾酮与紫罗兰酮相似，只是多了一个甲基，同样有三个几何异构体（双键位置不同）存在于鸢尾根部，再经发酵后鸢尾油得油5%，其中75%为γ-鸢尾酮，其余25%为α-异构体和β-异构体
8　肉豆蔻酸myristic acid	羧基，十四烷基酸	BP: 326℃ 肉豆蔻酸为白色蜡状结晶体，熔点54.4℃，肉豆蔻酸是鸢尾根油主要成分，含量约85%，肉豆蔻油中亦含有

表2-6　植物精油主要成分一览表（含氮硫化合物）（Ⅳ）

	名称	化学名称	特性
1	吲哚indole	氮杂辛烷	BP：253～254℃ 吲哚为鲜片状白色结晶体，熔点52～53℃，长期被光照易变黄色，吲哚存在于大花茉莉、小花茉莉油中，在苦橙油、甜橙油、柠檬油中亦有一定的含量，含量较少，但对精油的香韵起较重要作用
2	甲基吲哚methylindole	甲基氮杂辛烷 β-methylindole 3-methylindole skatole	BP：265℃ 甲基吲哚能溶于水和多种有机溶剂，主要存在于灵猫香中，在调香中被极度稀释后为定香剂之用
3	邻氨基苯甲酸甲酯 nthranilate	2-氨基-苯基甲酸酯	BP：256℃ 邻氨基苯甲酸甲酯在较低温为结晶体，常温下是无色，具橙花和葡萄特有香气，很多果香花香精油中都含有，但含量不大
4	二甲基硫醚 dimethyl mercaptan	对-二甲基硫醚 CH_3-S-CH_3	BP：37.5℃ 为较易挥发、具特殊气味的油状液体
5	大蒜辣素allicin	二烯丙基二硫	BP：248.6℃ 含有大蒜特有气味

表2-7　樟科樟属及其他属精油主要含量成分一览表

樟脑ccrmphor（1,7,7-三甲基-2-双环[2,2,1]庚酮）

植物名称	含量	部位（精油）	备注
樟Cinnamomum camphora	83.9%	枝叶1.0%	
	81.2%	枝叶1.1%	
	79.1%	木材3%～4%	台湾
	92.3%	根5%～6%	
湖北樟Cinnamomum bodinieri var. hupehanum	88.5%	枝叶	脑樟型
黄樟C. parthenoxylon	86.7%	枝叶0.5%～0.8%	大叶脑樟
芳樟C. camphora var. linaloolifera	80.0%	枝叶1.2%	脑樟型
尾叶樟C. caudiferum	60.4%	枝叶2.1%	脑樟型
油樟C. longepaniculatum	90.5%	枝叶2.8%	西双版纳
毛叶樟C. mollifolium	80.0%	枝叶1.25%～1.61%	脑樟型
细毛樟C. tenuipilum	85.7%	枝叶0.7%～1.0%	脑樟型
狭叶阴香（狭叶桂）C. burmannii f. heyneanum	48.7%	干叶0.7%	混合型
银木C. septentrionale	40.5%	干叶3.3%	广西南宁
天竺桂C. japonicum	30.3%	根	
三桠乌药Lindera obtusiloba	52.7%	枝叶0.9%～1.1%	
新樟Neocinnamum delavayi	41.0%	枝叶0.7%	

芳樟醇linalood（α–3，7–二甲基–3–辛二烯–1，7–醇）;（β–3，7–二甲基–3–辛二烯–1，6–醇）

植物名称	含量	部位（精油）	备注
樟Cinnamomum camphora	90.6%	枝叶0.3%～0.8%	左旋
芳樟C. camphora var. linaloolifera	92.7%	枝叶1.95%	西双版纳
黄樟C. parthenoxylon	82.8%	枝叶1.1%～1.4%	右旋
细毛樟C. tenuipile	97.5%	枝叶1.4%～2.1%	左旋
毛叶樟C. mollifolium	46.3%	枝叶1.37%	
阴香C. burmannii	57.0%	鲜叶0.72%～1.20%	芳樟醇型
土肉桂C. osmophloeum	83.3%	叶	
粗脉桂C. validinerve	43.8%	鲜叶0.21%	香叶醇17.4%
锈毛桂C. villosulum	59.5%	鲜叶0.25%～0.30%	广西大明山
假桂皮树C. tonkinense	21.6%	干叶0.47%	香叶醇37.3%
细毛樟C. tenuipilum	92.5%	枝叶1.5%～2.0%	

1,8–桉叶油素 1,8–cineole（1–甲基–4–异丙基–1,8–环己烷内醚）

植物名称	含量	部位（精油）	备注
樟Cinnamomum camphora	36.8%	枝叶1.2%～1.8%	西双版纳
	50.0%	枝叶0.75%	
尾叶樟C. caudiferum	54.0%	鲜叶2.5%	
云南樟C. glanduliferum	45.8%	鲜叶0.65%～0.75%	云南

（续表）

植物名称	含量	部位（精油）	备注
油樟 C. longepaniculatum	58.6%	叶1.2%	四川
毛叶樟 C. mollifolium	50%～60%	鲜叶0.13%～0.18%	
黄樟 C. parthenoxylon	55.1%	叶2.8%	
细毛樟 C. tenuipilum	54.1%	鲜叶1.2%～1.5%	
阴香 C. burmannii	50%～57%	枝叶0.3%～0.4%	油计树
毛桂 C. appelianum	37.6%	鲜叶1.1%	湖北咸丰
狭叶桂 C. burmannii f. heyneanum	33.0%	干叶0.71%	
香桂 C. subavenium	76.1%	枝叶0.24%	西双版纳

α.β-柠檬醛citral（α-3，7-二甲基-1-辛二烯-2，7-醛）；（β-3，7-二甲基-1-辛二烯-2，6-醛）

植物名称	含量	部位（精油）	备注
黄樟 Cinnamomum parthenoxylon	72.1%	枝叶0.6%～0.8%	姜樟
毛叶樟 C. mollifolium	74.0%	鲜叶0.7%～1.4%	柠檬醛型
湖北樟 C. bodinieri var. hupehanum	95.0%	鲜叶1.37%	
云南樟 C. glanduliferum	54.2%	鲜叶0.35%～0.5%	
细毛樟 C. tenuipilum	59%～75%	鲜叶1.0%～1.5%	柠檬醛型
阴香 C. burmannii	76.8%	叶	栽培种
假桂皮树 C. tonkinense	38.6%	鲜叶0.31%	云南麻栗坡
长柄樟 C. longipetiolatum	61.1%	叶0.53%	云南
山苍子（山鸡椒）Litsea cubeba	71.1%	果实3.0%～4.0%	广东

（续表）

植物名称	含量	部位（精油）	备注
清香木姜子 *L. euosma*	80.5%	果实2.5%～3.0%	
毛叶木姜子 *L. mollis*	70.0%	果实5.0%	
樟 *Cinnamomum camphora*	69.9%	鲜叶1.6%～2.0%	云南西双版纳

丁香酚系列 eugenol；methy eugenol（eugenyl methyl ether）（2–甲氧基–4–烯丙基苯酚）；（1，2–二甲氧基–4–烯丙基苯）

植物名称	含量	部位（精油）	备注
黄樟 *Cinnamomum parthenoxylon*	丁香酚甲醚71.5%	枝叶0.6%～0.8%	
锡兰肉桂 *C. zeylanicum*	t-丁香酚甲醚81.3%	枝叶1.5%～2.1%	西双版纳
阔叶樟 *C. platyphyllum*	t-丁香酚甲醚94.0%	鲜叶0.57%	湖北
毛叶樟 *C. mollifolium*	t-丁香酚甲醚83.3%	鲜叶0.75%	
岩樟 *C. saxatile*	t-丁香酚甲醚85.7%	叶1.10%	湖北利川
细毛樟 *C. tenuipilum*	丁香酚甲醚69%～89%	鲜叶1.2%～1.7%	栽培种
刀把木 *C. pittosporoides*	丁香酚64.9%	干叶1.5%	云南屏边
香桂 *C. subavenium*	丁香酚67.4%	干皮2.1%	西双版纳
假桂皮树 *C. tonkinense*	丁香酚73.4%	鲜叶0.13%	
银木 *C. septentrionale*	异丁香酚甲醚85.7%	叶1.1%	湖北利川
天竺桂 *C. japonicum*	丁香酚15.3% 丁香酚29.5%	枝叶 木材	
卵叶樟 *C. rigidissimum*	丁香酚甲醚28.6%	根1.4%	海南尖峰岭
银木 *C. septentrionale*	丁香酚甲醚85.7%	枝叶1.1%	湖北利川

黄樟油素 safrole（3，4-二氧亚甲基苯丙烯）

植物名称	含量	部位（精油）	备注
坚叶樟Cinnamomum chartophyllum	94.1%	根0.54%～1.22%	
八角樟C. ilicioides	83.1%	根0.21%	
长柄樟C. longipetiolatum	62.1%	根1.13%	
沉水樟C. micranthum	61.3%	木茎0.09%	
米槁C. migao	83.6%	木茎1.1%	
黄樟C. parthenoxylon	94.3%	根1.01%	
岩樟C. saxatile	94.7%	根1.46%	
少花桂C. pauciflorum	97.1%	根0.71%	
柴桂C. tamala	99.3% / 99.1%	木茎0.44% / 树皮2.3%	
卵叶桂C. rigidissimum	62.1%	根1.44%	
天竺桂C. japonicum	46.4%	根	
猴樟C. bodinieri	84.0%	枝叶0.3%	云南广南
假桂皮树（狭叶桂）Cinnamomum tonkinense	36.9%	树皮1.04%	
狭叶阴香C. burmannii f. heyneanum	97.5% / 98.8%	枝叶0.54% / 树皮0.81%	

桂醛 cinnamic aldehyde β-丙烯醛（3-苯基-1-丙烯-2-醛）

植物名称	含量	部位（精油）	备注
肉桂C. aromaticum	81.1% / 88.1%	枝叶0.35% / 干树皮0.83%～2.15%	广东肇庆
钝叶桂C. bojolghota	88.1%	清化桂皮0.8%～1.1%	云南河口

（续表）

植物名称	含量	部位（精油）	备注
土肉桂 *C. osmophloeum*	79.46%	叶0.28%~1.45%	台湾
聚花桂 *C. contractum*	t-桂醛90.96%	叶0.47%	云南

d-龙脑 borneol（1，7，7-三甲基-2-双环[2，2，1]庚醇）

植物名称	含量	部位（精油）	备注
阴香 *Cinnamomum burmannii*	51.2%	枝叶0.6%~1.1%	梅生树
樟 *C. camphora*	81%	枝叶0.8%	龙脑樟
细毛樟 *Cinnamomum tenuipilum*	45%	鲜叶1.43%	龙脑型
川桂 *C. wilsonii*	乙酸龙脑酯26.14% 龙脑16.8%	枝叶0.6%~0.8%	四川
天竺桂 *C. japonicum*	龙脑1.2%	鲜叶0.45%	西双版纳

特有和稀有倍半萜，倍半萜类（sesputerpene）

植物名称	含量	部位（精油）	备注
黄樟 *Cinnamomum parthenoxylon*	9-氧化橙花叔醇24.2% 橙花叔醇54.8%		广东、广西、福建、湖南、贵州、云南
岩樟 *C. saxatile*	橙花叔醇30.7% 乙酸金合欢酯12.1% 金合欢醇7.6%	枝叶0.19%	云南麻栗坡
滇南桂 *C. austro-yunnanense*	γ-榄香烯20.8%	枝叶	云南南部
猴樟 *Cinnamomum bodinieri*	橙花叔醇68.4% 金合欢醇13.4%	鲜叶0.4%	云南广南

（续表）

植物名称	含量	部位（精油）	备注
细毛樟 C. tenuipilum	金合欢醇70.03% 榄香烯56.4% 榄香脂素41.3% 肉豆蔻醚33.1%	根0.44%	云南南部和西部
卵叶樟 C. rigidissimum	苯甲酸苄脂82.8%	叶1.03%	西双版纳
钝叶桂 C. bejolghota	γ-榄香烯28.6% β-丁香烯10.1%	枝叶	

柠檬桉油（eucalyptus citriodora oil）

减压分馏（4 mmHg）

34理论塔板数

回流比（30～40）：1

馏分	收集温度/℃	薄层结果参考
1	80～84	可能为α-蒎烯
2	84～86	可能为α-蒎烯、月桂烯、β-蒎烯
3	86～89	可能为月桂烯、柠檬烯、2-菖烯
4	90～94	可能为1,8-桉叶油素和一羰基化合物
5	94～97	可能为萜类含氧衍生物芳樟醇
6	97～101	可能主要为香茅醛
7	101～105	可能为香茅醛、香芳醇、异胡薄荷醇
8	105～113	可能为香茅醇和秀叶醇
9	114～120	可能为石竹烯、倍半萜烯
10	120～128	可能为含氧倍半萜、β-桉叶醇
		经稀释后有白色粉末，重结晶纯化后鉴定为β-桉叶醇

柠檬桉油各馏分的鉴定

1馏分：	进行硅胶柱层析，分得一个较纯的单萜烯，经红外光谱对照和标样对照，鉴定为α-蒎烯
2馏分：	用硅胶柱层析，分得三个成分，经红外光谱对照和标样叠加对照，被鉴定为月桂烯、β-蒎烯和2-莰烯
3馏分：	经硅胶柱层析，分得三个成分，经红外光谱对照和标准样叠加对照，被鉴定为柠檬烯、2-莰烯
4馏分：	经硅胶柱层析，分得两个含量较高组分，经红外光谱和标准样叠加对照，被鉴定为1,8-桉叶油素和一个羰基化合物
5馏分：	经硅胶柱层析，分得其中一个含量较高组分，经红外光谱和制备3,5-二硝基苯甲酸酯对照，被鉴定为芳樟醇
6馏分：	经硅胶柱层析，分得一个较纯、含量很高羰基化合物，经叠加标样和制备2,4-二硝本腙衍生物，被鉴定为香茅醛
7馏分：	经硅胶柱层析，分得三个化合物，经红外光谱对照和衍生物制备，分别鉴定为香茅醛、香茅醇和异胡薄荷醇
8馏分：	经硅胶柱层析，分离出多个化合物，其中一个含量较高，经标样对照和制备衍生物3,5-二硝苯甲酸酯，被鉴定为香叶醇，还有倍半萜烯
9馏分：	经硅胶柱层析，分得多个化合物，含量较高组分，再用柱层析纯化后，经红外光谱对照，被鉴定为石竹烯
10馏分：	经硅胶柱层析，分得一白色结晶体，经重结晶纯化，经红外光谱对照和结晶熔点测试，被鉴为β-桉叶醇

11 釜底油脚同样也分离出β-桉叶醇，其量较10馏分高

  沉香木块超临界CO_2萃取后进行GC/MS分析（见图2-16），沉香精油气相色谱/质谱联用仪记录显示沉香精油的总离子流图（见图2-17）；用归一积分法可算出每一个净油在联用仪的碎片峰占各峰总和的比例。

  在质谱碎片峰最后较强的峰一般称为分子离子峰，在精油GC/MS分析中为分子量n。

在精油GC/MS分析中，经常会出现134、152、154、204、202、222等较强峰。

通过质谱分析检索其化学成分，通过NIST/EPA/MSDA，利用系统磁盘中计算机谱库进行检索，检索可能会出现4个相似碎峰，可根据实际情况选出相似度高的化合物。

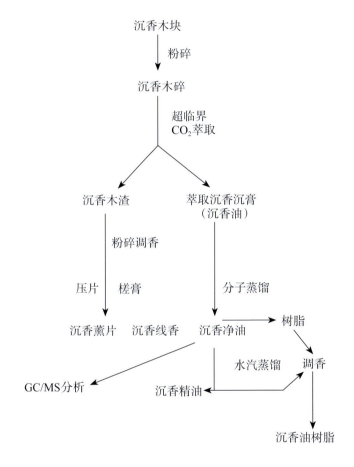

图2-16　沉香木块超临界CO_2萃取

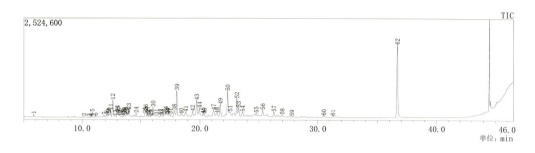

峰表 TIC

峰号	保留时间	峰面积	峰面积%	名称
1	5.829	203899	0.45	2-Butanone, 4-phenyl-
2	10.162	60297	0.13	Cyclohexanemethanol, 4-ethenyl-
3	10.602	63458	0.14	5.beta.,7.beta.H,10.alpha.-Eude
4	10.743	62350	0.14	Cyclohexane, 1-ethenyl-1-methyl
5	10.856	384390	0.85	.alpha.-Farnesene
6	11.139	86497	0.19	1H-Cycloprop[e]azulen-4-ol, dec
7	11.871	257662	0.57	2-Naphthalenemethanol, 1,2,3,4,
8	12.090	354002	0.78	1H-Cycloprop[e]azulene, decahyd
9	12.206	340213	0.75	5-Azulenemethanol, 1,2,3,4,5,6,
10	12.297	157407	0.35	5-Azulenemethanol, 1,2,3,4,5,6,
11	12.435	606650	1.34	5-Azulenemethanol, 1,2,3,4,5,6,
12	12.637	2838733	6.26	Cyclohexanemethanol, 4-ethenyl-
13	12.827	78277	0.17	Epiglobulol
14	13.006	592806	1.31	Aromadendrene oxide-(1)
15	13.058	134650	0.30	Naphtho[2,3-b]furan-2(3H)-one,
16	13.196	143030	0.32	3-Carene
17	13.277	41498	0.09	7-Tetracyclo[6.2.1.0(3.8)0(3.9)
18	13.454	80432	0.18	Cyclohexanemethanol, 4-ethenyl-
19	13.590	342466	0.75	4,7,10,13,16,19-Docosahexaenoic
20	13.642	65560	0.14	Spiro[4.5]decan-7-one, 1,8-dime
21	13.802	85585	0.19	Acetic acid, 1-[2-(2,2,6-trimet
22	13.873	161174	0.36	2H-Bisoxireno[2,3:8,8a]azuleno[
23	13.989	641735	1.41	1,3,6-Octatriene, 3,7-dimethyl-
24	14.629	467726	1.03	Andrographolide
25	15.378	272889	0.60	Androstan-17-one, 3-ethyl-3-hyd
26	15.446	595695	1.31	Cyclohexene, 2-ethenyl-1,3,3-tr
27	15.590	408321	0.90	Caryophyllene oxide
28	15.722	63355	0.14	Tricyclo[4.4.0.0(2,7)]dec-8-ene
29	15.877	118591	0.26	2(1H)Naphthalenone, 3,5,6,7,8,8
30	16.056	938657	2.07	Ledol
31	16.339	107081	0.24	Ledol
32	16.710	120786	0.27	9-Undecenal, 2,6,10-trimethyl-
33	16.808	60120	0.13	Ledol
34	17.127	353595	0.78	2(3H)-Naphthalenone, 4,4a,5,6,7
35	17.230	36631	0.08	Bicyclo[4.1.0]heptane-7-methano
36	17.322	142787	0.31	2-Naphthalenemethanol, decahydr
37	17.523	185615	0.41	1,8-Nonadiene, 2,7-dimethyl-5-(
38	17.861	1064617	2.35	5,5,8a-Trimethyldecalin-1-one
39	18.083	2773925	6.11	5,8,11,14,17-Eicosapentaenoic a
40	18.440	107148	0.24	Andrographolide
41	18.820	822474	1.81	.gamma.-Gurjunenepoxide-(2)
42	19.385	603987	1.33	1,3,6-Heptatriene, 2,5,6-trimet
43	19.724	1915557	4.22	Cyclohexanemethanol, 4-ethenyl-
44	19.954	865022	1.91	Andrographolide
45	20.247	250103	0.55	Cyclohexanemethanol, 4-ethenyl-
46	20.367	216624	0.48	Naphthalene, 2-hydroxy-7-[1-(([1
47	21.155	1028597	2.27	Andrographolide
48	21.442	463137	1.02	Ethanol, 2-(3,3-dimethylbicyclo
49	21.724	1827363	4.03	3-Oxatricyclo[20.8.0.0(7,16)]tr
50	22.345	3132506	6.91	Andrographolide
51	22.544	632135	1.39	Cyclohexanemethanol, 4-ethenyl-
52	23.120	2100845	4.63	5,6-Azulenedicarboxaldehyde, 1,
53	23.268	1298782	2.86	3,7-Cyclodecadiene-1-methanol,
54	23.615	607690	1.34	1-Hydroxy-6-(3-isopropenyl-cycl
55	24.782	391884	0.86	3-Pentanone, 1,5-diphenyl-
56	25.327	935508	2.06	Andrographolide
57	26.271	667804	1.47	Cyclodecacyclotetradecene, 14,1
58	26.978	304161	0.67	Norethynodrel
59	27.803	111474	0.25	Bicyclo[5.1.0]octan-2-one, 4,6-
60	30.492	269665	0.59	1-Penten-3-one, 1,5-diphenyl-
61	31.253	216111	0.48	Benzeneacetic acid, .alpha.-[[(
62	36.658	11103179	24.48	8-Naphthol, 1-(benzyloxy)-
		45364941	100.00	

图2-17 沉香精油的总离子流图和谱图检索

第三节　植物精油质量标准和理化常数的测定

植物精油各项理化性质是代表植物精油质量评价的重要依据。

一、测定样品取样方法

测定前先要取得具代表性的测定样品。由于被测定样品放置在不同的地方、不同的容器，数量或多或少，来源也不同，这样给取样带来一定困难。取出的样品一定是能代表整批需要测试的物料，取样工具需要规范化，取样方法要按统一规定，如果多个容器则每个容器必须取样并充分混合。分析样品精油应使之澄清，如果尚含有水或悬浮物需静置或离心，可再加入干燥剂，无水硫酸或焯烧过的无水硫酸镁，如油样颜色过深可用柠檬酸或酒石酸脱色，样品含有微量水分可根据标准方法（GB/T 606—2003）测定。

二、理化常数的测定

外观：从外观来观察如果是液体须澄清不能有悬浮物，固体则不能混有杂质，应是白色结晶或粉末结晶。

乙醇溶解度：不同种类有不同要求，一般是用不同乙醇浓度对样品有一定数量的溶解能力。

香气：香气代表各种精油的特征，特别是用于直接配制香精原料油和名贵鲜花的浸膏和净油，然而对于精油香气来说，是用感观来界定精油的优劣的，评价人员亦可以指出精油中精油香气的关键成分。目前GC/MS等已经能非常准确地测定精油的各个组分，但最终的精油产品多数是需要感官来体验香气的。

1. 物理测定

①折光指数。其测定有专门折光仪，指数随着温度不同而改变，目前国际上规定20℃为标准，光线波长为5.93毫米（参看ISO280）。

②旋光度。测量旋光度采用旋光仪，不少植物精油有旋光性，能使偏振光的偏振面向右或向左旋转，右旋用"＋"表示，而左旋则用"－"表示，这一方法有时甚至可估计旋光组分的含量。旋光同样是精油特征之一，如龙脑旋光度为左旋的中药称为"艾片"，龙脑旋光度为右旋的中药称为"梅

片"，两者药理性能完全不同，另外，如左旋芳樟醇香气为木香型，右旋芳樟醇为花香型，两者在配制香精时作用完全不同。旋光度也是体现精油纯度和质量的指标。

③比重或密度。一般有专门测定精油的比重瓶，比重同样也是表示精油质量的指标。

外贸和内销的精油均应标明上面三个物理指标参数，详细查看国际标准ISO3507。

2. 化学测定。

①酸值，这是精油的主要化学指标之一，它是指1克油样中和氢氧化钾的毫克数，一般来说植物精油含游离酸很少，但由于生产加工，精油含有一定的水，储存运输不当使精油发生氧化水解都会增加精油的酸值。

②酯值，是指精油所含酯类化合物的量，这是影响精油香气的重要化学指标之一。精油酯值是指中和1克精油中酯水解后释出酸所需要的氢氧化钾毫克数。RCOOR+KOH→RCOOK+ROH，每一种常见的植物精油上基本都标明准确酯值。（请参见国际标准ISO1241）

③乙酰化值，是指乙酰化后精油的酯值。这是估量精油中的含醇量，同样也是衡量精油香气的重要化学指标。这一方法比直接测定含醇量更恰当，主要是精油中含有未知醇类。（详见国际标准ISO1241）

由于醇类存在精油中有伯、仲、叔三种结构，则其测醇类试剂不尽相同。伯、仲醇用酸酐基本完全酯化，但叔醇则需要甲酸加入乙酸酐反应才可完成。用冷甲酸化后，测定酸值可参见国际标准ISO4090。

④羰值是指精油中含醛酮类化合物的数量，这也是评价精油香气质量的指标之一，羰值是指1克精油和盐酸羟胺肟反应中，释放出的盐酸所需要中和它的氢氧化钾的毫克数，反应式如下。

$$\begin{array}{c} R \\ R_1 \end{array} C{=}O \;+\; NH_2OH{\cdot}HCl \;\longrightarrow\; \begin{array}{c} R \\ R_1 \end{array} C{=}NOH \;+\; H_2O \;+\; HCl$$

⑤如果精油含酚量较多，可以用氢氧化钾将精油中的酚类转化成溶于水的酚钾盐，计算其上层不溶于水的量，把总量减去上层非水溶性油状部分则为含酚量。

⑥用气相色谱法测定精油各组分含量，这一方法随着分析仪器的发展和常规化，发展十分快，非常普遍。2012年实施精油高效液相色谱分析通用法的国家标准、2018年实施精油手性毛细管柱气相色谱分析通用法。

⑦同位素测定法：这一方法主要检测天然桂油中掺入一定量合成桂醛的问题，用传统GC/MS法一定是十分烦琐的，过去方法是用GC/MS找出合成桂油过程中的杂质，不同合成线有不同杂质，这样存在非常大的工作量，而且这些杂质含量极微。但使用同位素测定法，则由于天然桂醛碳原子和合成桂醛碳原子半衰期不同，非常容易测出来。

第二章 参考文献

[1] 安银岭.植物化学[M].哈尔滨：东北林业大学出版社，1996.

[2] 陈孝泉.植物化学分类学[M].北京：北京高等教育出版社，1990.

[3] 陈耀祖，涂亚平.有机质谱原理及应用[M].北京：科学出版社，2001.

[4] 陈业高.植物化学成分[M].北京：化学工业出版社，2004.

[5] 高锦明.植物化学[M].2版.北京：科学技术出版社，2012.

[6] 黄致喜，王慧辰.萜类香料化学[M].北京：中国轻工业出版社，1999.

[7] 林启寿.中草药成分化学[M].北京：中国科学出版社，1977.

[8] 刘米达夫.植物化学[M].杨本文，译.北京：科学出版社，1985.

[9] 宁永成.有机化合物结构鉴定与有机波普学[M].2版.北京：科学出版社，2002.

[10] 漆小泉，王玉兰，陈晓亚.植物代谢组学——方法与应用[M].北京：化学工业出版社，2011.

[11] 中国科学院上海药物研究所.中草药有效成分提取与分离[M].2版.上海：上海科学技术出版社，1981.

[12] 盛龙生，苏焕华，郭丹滨.色谱质谱联用技术[M].北京：化学工业出版社，2006.

[13] 孙中武.植物化学[M].哈尔滨：东北林业大学出版社，2001.

[14] 谭仁祥.植物成分分析[M].北京：科学出版社，2002.

[15] 吴立军.天然产物化学[M].北京：人民卫生出版社，2003.

[16] 徐任生.天然产物化学[M].北京：科学出版社，1993.

[17] 于德泉，杨峻山.分析化学手册：核磁共振波谱分析[M].2版.北京：化学工业出版社，1999.

[18] 朱亮锋，陆碧瑶，李宝灵，等.芳香植物及其化学成分（增订本)[M].海口：海南出版社，1993.

第三章　植物精油的提取工艺与设备

　　植物精油是指芳香植物的植株内各个部位（器官）所含有的挥发性物质、或称芳香成分，它们是一类次生物质。不同芳香植物种属分别存在于它们的根、茎、枝叶、树皮、果实、种子和全草中，例如樟科樟属中，樟（*Cinnamomum camphora*）的精油分别存在于茎、根和枝叶中，樟属阴香（*C.burmannii*）的精油存在于枝叶中；木犀科茉莉花（*Jasminum sambac*）的精油存在于花瓣中；唇形科薄荷（*Mentha canadensis*）的精油存在于全草中；芸香科柑橘属（*Citrus*）的精油主要存在于果皮和枝叶中；姜科砂仁属砂仁（阳春砂仁）（*Amomum villosum*）除全草含精油外其种仁也富含精油。

　　形成精油的器官是它们细胞原生的分泌物，而芳香植物器官的细胞都应该具有分泌精油的能力，这类组织亦称为"油胞"，这类细胞从解剖切片来分析，可以分为外部精油细胞分泌腺和内部精油细胞分泌腺，外部细胞存在于表皮组织，称为油腺，而内部细胞分泌腺则分别存在于裂生分泌细胞、分泌腔或分泌道等。不同种属所含精油分别以各种不同形式存在于腺毛、树脂导管、油胞、裂生分泌细胞等，如白兰叶、柑橘类果表皮中；也有与分泌树脂、黏液质等共存的，如枫香树脂、沉香；还有精油成分与糖结合形成苷存在，如苦杏仁苷、冬青苷等，它们水解后分别为苦杏仁油和冬青油等。我们提取芳香植物内所含的精油就是破坏它的腺毛、油胞、裂生细胞等，例如用蒸汽蒸馏提取精油就是加热使它们的细胞壁破坏，用溶剂和压榨等方法提取精油也是同一原理。

第一节　植物精油提取前预处理

提取植物精油的原料，即芳香植物各个部位，其中包括花、果、枝叶、木材、树根、树脂等。一般来说，采收来的原料应尽快处理，或及时运送到加工地点，尽量让采收的原料不在运输途中损坏。不同种类原料的处理方法也截然不同。

一、枝叶、草本

枝叶、草本采收后应尽快投入加工，如岩蔷薇薰衣草、互叶千层枝叶、白树枝叶以及草本类等，另一类原料如白兰叶、树兰叶、玳玳叶、薄荷、留兰香等应放置一段时间，但必须均匀摊开堆放好，使之通风、透气，避免发热。有的原料甚至要阴干一段时间，如桂叶必须放2～3个月才能用水汽蒸馏。也有些原料如藿香只有经过发酵蒸出的精油才符合标准。如白兰叶放置数天其出油可提高5%～20%（按采后鲜重计），干燥后其叶表面细胞孔扩大使油点易扩散。

二、鲜花

采收后的鲜花较易凋谢，损失香气，不同鲜花有不同处理方法，如桂花采收后可用食盐水泡浸，而小花茉莉则在花蕾时就采收，在一层层花架上铺开，通风、透气，待花蕾将要开放时再投入萃取。

三、地下茎类

地下茎类，如姜科的姜、山柰等，一般需要把姜切片，晾干才能蒸出姜油。

四、树根和茎木

树根和茎木等较为坚实，为了使贮存在木质内油腺的精油较容易被提取出来，应把原料破碎、切碎或打粉，如枫香、安息香、檀香、柏木、沉香木等。

五、种子

一些种仁含精油，其种子壳较为坚硬，用水汽蒸馏或溶剂提取不易获得收得率较高的精油，一般压破种子壳则能提高出油率和缩短加工时间。

六、发酵处理

一些芳香植物提取精油前，需把原料进行发酵处理才能提取出精油或浸膏，如鸢尾根、香荚兰豆等。

第二节 植物精油提取方法与设备

由于对象和要求不同，精油提取方法和设备均不相同，主要可分为两种：一种仅为获取少量精油样品进行分析研究，另一种为具备一定生产规模的大型工艺和设备。

一、研究用样提取方法和设备

1. 野外取样最简单的是利用吸附剂（大孔树脂）在一个无味PVC袋装上样品，然后用携式大气采取器进行抽气，进气口加接一大孔树脂管，过滤净化进入样品袋内的空气，样品为鲜花类则可由鲜花自然释放出香气被装在出气口的大孔树脂管吸收，如果样品为枝叶，则需要用外力揉磨样品袋的枝叶加速其释放香气。大孔树脂用精制乙醚或己烷脱吸附则可获得少量精油样，可进行初步GC/MS分析（图3-1）。在野外也可用精制混合脂肪（猪油和牛油混合）吸收鲜花释放的香气，最终可获得少量香脂，但要进行分析还需要除去脂肪，这道工序较烦琐，只能做一些初步评香，现在一般极少采用此法。

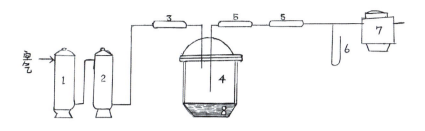

图3-1　大孔树脂吸附流程示意图

1、2.分子筛　3.大孔树脂管（参考用）　4.装被吸附原料的容器（野外可改用无味塑料袋）　5.大孔树脂管　6.空气流量计　7.抽气机（野外可改用野外大气采样器）　8.水

　　使用携带式整套水汽蒸馏装置到采样地附近，搭起整套水汽蒸馏设备，包括蒸馏锅、炉灶、冷凝器和驳接冷凝水的工作以及装好油水分离器（图3-2），还需解决柴火等十分繁重的工作，其取样时间较长，一天一般只能蒸馏2～3次，此方法一般多用于枝叶或草本，获得的样品较多，除用于成分分析外还可以做进一步研究之用。

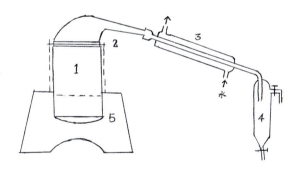

图3-2　野外蒸油设备

1.蒸馏器　2.水封　3.冷凝器　4.油水分离器　5.炉灶

　　2.用于实验室分析研究的精油，其提取方法比较简单，吸附方法与野外采样的相同，可能进气口要多加一支内装有分子筛的玻璃管，采样器可直接插实验室电源，水蒸气蒸馏则可安装玻璃设备，水汽发生器用电热蒸汽发生器提供蒸汽（图3-3）。

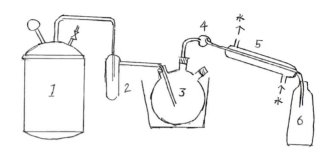

图3-3　实验室水汽蒸馏装置

1. 电蒸汽发生器　2. 水汽分离器　3. 水汽蒸馏三颈瓶　4. 防溅球　5. 冷凝器　6. 油水收集瓶

二、植物精油水汽蒸馏生产工艺和设备

由于生产精油的芳香植物种类繁多，客观上对精油产品种类需求不同，因此其生产工艺和设备也绝不相同。

1. 常压水汽蒸馏工艺与设备。这一方法是生产植物精油最常用的方法，它也有多种类似的不同方法与设备：水汽蒸馏原理，根据道尔顿原理，沸腾的定义是液体本身的蒸汽压力与外界大气压相同时就达到沸腾，液体本身达到外界蒸汽压时的温度，称为该液体的沸腾温度。加热两种混合液体时，两者蒸汽压之和等于大气压时，它们同时能达到沸腾，它们的混合蒸汽经冷却后能被蒸馏出来，混合液再经油水分离则可获得精油，至于被蒸出混合两相液体即油水的比率则取决于油与水在沸腾时它们蒸汽压力之比。由于水在常压下沸点为100℃，相对恒定，那么提高油相沸点温度，油相与水相混合液体中油相出油会增加。这样改善蒸馏工艺和设备将会改变油水蒸出比例，温度高则精油蒸出比率高（图3-4）。

一般水汽蒸馏按物料在蒸馏釜内堆放方式可以分为3种。第1种方法是水中蒸，被蒸物料与水混合后加热使水沸腾蒸出，此法可用于花类的水汽蒸，被蒸出的蒸出液经油水分离后水层可回到蒸馏釜中循环再蒸馏。第2种方法是隔水蒸馏，即物料在蒸馏釜中用筛缸与水隔开，加热水至沸腾后，蒸汽从蒸馏釜底部直接上升至被蒸物料，达到蒸馏目的。此法多用于中小型枝叶、草本等。同样可利用蒸出的水液回蒸馏釜再循环蒸馏。第3种方法是直接蒸汽蒸馏，蒸汽通过管道底部通入蒸馏釜底后由下而上进入被蒸物料进行蒸馏。这样蒸馏温度可以较高，油水蒸出比率比前两种方法高，蒸馏时间会缩短，但此方法需一个蒸汽发生器锅炉。这种方法蒸馏周期短，适合较大型的生产规模，同样可以利用回水（图3-5）。

（a）　　　　　　　　　　　　　（b）

图3-4　蒸汽蒸馏设备

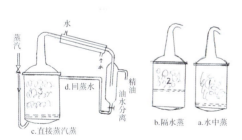 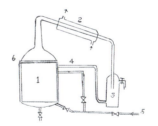

（a）三种水汽蒸馏方式　　　　（b）蒸汽蒸馏回水示意图

1. 蒸汽蒸釜　2.冷凝器　3.油水分离器

4. 蒸出水层回蒸馏釜导管　5.蒸汽　6.水封

图3-5　水汽蒸馏示意图

2. 减压水汽蒸馏法和加压水汽蒸馏法。基于水汽蒸馏时其外压降低则沸点也同时降低，蒸出温度也随之降低，这对温度敏感的芳香成分有利，特别是提取鲜花类的精油，但蒸出的精油与水中精油含量就降低。反过来外压增加，沸点提高则蒸出温度提高，则蒸出的精油与水中精油含量较高。

这对蒸木材、树根，如檀香、柏木、沉香等非常合适。但不论减压水汽蒸馏法还是加压水汽蒸馏法，其设备要求有专门的减压和加压整套设备，还需生产环境有防爆环境和措施（图3-6）。

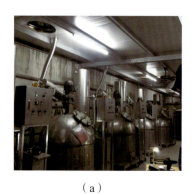 

（a） （b）

图3-6　高压蒸馏设备

3. 干蒸法。此法应用不多，主要针对某些特殊原料和客观特定要求，这种方法基本原理是隔绝空气，把碎物料放入钢制干蒸釜中，加热使温度达到物料中精油沸点则被蒸出。此法多用于杉木、柏木蒸馏，如柏木在干蒸过程中柏木脑（cedrenol）会随之脱水生成柏木烯（cedrene），这一组分是合成一种香原料甲基柏木酮（methyl cedrone）的原料。干蒸法用于杉木干蒸馏，得出黑色黏稠油状物称"黑油"，再经处理可获得结晶的柏木脑。采用干蒸法其出油率可达8%，其中柏木烯含量可达40%～70%，如果用一般水汽蒸馏法柏木烯只有1%～2%出油率（图3-7）。

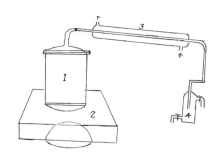

图3-7　干蒸法示意图

1.干蒸釜　2.炉灶　3.冷凝器　4.分油器

4. 水渗透蒸馏法。此方法与直接水汽蒸馏法相似，但不同之处是水汽是从蒸馏釜由上面下喷洒被蒸馏的物料，水汽是再经过二次加热才进入物料，蒸出混合液是从釜底流出，为此其蒸出温度较高、蒸馏液精油和水中精油含量较高，且经过热蒸汽同时迅速扩散到被蒸的物料，使蒸馏加快，被蒸出的混合液在蒸馏釜中停留时间缩短，使精油不论是香气、色泽还是出油率都高于常规水汽蒸馏（图3-8）。

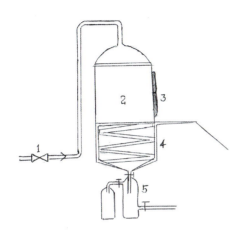

图3-8　水渗透蒸馏工艺流程示意图

1. 过热蒸汽　2. 水渗透蒸馏釜　3. 原料、废渣进出口　4. 冷却器　5. 油水分离器

5. 在水汽蒸馏过程或后续过程中还要进行如下事项。

1）投料蒸馏须注意被蒸物料堆放均匀，如鲜花、草本等受热后变软，需要分层同筛网相隔，叶则须加入小枝条，以免阻塞蒸汽扩散影响蒸馏。

2）蒸馏获得粗精油必须除去残留的水分，少量残存在精油中的水分会使精油变质，常用的吸水剂有无水硫酸钠、焙烧脱水硫酸镁，加热、高速离心等方法都可以达到除去残留水分的目的。

3）水汽蒸馏蒸出混合液经油水分离后，水层中可能尚有一定量极性芳香成分，如醛、酮、醇、酚类等，可复蒸一次或直接用低沸点溶剂萃取，水层加入电解质如食盐等效果更好，萃取液经无水硫酸钠脱水后浓缩除去溶剂，则获得极性芳香成分（图3-9～图3-10）。

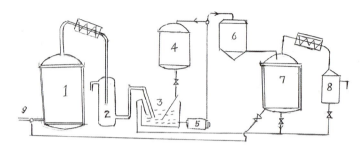

图3-9　复蒸萃取自动循环工艺流程示意图

1. 水汽蒸馏釜　2. 油水分离器　3. 储水池　4. 提取釜　5. 水泵　6. 高位槽　7. 复蒸釜　8. 分油器　9. 蒸汽

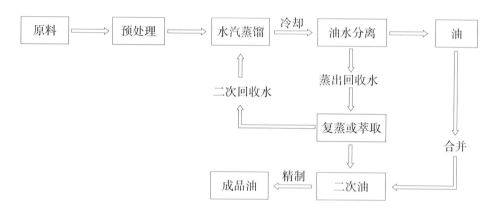

图3-10　水汽蒸馏提取精油工艺流程示意图

4）在水汽蒸馏过程中，冷却是很重要的一环，这是根据蒸出精油的性质来调整冷却温度，一般含单萜多的精油最终冷却温度须接近或达到外界温度。如果精油中含有在常温状态下为固体（如龙脑、樟脑）的成分，则冷却温度要比外界温度高10～20℃为好，避免倒塞冷却管道。而水汽蒸馏桂叶时须使冷却温度高于外界温度20～30℃，否则蒸出液会乳化，难以油水分离。如果是减压水汽蒸馏，其冷却温度为10℃以下，冷却面积须加大。

5）从树脂提取精油：植物精油大都分别存在于植物体内各个器官中，但有为数不多的精油存在于树脂中，如枫香树脂、沉香，还有松科的树脂中，马尾松等产的松脂提炼出的松节油是我国产量最大的植物精油，年产量达3万～4万吨，是一种用途很广的医药、食品重要原材料之一。其生产工艺流程如下（图3-11～图3-15）。

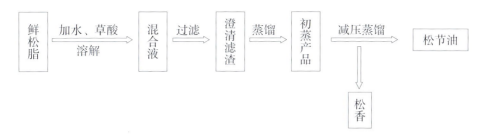

图3-11　松脂生产松香的过程

备注：生产出松脂应当尽快进行加工

图3-12　人工割松脂

图3-13　收集松脂

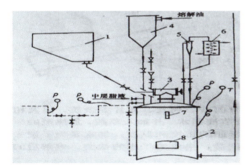

图3-14　松脂间歇熔解工艺流程图

图3-15　松香提取设备

1.加料斗　2.熔解釜　3.密闭加料阀
4.熔解没贮罐　　5.气液分离器
6.冷凝器　7.视镜　8.出渣口

第三节　溶剂浸提法

利用有挥发性、沸点低的有机溶剂，用萃取方法把贮存在芳香植物中的芳香成分提取出来，再通过蒸馏浓缩除去溶剂则获得含精油和蜡质、色素等的浸膏，再经低温乙醇处理后获得较纯净的精油，称为净油（pure oil）。

一、浸提原理

浸提亦称萃取（extract），溶剂浸提过程即溶剂经过渗透、溶解、扩散三阶段完成整个浸提或萃取过程。溶剂（或称溶媒）与芳香成分（溶质）相互分配，达到平衡后，溶媒中能溶解多少溶质，而还有多少东西留在溶质中，它们的比例称为分配系数。

$K=C_1/C_2$，C_1为溶媒浓度，C_2为溶质浓度，它们的比值K为分配系数。这个数字K代表溶剂萃取能力，一种溶质对不同溶媒其分配系数K是不同的，这是我们选择溶剂的一个有力依据。

二、溶剂的选择

用什么溶剂对提取芳香成分是相当重要的，溶剂的选择是基于芳香产品种类和客观要求。大致要求溶剂无色无味，不影响被浸提的芳香成分质量，即化学惰性，在常温下挥发性低，同时要求其沸点亦不高，容易被蒸去，对水溶解度极低。其选择性高，即对芳香成分溶解度高，而对杂质如脂肪、蜡质、色素、蛋白质、淀粉、糖类等溶解能力较低。化学惰性较高，基本不与芳香成分起化学反应，最后溶剂闪点要求较高，不易爆炸，不易燃烧，毒性较低。目前条件下，使用最多的溶剂为石油醚（沸点为65～75℃）。此外，也可选择用乙醇、苯、氯仿、二氯乙烷、乙腈等，也可用混合溶剂。

溶剂选择种类也影响到工艺和设备的确定。工业溶剂一般在提取前大都需要精制或处理，特别是被提取的芳香成分较为名贵。精制方法，一般多用复蒸，精制石油醚则可添加高沸石蜡油，或用较高效的分馏设备来精制。加入高锰酸钾酸性溶液，可以除去残留的石油烃芳类杂质。

三、浸提工艺与设备

工艺与设备相辅相成，可以用现成设备来选择工艺，也可设计好工艺来选购设备。

浸提工艺包括原料处理、浸提溶剂选择和芳香原料浸提、过滤出溶剂、溶剂蒸发回收、浓缩等过程，最终得到的产品为浸膏、香膏、油树脂、酊剂以及除蜡色素等杂质后所得的净油。

1. 按设备和溶质来分类，可分为静态浸提和动态浸提。静态浸提是指物料投入浸提釜中加入溶剂，一般每公斤加4～5升溶剂，待溶剂慢慢渗透扩散到物料，4～5小时后把溶剂放出再加入新鲜溶剂或回收溶剂。放出溶剂可马上蒸馏浓缩回收溶剂，回收溶剂可循环使用，但回收溶剂的水分一般比新鲜溶剂高，则必须添一些新鲜溶剂确保溶剂达到提取浓度要求。此方法已经改进为自动循环，浸提溶剂浸提后直接输送到蒸发浓缩器中，回收溶剂再经调整浓度后直接再输入浸提釜中不断循环。

此法所用设备简单，但溶剂使用量较大，物料静止会降低渗透和扩散速度，影响浸出效果和时间。物料不断翻动可使芳香物料较容易被溶剂扩散渗透，有助于加速芳香物料被浸提出来。另一种动态浸提方法是浸提釜内的物料与浸提溶剂相互按反向移动形成逆流扩散。一般浸提釜溶剂或回收溶剂在出料阀门前加入，浸提液则从加物料阀门流出，此操作是连续进行的。在加料阀门口附近由于溶剂最早接触物料，故这时浓度最高，在出料口阀门附近提取液浓度最低的，对物料来说情况刚好相反，这样在浸提釜中形成浓度差，而且在逆流连续液固接触中浸提和洗涤连续不断进行，则浸提效率高，提出率可高达90%，产品质量和香气均属上乘。这一方法生产能力大且适用性广，适用于鲜花如玫瑰、大花茉莉、小花茉莉、白兰花、晚香玉、黄兰花等的浸提，同时也适用于茎木材、枝叶、根、干块茎、块根以及粒状物料的浸提（图3-12、表3-1）。

表3-1　不同浸提方式的比较

	固定浸提	搅拌浸提	转动浸提	逆流浸提
提取方式	原料浸泡在溶剂中静止，但可以回流循环	原料浸泡在溶剂中，采用刮板搅动原料与溶剂同时转动	原料和溶剂一起在转鼓中相对运动	原料和溶剂作逆流方向移动以提高浸提效率
原料要求	适用于大花茉莉、晚香玉、紫罗兰等娇嫩的鲜花	适用于桂花、米兰等小型粒状鲜花	适用于白兰、小花茉莉、玫瑰等花瓣类鲜花	适用性较广，多用于量较大的鲜花浸提
生产效率	较低	较高	高	最高
浸提效率	60%～70%	80%左右	80%～90%	90%左右
产品质量	由于原料静止，对易损鲜花有一定好处，浸泡杂质较少	由于刮板搅动不太伤原料，原料损坏少，故杂质也不多	原料损伤较多，杂质较多	浸提较充分，浸提比较完全，但杂质也较多

2. 浸提温度与浸提时间。温度可影响浸提速度和浸提产品优劣。温度高浸提速度快，但浸出物杂质多，同时增加能耗，一般浸提鲜花，特别是较名贵鲜花多采用室温浸提，而对根茎木质类如檀香、沉香、鸢尾等，为了能把高沸点芳香成分提出，增加溶剂的渗透和扩散，一般将浸提温度控制在溶剂沸点以下。

需严格把握浸提过程中的浸提时间，不同原料浸提时间不尽相同，一般是掌握浸提"终点"，即是浸提液的浓度与原料液香成分呈现动态平衡（即浸提液香成分浓度不再增加），此时为浸提"终点"。在实践中一般浸出率达到80%～95%，两者达到平衡，但不同原料也不尽相同，如浸提大花茉莉70%～75%，两者就基本达到平衡，浸提时间长短当然对生产周期、能耗、溶剂耗量有影响，而且也会增加产品的杂质等。

3. 提取液的蒸发与浓缩。提取液经过蒸发后成为浓缩液，但为了使浓缩液中的芳香成分保全免受破坏，则需要尽快把残存溶剂清除，必须进行减压蒸馏的二次浓缩，为了确保获得净油的香气，在减压浓缩过程中一定要掌握真空度、蒸馏温度以及冷却系统。就符合鲜花浸膏的要求来说，真空度80～84kPa，用喷射泵减压较为安全，加热温度较室温高10～15℃，减压蒸馏冷却温度十分重要，最适宜是用干冰或加食盐或氯化钙的冰水（在−10℃以下）来冷却，而收集器则也要有5～10℃冷却保温，减压蒸馏釜宜能旋转，所得的"粗产品"将根据市场和客观要求配制成不同浓度的浓缩液、酊剂、浸膏和香膏等（图3-16）。

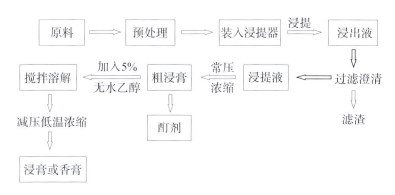

图3-16　浸膏制备工艺流程示意图

四、净油的制备

净油（pure oil）是浸提芳香成分的最终产品，其香气是芳香植物的芳香成分的高度浓缩特征和体现，经多个步骤获得的浸膏是一种较黏稠膏状物，这是浸提过程中溶解在溶剂中的香成分、蜡质（花蜡）色素等，为此必须除去蜡质和其他杂质，基于这些蜡质不溶于低温乙醇，但能溶于温度较高的乙醇，而精油不论温度怎样变化，都能较好溶解在乙醇中，根据这一现象我们在室温或更高的温度下加入一定量的95%精制（12～15倍）乙醇，把浸膏全部溶解后，慢慢把混合物温度降低，此方法可以分步冷冻或一次冷冻（图3-17）。

分步冷冻则先把混合物冷冻到−10℃然后减压过滤，过滤要用−10℃、95%乙醇洗涤至滤渣无香味，然后再冷冻至−26℃，同样方法减压过滤和洗涤滤渣至无香气，一次性除蜡质即一次把混合物冷冻至−25℃进行除蜡质。冷冻时间一般2～3小时，一般来说这种方法只能除去85%～90%的蜡，为获更优质的净油，还要重复除蜡两次以上。

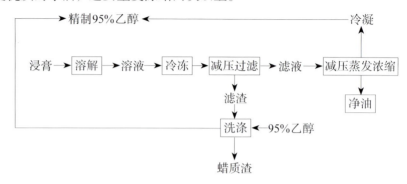

图3-17　净油制备工艺流程图

除蜡过滤和洗涤后的乙醇滤液需要蒸发浓缩回收，可以分两次进行，首先在减压下73.3～80kPa、温度（水浴）45～55℃，浓缩物料的大部分乙醇被蒸出后可把蒸馏真空提高到93.3kPa，使残存的乙醇基本蒸出，一般乙醇残留量在0.5%以下。这种获得净油的方法十分烦琐，已经开始用分子蒸馏，能直接获得纯度极高的净油，见本章第七节。

五、浸提设备

选择什么样的浸提设备，其依据是多方面的，包括生产规模、浸提的品种是单一种类还是多种类等。

1. 固定式浸提设备。这一设备比较简单，通常使用1～4台，2000～4000升有夹套不锈钢釜，备下口出渣用，上盖可装入料口和冷凝回流冷却器，釜底装有浸出液出口，出口管道连接蒸发器回收溶剂。如果所需物料能翻动，则在提取釜顶装上一搅拌棒直入釜中。

2. 刮板式浸提设备。其工作原理为将物料浸泡在有机溶剂中，用刮板式搅拌器使物料和溶剂都一起转，使物料始终都完全在动态下与溶剂保持接触，转动速度不需要太快，每分钟两转左右，浸提完成经过滤后可把浸提液直接送到蒸发器回收溶剂，回收的溶剂调整浓度后可再次使用（图3-18）。

3. 浮滤式浸提设备。其工作原理：安装在提取釜底的涡轮搅拌器的转动叶轮与固定的导轮及挡板，其作用是使物料与溶剂能迅速充分地混合接触，达到高效率扩散渗透的目的，利用安装在浮筒中心的吸滤管，上下浮动，使滤片保持在液面上，使浸提液与物料残渣分离，提取液抽出速度较快，在这一过程中澄清液和滤渣不易与残渣微粒接触，特别适合粉末状和树脂状物料的提取，浸出液可直接送至蒸发回收处。该设备在转动密封原件方面的加工精度要求较高（图3-19）。

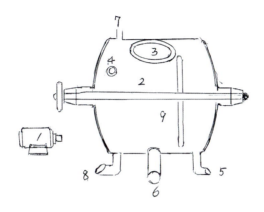

图3-18　刮板式浸提设备

1. 马达　2. 转轴　3. 加料口　4. 蒸汽入口　5. 废液出口　6. 排渣口　7. 溶剂入口　8. 浸提液出口　9. 刮板

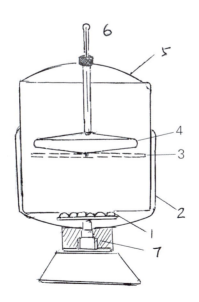

图3-19　浮滤式浸提设备

1. 涡轮搅拌器　2. 夹套　3. 过滤片　4. 浮筒　5. 快干盖　6. 中心吸虑管　7. 动力

4. 平转式逆流连续浸提设备。该设备工作原理和浸提过程是连续进行的，根据渗滤原理，溶剂是喷淋在物料表层，由上而下一层层浸润物料，把芳香成分从物料中溶解浸出后流至底层，经过滤流入储备槽中。再送入蒸发回收溶剂设备中，该设备处理量大，由于装物料斗不动，溶剂从上而下喷淋不会损伤物料（鲜花），保证以后产品的质量。喷淋溶剂尾气含量较高应注意回收。

5. 泳浸桨叶式连续浸提设备。该装置较特别，设备为一不锈钢圆柱，长11～12米，直径约1米，安装时约倾斜8°，底部装有垂直进料高位槽，圆柱形浸提釜内装有一根能转动的带式螺旋叶主轴，物料经高位槽进入圆柱形卧式倾斜提取釜前，首先被高浓度提取溶剂喷淋，再进入圆柱形卧式浸提釜后，物料被带式螺旋叶慢慢推向顶部，溶剂则从顶部向底部流动形成逆向对流，从而迅速渗透溶解物料中芳香成分。最浓的浸出液又先喷淋从高位槽投料口投入的新鲜物料，喷淋后的浓浸提液则被送到蒸发回收装置中，这一装置投料量大，可连续生产，生产周期短，效率高，最适合提取玫瑰、茉莉、栀子、墨红等鲜花中的精油（图3-20）。

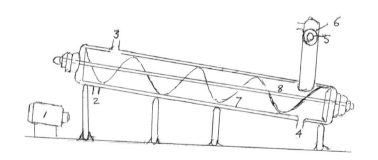

图3-20　泳浸桨叶式连续浸提设备

1. 马达　2. 废料出口　3. 溶剂入口　4. 浸提液出口　5. 鲜料入口　6. 进料漏斗
7. 带式螺旋液　8. 转轴

6. 热回流浸提装备。该装置与传统通用浸提装置相同，圆筒形带夹层的不锈钢浸提釜，底部为一倾斜椎形体，立式，上盖安装一投料入口，顶中央装有搅拌棒，出料口在釜底，物料从顶部进料口投入，溶剂亦可从进料口放入，并有回流冷凝器，加温时可从夹套通蒸汽加温，同时可用搅拌棒搅动物料，这一装置最适合岩蔷薇、鸢尾、灵香草、云木香以及一些中药材（如田七、桂圆）中精油的提取等。提取是非连续进行，废渣出料口是用气动开底出料盖（图3-21）。

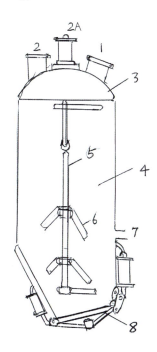

图3-21　热回流浸提装备

　　1. 加料口　2. 安装冷凝器口　2A. 安装搅拌器口　3. 顶盖　4. 提取釜　5. 转轴
6. 搅拌桨　7. 蒸汽出口　8. 出料口

第四节　榨磨法

　　榨磨法主要用于提取柑橘类果皮精油，此方法生产的精油主要用于食品、化妆品、医药等方面，且产量非常大。提取柑橘类果皮精油可分两大类，一是压榨法，二是整果磨榨法。

一、前处理

不管用什么方法，其目的是把储存在外果皮层中的油囊破碎才能获得其中的精油，果皮过软和过硬均不利于使油囊破裂，同时果肉与果外皮层还存在厚厚的海绵状内果皮，它的存在往往会吸附果皮精油，并使那些果胶随精油一起被收集，影响提取率，最常用的方法是把果皮浸泡在水或石灰水中，这样不但使果皮软硬适中，而且可使内皮层含大量水，降低海绵状内果皮对精油的吸附。石灰水也是使内皮层的果胶钙化使之不溶于水，避免精油中混入果胶影响精油分离，要注意在投料前一定要用水冲洗以除浸泡剂，由于柑橘类精油对热较敏感，故一般加工过程在室温下进行。

二、压榨法

压榨法主要是针对非球形果皮和零散的柑橘，把果皮处理后进行压榨，多采用螺旋压榨，上下压板锥形均有钉刺，且有孔道，压榨后流出的混浊液中有精油、水、少量果皮碎粒、果胶，需要通过离心或静置获得柑橘精油，也有把压榨流出的混浊液用海绵吸收，待海绵吸收饱和后，取出用挤压法把油和水分压出，这样果皮碎了，果胶则留在海绵内，将挤压出的油水液体静置进行油水分离，此法多为手工操作和间歇方法获取精油。

三、整果磨榨法

这是针对整果，果的形状为球形、大小相对一致，如甜橙、酸橙、广柑。磨盆为一平面或具轻微锥状形、漏斗形底盆，盆表面有小孔和钉刺，上压转盒形状与底盆形状一致，并有钉刺，底盆和上转盒距离可调节，视物料大小而定，转动可上盆转动下盆不转动，也可两者逆向转动，加快提取速度，在转动过程中需不断喷淋水，加工前同样根据实际情况进行上述操作。经榨磨后一般果还可加工利用，如压榨果汁、加工成蜜饯等。磨榨液是从磨盆底部流出，过滤后可静置分层取精油，为提高收得率和获得高质量的柑橘果皮精油，现大型生产厂家均用连续进料式高速离心处理流出混合液，加工果皮精油后，海绵的内皮层还可生产非常有用的食用果胶（图3-22～图3-24）。

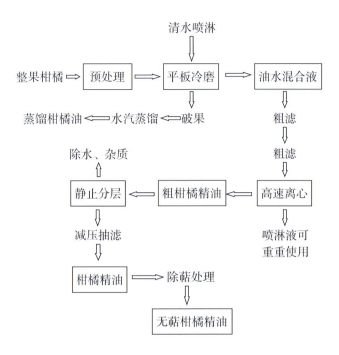

图3-22 柑橘类整果冷磨精油生产工艺示意图

注：预处理包括水洗、泡浸等

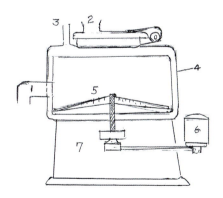

图3-23 平板式榨磨机示意图

1.出渣口 2.进料口 3.喷淋口 4.机壳 5.磨盘 6.转动马达 7.机座

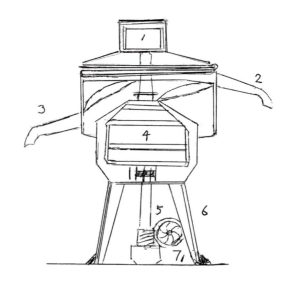

图3-24　碟式高速离心分离机示意图

1.加料口　2.出油口　3.出水口　4.转鼓　5.主轴　6.机座　7.涡轮

第五节　吸附法

一、树脂吸附法

由于植物精油是一挥发性化学成分，会自然散发出一定香气，当然对某些芳香成分如枝叶可通过机械法拢碎加速其释放香气。用表面活性高的物质如活性炭、硅胶等可吸附自然放出的芳香成分，因此很容易用低沸点溶剂就能解吸回收芳香成分，再除溶剂则获得优质鲜花精油。这一吸附过程是一物理吸附过程（即所谓"范德华力"），吸附剂并没有与吸附的芳香成分起化学作用，同时加入有机低极性溶剂则非常容易解吸。大孔树脂吸附量大，吸附能力强，化学惰性比上述两类吸附剂优，还可重复再生使用，它是由聚丙烯微粒堆积而成的表面积非常大的吸附剂。它有多种型号，大孔树脂其中一种商品名为XAD系列。针对吸附有机气体为XAD-4，针对从水中收集微量有机物为XAD-2，在环保上使用的多为XAD-7，这类吸附剂如果利用超临界CO_2萃取效果更优，不过设备成本费用会较高。

二、脂肪吸附法

吸附法是利用精炼脂肪如猪油、牛羊油或两者混合加热混溶后制成脂肪基，在能够对流半封闭透明（50cm×40cm×5cm）玻璃框中放多层涂有两面脂肪基的玻璃板，进入空间的空气一定要过滤净化，同时不断与外界保持湿度、温度一致。然后把加工物料轻轻铺在涂有脂肪基的玻璃板上，吸附24小时取出物料后，再铺上新鲜的物料，每24小时更换一次，直至脂肪基完全吸附饱和为止，此法对释放香势强的鲜花如大花茉莉、小花茉莉、晚香玉较为合适，对茉莉花来说一个周期需要约30天，吸附饱和后刮下的脂肪基即为产品，统称为香脂。香脂为天然香料中的名贵产品，可直接用于高级化妆品中（图3-25）。

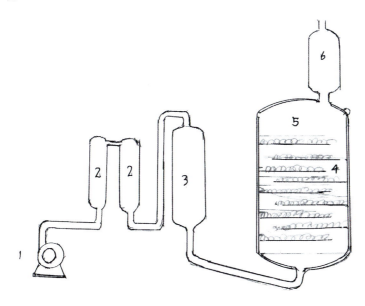

图3-25　动态吸附装置示意图

1.鼓风机　2.空气净化装置　3.增温器　4.鲜花　5.吸附器　6.吸附剂

第六节　超临界 CO_2 流体萃取法

超临界流体萃取是比较崭新的萃取方法，特别应用到芳香成分的萃取，其主要原理是在特定温度和压力下使某一流体具有接近液体密度、接近气体低黏度和扩散速度等的特性，使其具有很强的溶解度和渗透能力，能迅速使溶质和溶媒达到平衡（图3-26）。

（a）　　　　　　　　（b）　　　　　　　　（c）

图3-26　超临界萃取设备

CO_2 在常温下是气体状态，无毒，具有化学惰性，不易燃烧和爆炸。如果把温度降到-56℃、压力为0.51MPa时，二氧化碳处于气、液、固三相共存的状态，这时其黏度与密度已经非常接近液体状态，很容易渗透扩散到物料中，使两者达到动态平衡，这样基本上完成一次萃取过程。完成后先把耐压萃取釜降压至与室外一致的压力，这就等于把二氧化碳释放。萃取釜同时把温度慢慢升高至40~55℃，釜内的二氧化碳全部排除，留在釜中的主要是被萃物料的芳香成分，蜡质色素还可能存在少量水分，这是物料中水分带来的。另外由于超临界二氧化碳不能转变成液态后再转变成气态，故蒸发萃取后的二氧化碳更容易被排除干净（图3-27）。

一般二氧化碳萃取所得的粗浸膏，可用分子蒸馏除去蜡质和其他杂质，得到纯正的净油。

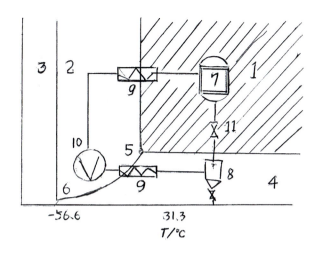

图3-27 二氧化碳超临界状态下萃取工艺流程示意图

1.超临界状态区 2.液态状态区 3.固态状态区 4.气态状态区 5.接水瓶 6.气液固三相点 7.萃取釜 8.分离器 9.冷却器 10.压缩机 11.阀门

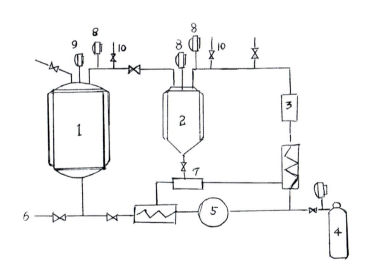

图3-28 二氧化碳萃取工艺流程示意图

1.萃取釜 2.分离器（萃取物） 3.放出CO_2 4.二氧化碳钢瓶 5.压缩机 6.原料或半成品 7.换热器 8.压力表 9.温度表 10.安全阀

第七节　分子蒸馏法

分子蒸馏法亦称短径蒸馏，它与我们常用的蒸馏原理完全不同。常用蒸馏按道尔顿原理是把被蒸馏的物料加热（加速分子运动）到其蒸气压与外界大气压相等时沸腾，其蒸气逸出过程就是蒸馏过程。而分子蒸馏的工作原理：使用的设备为短径蒸馏釜（或称短径蒸发器）（short-path evaporator），基本工作条件是高真空（0.1～10Pa），首先是最大限度排走蒸发器内的残存空气，减少空气对逸出物料分子的冲撞。这样物料（芳香成分）分子就能保持在较低温度下，无阻碍地逸出蒸发从而迅速达到收集冷却面，这种蒸发器的蒸发面到冷却收集面的距离应较物料分子的平均自由程短，故称短径蒸发。由于真空度高，物料中挥发性成分的沸点可大大降低，蒸发速度同时也提高，故能生产出优质的净油。这一方法多用于高级香原料生产加工如岩兰草精油、杂薰衣草净油等，这类产品的专门名词为U.V.产品。

分子蒸馏最根本的就是高真空，早期高真空是分级进行的，先用机械泵使其达到一定真空度后再用扩散泵，早期用汞扩散泵，后来改用油扩散泵。已经有不少工厂改用更先进的分子泵，这可大大降低各类扩散泵的内外反复污染，而前置泵也改用水环泵，减少环境污染（图3-29～图3-32）。

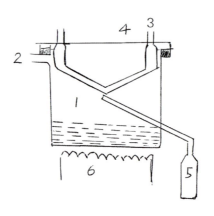

图3-29　静止式分子蒸馏设备

1.分子蒸馏釜　2.接真空泵组　3.冷却水　4.冷凝器　5.接收器　6.加热器

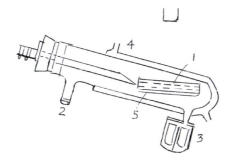

图3-30　静止盘式分子蒸馏设备

1.蒸发盘　2.接油扩散泵　3.可转动收集器　4.冷却水　5.加热器

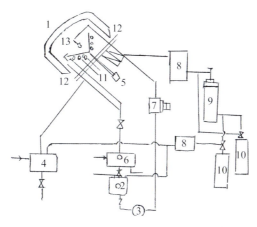

图3-31　离心式分子蒸馏工业化流程示意图

1.分子蒸馏器　2.原料罐　3.进料泵　4.馏出物罐　5.蒸发器马达
6.蒸余物罐　7.蒸发器　8.冷阱　9.油扩散泵（分子泵）
10.油旋转泵（水循环泵）　11.电加热棒　12.冷却水　13.进料

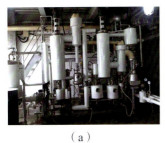

（a）　　　　　　　　　　（b）　　　　　　　　　　（c）

图3-32　分子蒸馏设备

第八节　微波提取技术

微波是一种电磁波，其波长介于红外和无线电波之间，一般作为热源的微波炉波长为12.2cm（2.45GHz）。微波提取植物精油过程如下：微波加热导致植物细胞（指含有精油的细胞）如油胞、油腺、树脂导管等内的极性物质（细胞液）吸收微波能量，从而产生大量热量，促使细胞内温度迅速上升，液态细胞液（水分）汽化膨胀产生压力把细胞膜和细胞壁冲破，继续加热使细胞表面出现裂纹和孔洞，使提取液体（溶剂）直接进入细胞内部进行渗透和扩散，并溶解细胞内的挥发物质（精油），这样在微波照射下迅速加热至沸点，在密封容器（提取器）中这时温度应较溶液正常沸点高，这样促使精油成分较快被萃取出来。应用微波这一技术提取植物精油，其优点在于速度较快，减少热破坏，对含水分较多的被萃取植物原料更为有利，因为水的矩较大，产热较快，这样对新鲜植物原料含水分较高更有利。另外，微波在提取植物精油时是否会使精油某些成分产生重排或异构化，有不同报道，但把握好时间、溶剂、温度等影响因素，所提取的精油成分相对比较稳定。目前这一技术在精油提取方面报道较多。

第三章　参考文献

[1] 安鑫南.林产化学工艺学[M].北京：中国林业出版社，2002.

[2] 金琦.香科生产工艺学[M].哈尔滨：东北林业大学出版社，1994.

[3] 贺近恪，李启基.林产化学工业全书（第三卷）[M].北京：中国林业出版社，1997.

[4] 孙凌峰，汪洪武，汤敏燕.黄樟素的天然来源及其在合成香料中的利用[J].香料香精化妆品，1998(2)：14-18.

[5] 黄彪，刘雁，陈彦，等.超临界二氧化碳提取肉桂精油的研究[J].林产化学与工业，2003(1)：59-62.

[6] 庞建光，张明霞，韩俊杰.植物精油研究及应用[J].邯郸农业高等专科学

校学报，2003，20（1）：26-29.

[7] 邹超贤，梁丽明，陈衍松，等.高效节能提取肉桂油新工艺的研究[J]. 林产化工通讯，1995（5）：11-15.

[8] 周雪敏，朱科学，房一明，等.黑胡椒油树脂乙醇浸提工艺研究[J]. 中国调味品，2016，41（10）：1-11.

[9] 李平月，聂青玉，陈鲁，等.超声法制备胡椒油树脂工艺研究[J].食品科技，2010，35（2）：60-62.

[10] 柳中，李银聪.索氏法提取白胡椒油树脂的工艺研究[J].食品与发酵科技，2011，47（1）：65-68.

[11] 戴猷元.新型萃取分离技术的发展及应用：超临界流体萃取技术[M].北京：化学工业出版社，2007：129-144.

[12] 李毓敬，朱亮锋，陆碧瑶，等.天然右旋龙脑新资源——梅片树的研究[J].植物学报，1987，29（5）：527-537.

[13] 陆碧瑶，李毓敬，朱亮锋.樟属芳香植物资源[J].植物杂志，1985（6）：18-19.

[14] 陆生椿，黄秀丽，卢剑飞，等.茉莉花香成分研究及其应用试验[J].广州轻工，1985（3）：1-5.

[15] 罗金岳，安鑫南.植物精油和天然色素加工工艺[M].北京：化学工业出版社，2005.

[16] 丘雁玉，李飞飞，邓超宏，等.广东省3种野生香茅属植物精油的化学成分含量分析[J].植物资源与环境学报，2009，18（1）：48-51.

[17] 朱亮锋，陆碧瑶，李宝灵，等.芳香植物及其化学成分（增订版)[M].海口：海南出版社，1993：358.

[18] 朱亮锋，陆碧瑶，罗友娇.茉莉花头香化学成分的初步研究[J].植物学报，1984，26（2）：189-194.

[19] 朱亮锋，陆碧瑶，徐丹，等.阳春砂仁叶油和广宁绿壳砂仁叶油化学成分初步研究[J].广西植物，1983，3（1）：43-47.

第四章 植物精油及其成分的用途

　　植物精油在古今中外的历史上都与人类息息相关。在访问非洲南部时，我们较深入地了解到远在非洲的原住民所用的草药绝大多数都是含有芳香气味的植物，除煎煮成汤剂服用外，最主要还是熏蒸和外敷。我们祖先早在两三千年前就已经用菊科蒿属精油熏蒸洁身防病治病，一千多年前就利用茉莉花熏制茶叶供饮用。至于被用作食品的调味调香料以及美容养生是近代蓬勃发展起来的关系到民生的一门产业。

第一节　植物精油的活性成分与医疗药物

　　植物精油很多成分存在生物活性以及药性。实践表明，它们大多具有芳香健胃、解表发散、祛痰、止咳、平喘、抑菌、消炎、镇痛、止瘙痒、止血等功效。用精油直接配制的各种风油精类，在我国南方以至东南亚和不少华人聚居的欧美地方已十分普遍，这类风油精最大特点是可随身携带，使用方便，大多数只要外涂一两滴即可，对肌肉疼痛、蛇虫咬伤、感冒不适、舟车晕船、瘙痒、头晕头痛、伤风鼻塞等有一定缓解效果，且多为家里和旅行外出常备的药物之一。常见到的风油精有香港和兴白花油，新加坡生产斧标驱风油，马来西亚生产的虎标万金油系列，广州生产的穗草油、罗浮山生产的百草油、追风油和产于印度尼西亚的千层属白树（Melaleuca leucadeadron var. cajaputi）的白树油，其主要含1,8-桉叶油素，以及名目繁多的药用植物精油。总体来说这些药用植物精油离不开薄荷油、桉叶油（蓝桉油）、樟脑、水杨酸甲酯（冬青油）、薰衣草油、迷迭香油、丁香油等。

另外《中国药典》（2020版）所列出的中成药中添加植物精油作为主料或副料的也不少，如十香丸系列、保济丸、藿香正气丸、七制香附丸、十五味沉香丸、八味檀香散、九味羌活丸、川芎茶调丸、小儿香橘丸、小儿解表颗粒、云香祛风止痛酊、木香顺气丸、止咳橘红丸等等。在西药方面，酊水剂中也少不了樟脑和薄荷。

植物精油适合居家、健体、美容、治疗亚健康的芳香疗法。这整个产业是在有关专业医疗人员指导下进行的，内容包括嗅吸、口服、按摩、香薰、沐浴等，常使用的植物精油多达70多种，现选择30多种我们常见的植物精油进行介绍，详见表4-1。

表4-1　30多种常见植物精油用于芳香治疗

名称	主成分	特性
1. 澳洲互叶白千层精油 *Melaleuca alternifolia*	4-松油醇、松油烯、1,8-桉叶油素	抑杀细菌、病毒、真菌和寄生虫
2. 罗勒精油 *Ocimum basilicum*	丁香酚	止痛，改善痛经，治疗肠道病、腹胀、便秘，消除烦躁、忧郁
3. 佛手柑皮精油 *Citrus medica* var. *sarcodactylis*	柠檬烯、芳樟醇、乙酸酯	可治白癜风，杀菌消毒，放松心情，健康肌肤
4. 玫瑰木油（蔷薇管花樟）*Aniba rosaeodora*	芳樟醇、香叶醇、乙酸芳樟酯	改善皮肤粗糙，抗皱，延缓皮肤衰老，消除烦躁
5. 柏木油（福建柏木油）*Fokienia hodginsii*	橙花叔醇、侧柏烯	改善前列腺充血
6. 锡兰玉桂 *Cinnamomum zeylanicum*	丁香酚、肉桂醛	杀菌，治疗尿道炎，对哮喘性气管炎有疗效，杀灭人体寄生虫，消除疲劳
7. 岩蔷薇种子精油 *Cistus ladaniferus*	α-蒎烯	止血，可治流鼻血、痔疮出血，改善失眠
8. 柠檬果皮精油 *Citrus limon*	柠檬烯、柠檬醛	利肝清肝，净化消化系统，利尿，改善血液循环

（续表）

名称	主成分	特性
9. 香茅（亚香茅） *Cymbopogon nardus*	香茅醇、香叶醇	驱虫，杀菌消炎，清新空气，消除异味
10. 乳香（树脂） *Boswellia carterii*	α-蒎烯、桧烯、柠檬烯、香叶烯	抑制沮丧情绪，增强免疫力
11. 龙蒿（花精油） *Artemisia dracunculus*	甲基萎叶酚	治疗呼吸系统过敏，缓解风湿疼痛
12. 柠檬桉叶精油 *Eucalyptus citriodora*	香茅醛、香茅醇	消除真菌，驱除蚊蝇，缓解痉挛
13. 蓝桉叶精油 *Eucalyptus globulus*	1,8-桉叶油素	治疗肺炎，舒畅咽喉
14. 桉树（大叶桉） *Eucalyptus robusta*	1,8-桉叶油素、对伞花烃	治疗感冒、流感、气管炎
15. 平铺白珠树 *Gaultheria procumbens*	水杨酸甲酯	改善由炎症引发的疼痛，消炎、退热、利尿
16. 天竺葵（香叶） *Pelargonium graveolens*	香茅醇、香叶醇	治疗低血糖及相关疾病，改善身体疲乏，治疗脱发、痤疮，改善皮肤衰老，紧致皮肤
17. 姜（块茎） *Zingiber officinale*	姜油烯、莰烯、1,8-桉叶油素	缓解各种恶心症状，缓解身体疲劳、肌肉紧张，治疗脱发，对关节炎、肌肉疼痛治疗效果明显
18. 公丁香花蕾精油 *Eugenia caryophyllata*	丁香酚、乙酸丁香酯	缓解各种牙痛、口腔溃疡，具杀菌、消毒之效，可消除疲劳
19. 土木香花精油 *Inula graveolens*	乙酸龙脑酯	治疗黏液分泌过多，对多种炎症有效，如阴道炎、膀胱炎等
20. 月桂（开花枝桠） *Laurus nobilis*	1,8-桉叶油素、芳樟醇、α-松油醇、松油烯醇、木香内酯	杀菌，止痛，可治流感、风湿
21. 宽叶薰衣草精油（开花时）*Lavandula latifolia*	1,8-桉叶油素、芳樟醇、樟脑	对重度烧伤、蛇咬蜂叮、蝎子、蜈蚣咬等有一定效果

（续表）

名称	主成分	特性
22. 薰衣草 *L. angustifolia*	乙酸芳樟酯、芳樟醇、1,8-桉叶油素、薰衣草醇	治疗偏头痛、眩晕、失眠、神经紧张、心悸、心律不齐、脉管炎
23. 柠檬草精油（枫茅） *Cymbopogon citratus*	α，β-柠檬醛、芳樟醇	镇静，安抚情绪，消炎、杀菌、消脂，治疗蜂窝性组织炎
24. 黄连木开花期枝叶精油（中国黄连木） *Pistacia chinensis*	β，α-蒎烯、α，β-罗勒烯、柠檬烯	治疗各种充血、静脉曲张，对耳鸣有一定效果
25. 山苍子果实精油 *Litsea cubeba*	α，β-柠檬醛、柠檬烯	抑制沮丧情绪，减压、安眠、消炎
26. 辣薄荷全株精油 *Mentha piperita*	薄荷脑、薄荷酮、辣薄荷醇	止痛，改善呼吸不畅，治消化疾病
27. 匙叶甘松 *Nardostachys jatamansi*	β-古芸烯、白菖油萜烯	呼吸镇静剂，头发滋养剂，刺激头发生长
28. 苦橙花精油 *Citrus aurantium*	芳樟醇、α-松油醇、橙花醇	排解不良情绪，调节神经
29. 牛至花期枝叶精油 *Origanum vulgare*（美洲）	香芹酚、麝香草酚、芳樟醇	消毒杀菌，改善病毒性炎症，增强免疫力，治疗耳鼻喉病
30. 葡萄柚果皮精油 *Citrus paradisi*	柠檬烯、芳樟醇、橙花醇、α，β-柠檬醛	杀菌消毒，净化室内空气，紧致皮肤，如面部、大腿等
31. 檀香（木）精油 *Santalum album*	α，β-檀香醇、檀香醇异构体	对各种炎症有效，紧致肌肤
32. 巴山冷杉针叶精油 *Abies fargesii*	β-蒎烯、龙脑、乙酸龙脑酯	治感冒，缓解流鼻涕症状，治疗气管炎，抗疲劳
33. 柠檬果皮精油 *Citrus limon*	柠檬烯、α，β-柠檬醛、香叶醇、4-松油醇	改善人体血液循环，杀菌消炎，消除异味，清新空气

第二节　植物精油用于食品

　　植物精油用于食品，主要是以食品添加剂的形式出现，其中柑橘类果皮精油广泛应用于饮料、饼干、糕点和冰激凌中，可以增加食品风味，用量非常大；另一类植物精油直接用于调味料，如花椒油、八角茴香油、胡椒油、孜然等辛辣型调味料。部分制成油树脂形式使用。

　　由于植物精油具有抗细菌、抗真菌和抗氧化等多种潜在的特性，它在食品工业中得到了广泛的应用。植物精油作为天然食品防腐剂的功效已经在许多食品中进行了测试，如烘焙食品、水果和蔬菜以及肉类产品。从牛至、丁香、百里香和柑橘中提取的精油对细菌和酵母具有较强的抑制活性，在乳制品中添加精油既增强了产品的香气，又赋予其抗菌特性，使其更容易被消费者接受。在水果和蔬菜中使用植物精油有助于延长其保质期。与水果和蔬菜有关的主要问题是其真菌腐败，有研究已经表明在水果中使用植物精油可以减少微生物数量。尽管植物精油具有一系列有益的特性，但其易挥发性和疏水性使其难以直接用于食品体系。纳米技术已被报道为一种可以保护活性化合物不被降解并提高其在食品系统中的功效的植物精油递送系统的解决方案。纳米封装保护植物精油免受光线、水分、pH值等恶化因素的影响，在加工和储存过程中也是如此。它还有助于在水介质中溶解亲脂性化合物并将其释放到目标位置。

第三节　植物精油用于化妆品、护肤品以及洗涤用品

　　绝大多数植物精油是调配成香精添加于膏霜类、洗涤香波、高级香水、花露水等各类化妆、护肤产品。如果缺少植物精油，可能就不会有这么多化妆品、护肤品、沐浴露等产品与消费者见面了。还有精油调配的一种产

品就是烟用香精，这也是用量非常大的精油调配产品，它是卷烟工业中非常重要的辅料。目前国内使用烟用天然香料主要有白兰叶油、广藿香油、八角茴香油、姜油、灵香草油、肉桂油、小豆蔻油、薰衣草油、桂花净油、茉莉花净油、香叶油、丁香油、沉香油等，它们除了可提高香烟的香味外，对降低焦油、尼古丁后提高烟草香浓味同样能起主要作用。还有一类就是添加到露酒的酒类香精。

第四节　植物精油及含精油的芳香植物在抑菌杀虫上的作用

植物精油被认为对微生物具有一定的抑制或抑杀作用，但由于植物精油的化学成分非常复杂，每一个化学成分的抑菌效果截然不同。据研究，植物精油主要成分为萜类化合物，其次为芳香族化合物，它们不含氧的萜烯，抑菌能力较差，而含氧衍生物中羰基化合物较羟基化合物好，基团前有双键特别是不饱和或共轭双键，其抑菌效能较高，碳链短链较长链抑菌效能高，详见表4-2。

研究表明，植物精油成分醛类如桂醛、柠檬醛、茴香醛，酚类如麝香草酚、丁香酚、香芹酚，酮类如香芹酮、薄荷酮、辣薄荷酮，醇类如芳樟醇、香茅醇、薄荷醇（脑）等均具较强的抑制真菌生长及抑制真菌毒素合成的能力。另外4种植物精油成分4-松油醇、丁香酚、1,8-桉叶油素、麝香草酚对镰刀菌属、曲霉菌属、青霉菌属均具较明显抑制生长或抑杀功效，其中不少对霉菌达到100%抑制率。

关于植物精油及其活性成分抑制真菌作用机制，以及对抑菌效能较强的精油成分柠檬醛研究表明，它能通过损伤黄曲霉菌的质膜，使其失去选择通透性而进入细胞内对其他细胞器官产生影响，受破坏的霉菌质膜，再使细胞内大分子空间结构改变，新陈代谢紊乱，从而抑制霉菌的生长。

表4-2　植物精油成分抑菌试验

样品	黑曲霉 Aspergillus niger	西氏曲霉 A. Sydowi	土曲霉 A. terreus	产黄青霉菌 Penicillium chrysogenum	宛氏拟青霉 Paecilomyces variotii	球毛壳霉 Chaetomium globosum	甘露芽孢霉 Cladosperium herbarum	木霉 Trichoderma sp.
依兰油烯 ylangene	--	--	--	--	--	--	--	--
蛇麻烯 humulene	--	○	--	-	-	-	-	-
β-榄香烯 β-elemene	--	○	--	-	-	○	○	-
龙脑 borneol	-	○△	-	○	○	++	○	○△
十二碳醛 dodecanal	-	+	-	○	+	++	○	○
乙醛 aldehyde	-	○△	-	○△	○	+△	+△△	-
枞油烯 sylvestrene	○	○	+	+△	+	++	+	+
β-水芹烯 β-phellandrene	○	○	○	○	+	++	+	+
乙酸正辛酯 octyl acetate	○△	++	+△	○	○	++	+	○
α-派烯 α-pinene	+	++	○	+	+	++	+	+
辛醛 octanal	○	++	+	+△	+△	+△△	+	++
α-松油醇 α-terpineol	++	+△	○	+△	+△	++	+	+
芳樟醇 linalool	++	++	+△	++	++	++	++	++
4-甲基-6-乙酸基己醛 4-methyl-6-cetoxyhexanal	+△	++△△	+	++	++	++	++	++
柠檬醛 citral	++	++	++	++	++	++	++	++
正辛醇 L-octanol	++	++	++	++	++	++	++	++

注：++表示未见霉菌生长；+表示抑菌生长；-表示抑菌圈在5mm以上；○表示抑菌圈为2～4mm；—表示抑菌圈小于2mm，但未在滤纸上生长；--表示霉菌已在滤纸上生长；△表示抑制生长；△△表示只长个别菌落。

表4-3　具有广谱抑真霉菌活性的植物精油及其活性成分

名称	来源	抑制的真霉菌
野蔷薇 *Rosa multiflora*	蔷薇科蔷薇属	茄病镰刀菌、尖孢镰刀菌、中珠镰刀菌等12种真霉菌
孜然 *Cuminum cyminum*	伞形科孜然芹属	烟曲霉菌、尖孢镰刀菌、青霉菌等17种真霉菌
小茴香 *Foeniculum vulgare*	伞形科茴香属	烟曲霉菌、尖孢镰刀菌、青霉菌等17种真霉菌
圣罗勒 *Ocimum sanctum*	唇形科罗勒属	黄曲霉菌、烟曲霉菌、桔青霉菌等13种真霉菌
竹叶花椒 *Zanthoxylum armatum*	芸香科花椒属	黄曲霉菌、黑曲霉菌、桔青霉菌等12种真霉菌
百里香 *Thymus vulgaris*	唇形科百里香属	黄曲霉菌、尖孢镰刀菌、黑曲霉菌等9种真霉菌
薄荷醇 menthol	薄荷油	皮肤癣菌、念珠菌、黄曲霉菌、黑曲霉菌等真霉菌
百里香酚 thymol	百里香精油	黄曲霉菌、赭曲霉菌和桔青霉菌等11种真菌
丁香酚 eugenol	丁香油	禾谷镰刀菌、炭黑曲霉菌和青霉菌等10种真霉菌
香芹酮 carvone	香芹精油	禾谷镰刀菌、曲霉菌、黑曲霉菌等真霉菌
柠檬醛 citral	山苍子油	黄曲霉菌、桔青霉菌、禾谷镰刀菌等5种真霉菌

第五节　植物精油对抑杀害虫的作用

采用66种植物精油对3种主要储粮仓内害虫赤拟谷盗、玉米象、谷蠹，在0.2%剂量的植物精油下进行繁殖抑制试验。其中齿叶黄皮、芸香、猪毛蒿、八角茴香、草八角、地檀香、香楠、沉水樟、狭叶阴香、肉桂油完全能抑制赤拟谷盗F1代的发生，0.2%浓度的上列植物精油能抑制赤拟谷盗F1繁殖率达到94.36%，同样0.1%浓度的上列植物精油对谷蠹F1繁殖达到基本完全抑制，对玉米象致死率高达85%，这说明上列植物精油对三种储粮仓害虫不论繁殖或毒杀都存在非常有效的抑制功效。研究报道从芸香科黄皮属齿叶黄皮枝叶精油中获得一个主含93%的爱草脑对7种霉菌具有强的抑菌活性，但对玉米象（*Sitophilus zeamais*）、赤拟谷盗（*Tribolium castaneum*）、黄粉虫（*Tenebrio molitor*）、谷蠹（*Rhizopertha dominica*）在浓度0.2mg/L时抑杀率达到100%，这是一类使用植物精油预防和抑杀粮储害虫的天然药物，由于植物精油均为挥发性天然产物，同时其安全性是经过测试的，因此基本不存在残留问题（表4-4）。

表4-4　植物精油的活性成分的安全评价

常见精油的活性成分	评价方法	实验对象	浓度 LD50/ （g·kg⁻¹）	评价 （毒性级别）
木橘 *Aegle marmelos*	经口急性毒性	小鼠	23.660	无毒
百里香 *Thymus vulgaris*	经口急性毒性	小鼠	17.500	无毒
丁香 *Syzygium aromaticum*	经口急性毒性	小鼠	55.230	无毒
肉桂 *Cinnamomum aromaticum*	经口急性毒性	小鼠	5.038	无毒
枫茅 *Cymbopogon winterianus*	经口急性毒性	小鼠	3.500	低毒
薄荷 *Mentha canadensis*	经口急性毒性	小鼠	2.000	低毒
印棟素 *Azadirachtin*	经口急性毒性	小鼠	13.000	无毒
马鞭草酮 *Verbenone*	经口急性毒性	小鼠	8.088	无毒
1,8-桉叶油素 1,8-cineole	经口急性毒性	小鼠	7.220	无毒
丁香酚 *Eugenol*	经口急性毒性	小鼠	3.000	低毒
肉桂醛 *Cinnamaldehyde*	经口急性毒性	小鼠	2.220	低毒

第六节　植物精油用于工业原料

在植物精油中产量最大、应用范围最广的当属主含 α-蒎烯和 β-蒎烯的松节油。我国松节油主要从马尾松脂经蒸馏所得，称为GT松节油，这有别于北美洲的主要国家利用针叶树木材生产纸浆回收的MPT松节油，我国每年生产松节油4.5万～5万吨，除本国使用外大部分出口，松节油除少量药用外，大部分是用于合成其他化工产品如合成樟脑、龙脑以及其他合成香料、黏结剂、溶剂等。

松节油主要成分为 α-蒎烯、β-蒎烯、月桂烯、松油烯、α-水芹烯，它们都是被允许作为合成食用香料的原料，如合成松油醇、合成松油醇酯、合成芳樟醇、合成香叶醇和橙花醇、合成薄荷醇、合成龙脑、合成乙酸龙脑酯、合成乙酸橙花酯和乙酸香叶酯、合成香叶基丙酮、合成乙酸异丙基甲苯、合成 N-烷基酰亚胺农药增效剂，还可合成昆虫保幼激素或昆虫生长调节剂、昆虫引诱剂、昆虫驱避剂、昆虫拒食剂。

另外松节油外用对皮肤具有增进局部血液循环、缓解肿胀和轻度止痛作用，调制成乳膏乳液，对于缓解运动疼痛、疲劳效果显著。松节油与鱼肝油复配外涂可止血，还可加速伤口愈合。松节油还可代替有毒二甲苯用于冰冻切片剂。

另一些植物精油成分，虽然产量远远不及松节油，但其结构特殊，能作为一些特有的产品合成原料，如黄樟油素（safrole），它在植物中含量大于40%者有34种，其中存在于根茎中的有20种，主要是樟科樟属黄樟、狭叶桂、少花桂、香樟、沉水樟等；存在于枝叶中的有10种；存在于草本中的有4种。由此看来可以用于连续生产黄樟油素的有14种，可建立种植生产基地，不再用挖根伐树来获取黄樟油素。黄樟油素是合成洋茉莉醛、大茴香脑、香荚兰素最理想的天然原料。

第七节　植物精油的防腐保鲜作用

具有抑制霉菌成分的植物精油在防腐、防霉方面所发挥的作用值得注意。首先这一类植物精油的安全性，比起化学合成的防腐保鲜剂更容易为人们所接受，利用柠檬醛、爱草脑、大茴香醛等生产出一种保鲜膜，用作水果保鲜膜，防止指状青霉、意大利青霉入侵的效果十分好。

第八节　在中药材中的精油植物（芳香植物）

《中国药典》（2020年版）第一部分的有关中药材中，含有精油成分的中药有80～90种，如果加上民间草药能超过100种，但它们的药性大致相似，如丁香性味为辛、温，归脾、胃、肺、肾经；人参性味为甘、微苦、微温，归脾、肺、心、肾经；九里香性味辛、微苦、温，有小毒，归肝、胃经。它们大多用于行气止痛、风湿痹痛、跌打肿痛、虫蛇咬伤、脘腹胀痛、芳香化浊、发表解暑、理气宽中、燥湿化痰、活血散瘀、解毒、肿痛、疟疾、散瘀解毒消痛、外伤出血、散寒止痛、月经过多、保气养阴、清热生津等等，详见表4-5。

表4-5　芳香植物中的中药材

名称	主要精油成分	特性
1. 丁香（干花蕾） *Syzygium aromaticum*	丁香酚	补肾助阳，可治脾胃虚寒、肾虚阳痿、食少吐泻
2. 人参（块根） *Panax ginseng*	α，β-榄香烯、β-香树烯、β-石竹烯	大补元气、生津养血、安神益智，可治血气亏虚、肺虚喘咳、惊悸失眠

名称	主要精油成分	特性
3. 九里香 *Murraya paniculata*	t-石竹烯、α-姜黄烯、γ-柏木烯、蛇麻烯	行气止痛、活血散瘀、风湿痹痛，外治牙病、虫蛇咬伤、跌打肿痛
4. 土木香 *Lnula helenium*	木香内酯、木香烯、鹰爪甲素	健脾和胃、行气止痛、安胎、驱虫，可治呕吐泻痢
5. 土荆芥 *Chenopodium ambrosioides*	乙酸4-松油醇酯、α-松油烯、驱蛔素	杀虫、疗癣、止痒（外用）
6. 山柰（沙姜） *Kaempferia galanga*	cis-t-对甲氧基桂酸乙酯、桂酸乙酯、大茴香醚	行气温中、消食、止痛，可治脘腹冷痛、饮食不消
7. 华南忍冬（山银花） *Lonicera confusa*	顺反-氧化芳樟醇、苯乙醇、罗勒烯	清热解毒、疏散风热，可治热毒血痢、风热感冒，用于肿痛疗疮
8. 川木香 *Vladimiria souliei*	木香内酯、倍半萜烯	行气止痛，可治脘腹胀痛、肠鸣腹泻、里急后重
9. 川芎全草（藁本） *Ligusticum sinense*	新蛇床内酯、榄香脂素、β-水芹烯、t-罗勒烯	活血行气、祛风止痛，可治胸痹心痛、跌打肿痛、月经不调、头痛、风湿痹痛
10. 姜（含干姜） *Zingiber officinale*	姜油酮、6-姜辣素	抑菌驱风、温中散寒，可治寒饮喘咳
11. 土香圆叶（香橼叶） *Citrus medica*	α，β-柠檬醛、柠檬烯	理气化痰、活血止痛，可治咽喉肿痛、跌打伤痛
12. 广藿香 *Pogostemon cablin*	广藿香薁醇、广藿香醇、丁香酚、α-广藿香烯	镇呕、芳香健胃、解热、发表解暑、芳香化浊
13. 小茴香 *Foeniculum vulgare*	t-茴香脑、葛缕酮、芹菜脑、柠檬烯	祛风健胃、散瘀，可治少腹冷痛、经寒腹痛
14. 天然右旋冰片（梅片） *Borenol*	d-龙脑	开窍醒神、清热止痛，可治中风痰厥、目赤口疮、惊厥
15. 木香 *Radix Aucklandiae*	木香内酯、去氢木香内酯	行气止痛、健胃消食、治泄泻腹痛、泻痢后重
16. 片姜黄（含姜黄） *Curcuma wenyujin*	芳姜黄烯、姜黄烯、姜油烯、莪术酮	行气止痛、抗菌降压、通经止痛，可治风湿肩臂疼痛

（续表）

名称	主要精油成分	特性
17. 甘松（根茎） *Nardostachys jatamansi*	甘松新酮、马兜铃烯、甘松醇	理气止痛，开郁醒脾，外用祛湿消肿
18. 小蓬草 *Conyza canadensis*	左旋龙脑	开窍醒神、清热止痛，可治中风痰厥、目赤口疮
19. 艾叶（野艾蒿） *Artemisia lavandulaefolia*	左旋龙脑、樟脑、1,8-桉叶油素	温经止血，散寒止痛；外用祛湿止痒
20. 橘红（化州橘红） *Citrus grandis* 'Juhong'	柚皮苷、柠檬烯、柠檬醛	理气宽中、燥湿化痰，可治食积伤酒，用于咳嗽痰多
21. 月季花 *Rosa chinensis*	柠檬烯、香茅醇、乙酸芳樟酯、芳樟醇	活血调经、疏肝解郁，可治月经不调、痛经、闭经、腹肋胀痛
22. 石菖蒲（菖蒲） *Acorus calamus*	β-细辛醚、白菖烯、细辛酮、白菖醇	开窍健脾、醒神益智、化湿开胃，可治健忘失眠
23. 白芷 *Angelica dahurica*	欧前胡素、1,8-桉叶油素	解表散寒、宣通鼻窍、消肿排脓，可治鼻塞流涕、牙痛、疮疡肿痛
24. 白术（根茎） *Atractylodes macrocephala*	苍术酮、茅术醇、白术内酯A、1,8-桉叶油素、香芹二酮	补中益气、燥湿利尿、止汗安胎，可治消化不良、腹胀泄泻、水肿自汗
25. 母丁香（干果实） *Syzygium aromaticum*	丁香酚	温中降逆、补肾助阳，用于脾胃虚寒、食少吐泻、肾虚阳痿
26. 当归（块根） *Angelica sinensis*	阿魏酸、藏红花醛、柠檬烯、2,4,6-三甲基苯醛、佛手烯、柠檬烯、藁本内酯、2,4-二甲苯甲醛	降压利尿、补血活血、调经止痛，可治眩晕心悸、闭经痛经、虚寒腹痛、风湿痹痛、跌打损伤
27. 肉豆蔻（干种仁） *Myristica fragrans*	去氢异丁香酚、1,8-桉叶油素	温中行气、涩肠止泻，可治久泻不止、食少呕吐
28. 肉桂 *Cinnamomum aromaticum*	肉桂醛、乙酸肉桂酯、肉桂酸	补火助阳、散寒止痛，可治腰膝冷痛、肾虚作喘、痛经闭经、吐寒吐泻

（续表）

名称	主要精油成分	特性
29. 芫花 *Daphne genkwa*	芳樟醇、乙酸香叶酯、α-蒎烯	健脾胃、泄水通饮，可治胸腹积水、气逆咳喘，外用可杀虫疗疮
30. 苏合香（树脂） *Liquidambar orientalis*	肉桂酸、乙酸肉桂酯	开窍、辟秽、止痛，用于中风痰厥、腔腹冷痛
31. 西洋参 *Panax quinquefolius*	人参皂苷Rg1、Rb，Re精油	清热生津，用于气虚阴亏、虚热烦倦、咳喘痰血、内热消渴
32. 安息香（树脂） *Styrax tonkinensis*	苯甲酯、苯甲酸酯类、安息香酸、安息香酸酯类	行气活血、开窍醒神，用于中风痰厥、中恶昏迷、心腹疼痛
33. 红豆蔻（种子）（大高良姜） *Alpinia galanga*	1,8-桉叶油素、樟脑、α-蒎烯	散寒燥湿、醒脾消食，用于食积胀满、饮酒过多
34. 豆蔻（种子） *Amomum kravanh*	1,8-桉叶油素、4-松油醇、α-松油醇	化湿行气、温中止痛、开胃消食，可治腔腹胀痛、寒湿呕逆、湿温初起
35. 牡荆（叶） *Vitex negundo* var. *cannabifolia*	1,8-桉叶油素、4-松油醇、β-石竹烯、β-荜澄茄烯	祛痰、止咳平喘，用于咳嗽痰多
36. 佛手（果皮精油） *Citrus medica* var. *sarcodactylis*	柠檬醛、橙花醇、橙皮苷、α-佛手烯、柠檬烯	疏肝理气、和胃止痛，可治咳嗽痰多、腔肋胀痛
37. 羌活 *Notopterygium incisum*	羌活醇、异欧前胡素、芳樟醇、β-水芹烯、对伞花烃、桂酸甲酯	解表散寒、祛风除湿、止痛，用于风寒感冒、肩背酸痛、风湿痹痛
38. 玫瑰花 *Rosa rugosa*	香叶醇、芳樟醇、金合欢醇、香树烯	行气解郁、和血、止痛，用于肝胃气痛、食少呕恶、月经不调、跌打伤痛
39. 郁金 *Curcuma aromatica*	1,8-桉叶油素、芳樟醇、牻牛儿酮、龙脑、莪术酮、樟脑	活血止痛、行气解郁、利胆退黄，用于闭经痛经、乳房胀痛、黄疸尿赤
40. 青蒿 *Artemisia carvifolia*	青蒿素、蒿酮、芳樟醇	清虚热、除骨蒸、解暑热、截疟、退黄，用于疟疾寒热、湿热黄疸、阴虚发热

（续表）

名称	主要精油成分	特性
41. 土沉香（树脂） *Aquilaria sinensis*	沉香四醇、5,6,7-三羟基色酮、螺旋沉香醇、β-呋喃沉香、呋喃白木香醇	行气止痛、温中止呕，可治胃寒呕吐、肾虚、气逆、喘急
42. 卷柏 *Selaginella tamariscina*	穗花衫双黄酮、4-松油醇、乙酸龙脑酯	活血通经，用于闭经痛经、症瘕痞块、跌打损伤
43. 侧柏 *Thuja orientalis*	槲皮素、罗汉柏烯、柏木烯醇、3-蒈烯	凉血止血、化痰止咳、生发乌发，用于吐血、咯血、便血、崩漏下血、血热脱发
44. 金银花（忍冬花） *Lonicera japonica*	木犀、草苷、芳樟醇、橙花醇、苯甲醇、异橙花醇、α-松油醇	清热解毒、疏散风热，用于肿痛疔疮、喉痹、丹毒、风热感冒、温病发热
45. 荆芥 *Schizonepeta tenuifolia*	胡薄荷酮	解表散风、透疹、消疮，用于感冒、头痛、麻疹、疮疡初起
46. 草豆蔻 *Alpinia hainanensis*	山姜素、小豆蔻明、桤木酮、金合欢醇、1,8-桉叶油素、肉桂醛	燥湿行气、温中止呕，用于寒湿内阻、不思饮食、嗳气呕逆、脘腹胀痛
47. 砂仁（含绿壳砂仁） *Amomum villosum*	乙酸龙脑酯、龙脑、橙花叔醇、樟脑	化湿开胃、温脾止泻、理气安胎，用于湿浊中阻、脘痞不饥、脾胃虚寒
48. 降香（心材） *Dalbergia odorifera*	苯甲酸甲酯、芳樟醇、苯乙醇、柠檬烯	化瘀止血，用于吐血呕血、外伤出血、肝郁胁痛、跌打损伤、呕吐腹痛
49. 草果 *Amomum tsaoko*	1,8-桉叶油素、2-甲基桂醛、乙酸香叶酯、α,β-柠檬醛	燥湿温中、截疟除痰，用于寒湿内阻、痞满呕吐、疟疾寒热、瘟疫发热
50. 香橼 *Citrus medica*	柠檬烯、α，β-柠檬醛、香茅醛、乙酸香叶酯	疏肝理气、宽中、化痰，用于肝胃气滞、胸胁胁痛、痰多咳嗽、呕吐噫气
51. 香薷 *Mosla chinensis*	α,β-蒎烯、柠檬烯、乙酸龙脑酯、α-石竹烯	发汗解表、化湿中和，用于暑湿感冒、恶寒发热、头痛无汗、水肿、小便不利

（续表）

名称	主要精油成分	特性
52. 松节油（树脂精油） *Pinus tabuliformis*	α-蒎烯	祛风除湿、通络止痛，用于风寒湿痹、历节风痛、跌打伤痛
53. 细辛 *Asarum sieboldii*	细辛醚、β-金合欢烯、乙酸龙脑樟	祛痰止咳、散瘀消肿、温经散寒
54. 姜黄（块根） *Curcuma longa*	姜黄素、姜黄酮、姜黄烯、1,8-桉叶油素、姜油烯、牻牛儿酮	破血行气、通经止痛，用于胸肋刺痛、胸痹心痛、痛经闭经、症瘕、跌打肿痛
55. 莪术（根茎） *Curcuma phaeocaulis*	姜烯、莪术酮、芳姜酚、姜黄酮、樟脑、龙脑	行气破血、消积止痛，用于症瘕痞块、淤血经闭、胸痹心痛、食积胀痛
56. 高良姜 *Alpinia officinarum*	高良姜素	湿胃止呕、散寒止痛，用于脘腹冷痛、胃寒呕吐、嗳气发酸
57. 菊花 *Dendranthema morifolium*	木犀草苷、绿原酸、龙脑、樟脑、1,8-桉叶油素	散风清热、平肝明目、清热解毒，用于风热感冒、头痛眩晕、目赤肿痛
58. 紫苏叶 *Perilla frutescens*	α，β-柠檬醛、香叶醇、芳樟醇、香茅醇	解表散寒、行气和胃，用于风寒感冒、咳嗽呕恶、鱼蟹中毒、妊娠呕吐
59. 蔓荆叶（单叶蔓荆） *Vitex trifolia* var. *simplicifolia*	蔓荆子黄素、1,8-桉叶油素、β-石竹烯	疏散风热、清利头目，用于风热、感冒、头痛、齿龈肿痛、目赤多泪
60. 薄荷 *Mentha canadensis*	薄荷脑、薄荷酮、胡薄荷酮	疏散风热、清利头目、利咽透疹、疏肝行，用于风温初起、头痛、目赤喉痹、口疮麻疹
61. 藁本（川芎） *Ligusticum sinense*	阿魏酸、新蛇床内酯、γ-依兰油烯、β-石竹烯、t-罗勒烯、薰衣草醇	祛风、散寒、除湿、止痛，用于风寒感冒、巅顶疼痛、风湿痹痛
62. 檀香（木） *Santalum album*	α-檀香醇、α-异檀香醇、β-檀香烯、α-檀香烯	行气温中、开胃止痛，用于寒凝气滞、胸膈不舒、呕吐食少、脘腹疼痛

（续表）

名称	主要精油成分	特性
63. 茵陈（茵陈蒿） *Artemisia capillaris*	茵陈炔酮、丁香酚、茵陈炔、丁香酚甲醚	清利湿热、利胆退黄，用于黄疸尿少、湿温暑湿、湿疮瘙痒
64. 荜澄茄（山苍子） *Litsea cubeba*	α-柠檬醛、芳樟醇、6-甲基庚烯酮	温中散寒、行气止痛、用于胃寒呕逆、脘腹冷痛、寒疝腹痛、小便浑浊、寒湿郁滞
65. 乳香（树脂） *Boswellia carteri*	α-蒎烯、桧烯、柠檬烯、香叶烯	活血行气止痛，消肿生肌
66. 龙蒿 *Artemisia dracunculus*	甲基娄叶酚	治暑湿发热、虚劳等
67. 青花椒（花椒） *Zanthoxylum schinifolium*	β-罗勒烯-X、爱草脑、柠檬烯、1,8-桉叶油素	散寒、祛温、止痒、驱虫、温中助阳
68. 藏龙蒿 *Artemisia waltonii*	丁香酚甲醚、榄香脂素、茵陈炔、茵陈炔酮	杀菌防腐
69. 乳香 *Boswellia carteri*	α-蒎烯、桧烯、香叶烯、柠檬烯	活血定痛、消肿生肌，用于胸痹心痛、痛经经闭、产后瘀阻、筋脉拘挛、跌打损伤、痈肿疮疡
70. 细辛 *Asarum sieboldii*	细辛醚、马兜铃酸	解表散寒、祛风止痛、通窍、温肺化饮、风寒感冒、头痛、牙痛
71. 花椒 *Zanthoxylum bungeanum*	1,8-桉叶油素、芳樟醇、α-松油醇、辣薄荷酮	止痛、助消化、牙痛、腹痛、止泻、杀虫
72. 枫香（树脂） *Liquidambar formosana*	α-蒎烯、β-蒎烯、莰烯	痰饮喘咳、鼻塞流涕

另外含有精油的植物除被用于中药材外，还可以直接用于调味料，如八角茴香、花椒、胡椒、孜然、肉桂皮、小茴香、丁香等，它们主要是被用作辛辣调味，还有兰科的香荚兰豆主要用作酒香精、烟香精等。

第九节　植物精油主要成分萜类的应用

　　植物精油的化学组成中，萜类中的单萜和倍半萜占了一大部分，其中单萜主要被用于药用香料和轻工化工业上，而倍半萜的化学结构较为复杂，因此它们具有广泛的生物活性。不少倍半萜含氧衍生物具有强烈和持久的香气，广泛用于食品、化妆品和医疗药物。如青蒿素以及它们的各类衍生物对各种疟疾如间日疟疾、恶性疟疾、脑型疟疾以及危重型疟疾颇具疗效，青蒿素衍生物、青蒿素甲醚、青蒿素琥珀酸酯等已经正式为临床用药。

　　1. 具有抗肿瘤作用的天然倍半萜种类最多。其中又多为倍半内酯类，据不完全统计，有细胞毒活性的就有上百种。据报道，α-亚甲基-γ-丁内酯或α-亚甲基环戊酮是使这类倍半萜类化合物具有抗癌生物活性的有效基团。除内酯类倍半萜外，如人参精油中富含的β-榄香烯（β-elemene）对多种肿瘤具抑杀活性的作用，已被用于治疗脑瘤、肺癌、肝癌、食道癌、子宫癌、白血病等等，其药理作用被认为是干扰肿瘤细胞生长代谢、抑制肿瘤细胞的增殖，此化合物已被用于临床治疗。

　　2. 驱虫杀虫作用。多种倍半萜成分可驱杀人体内的寄生虫，山道年是长期被使用的一种针对蛔虫的天然药物。土木香内酯和菊科土木香（*Lnula helenium*）根有驱虫成分鹰爪甲素。根中分离出一种倍半萜，对鼠疟原虫生长具有强抑制作用，它和青蒿一样具有过氧键的倍半萜化合物。

| costunolide | eremanthine | santonin | artemisinin |
| 木香烃内酯 | 巴西菊内酯 | 蛔蒿素 | 青蒿素 |

　　3. 昆虫拒食性、驱避及引诱作用。昆虫拒食剂是倍半萜的衍生物如卫茅科苦树（Celastrus angulatus）根皮分离出苦树皮苦素Ⅰ（angulatin Ⅰ）等

多种类似化合物的二氢呋喃沉香多元醇类衍生物，对多种昆虫有拒食或毒杀作用，而苦树叶粉民间亦有用于菜虫拒食剂。从一种异水霉菌（*Allomyces sabviscula*）得到一个倍半萜双羟基化合物雌诱素，可作为某些昆虫及低等动物的引诱剂，其生理作用十分强。

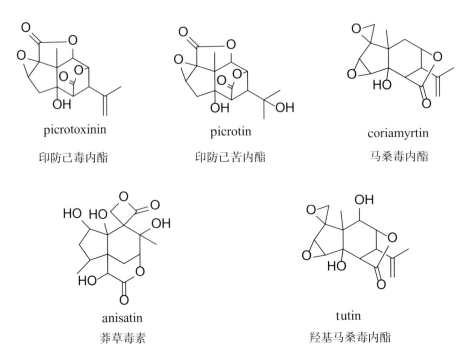

sirenin

雌诱素

4. 对神经系统的生理作用。这类具有神经系统生理活性的萜类多为倍半萜内酯，如从印度防己果实（*Amamirta cocculus*）获得的印防己毒内酯（picrotoxinin）和印防己苦毒素（picrotin），还有存在于日本产毒空木（*Coriaria japonica*）和我国产马桑（*C. sinica*）中的马桑毒内酯（coriamyrtin）和羟基马桑毒内酯（tutin），还有八角科莽草（*Illicium anisatum*，毒八角）果、叶、树皮中含有的莽草毒素（anisatin），它们对中枢神经系统均有兴奋作用，与剧毒的番木鳖碱一样具有很强的致惊厥作用。

picrotoxinin

印防己毒内酯

picrotin

印防己苦内酯

coriamyrtin

马桑毒内酯

anisatin

莽草毒素

tutin

羟基马桑毒内酯

5. 调节植物生长作用。对植物生长的调节作用包括抑制植物生长和促进植物生长两个相互作用，脱落酸（abscisic acid）早期是在脱落的棉铃中分离得到的，是一种促进落叶的激素，并可以控制植物落叶以及抑制种子和球根顶芽发芽。马齿苋科半支莲（*Portulaca grandiflora*）含有的半支莲醛（portulal），亦是一种植物生长素，对萝卜生根有阻碍作用。另外从棉花根部渗出的分泌物独脚金醇（strigol），是一种倍半萜衍生物，能促进玄参科种子发芽。

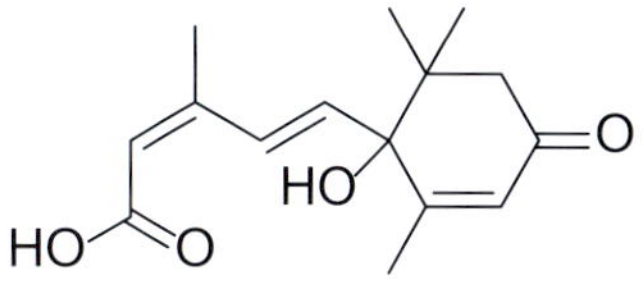

abscisic acid

脱落酸

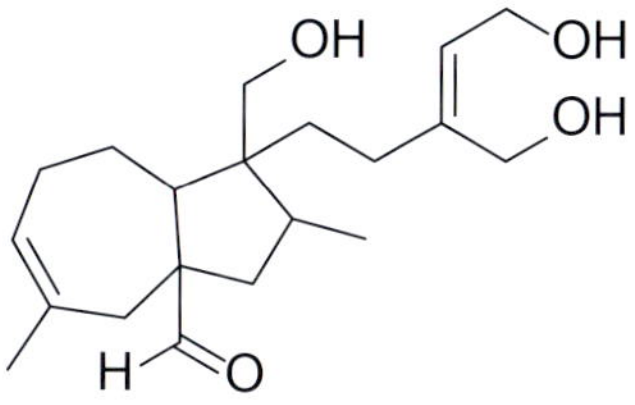

portulal

半支莲醛

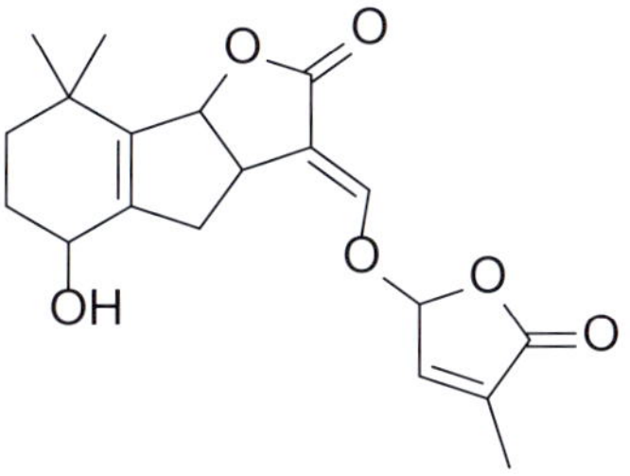

strigol

独脚金醇

6. 抑菌活性。本书在第四章第四节中主要介绍一些植物精油和植物精油成分中的单萜抑杀微生物的概况，这一部分主要介绍精油中倍半萜抑菌活性，其中菊科银胶菊（*Parthenium hysterophorus*）及豚草属植物（*Ambrosia maritma*）存在的倍半萜和它的衍生物 parthenin 和 ambrosin，具抗革兰阳性菌作用。从凹顶属（*Laurencia*）提取出的对凹藻醇（oppositol），是一种含溴的倍半萜醇，能较好地抑制金黄色葡萄球菌活性。它们都是薁类的骨架。

7. 可用于食品、化妆品的天然抗氧化剂——迷迭香精油。唇形科草本植物迷迭香（*Rosmarinus officinalis*）抗氧化能力十分强，而且完全无毒。迷迷香精油原产于地中海沿岸，在世界各地广泛被引种，而且还具有较强的杀菌、消炎、抗病毒及肿瘤等生物活性。

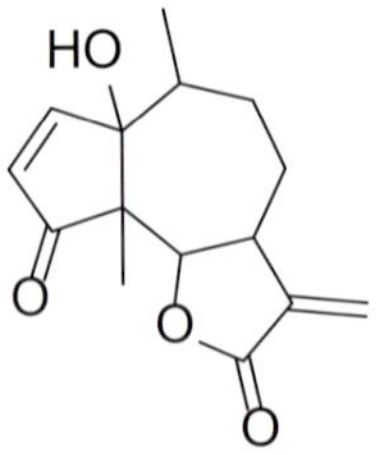

parthenin
银胶菊素

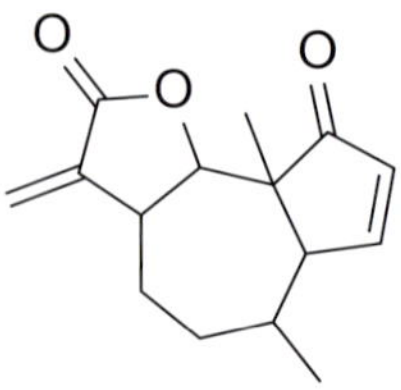

ambrosin
豚草素

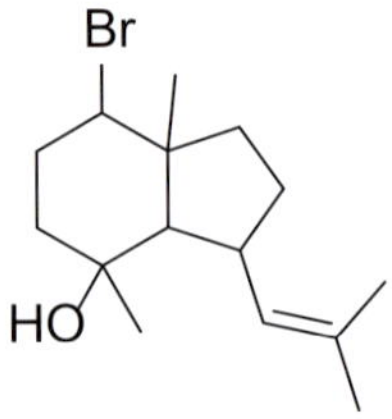

oppositol
对凹藻醇

第四章　参考文献

[1] 国家药典委员会.成方剂和单味制药[M]//中国药典：第一部分 药材和饮片、植物油脂和提取物.北京：中国医药科技出版社，2020.

[2] 菲斯蒂.精油圣经[M].刘曦，张昕，余春红，等译.上海科技文献出版社，2011.

[3] 哈成勇.天然产物化学与应用[M].北京：化学工业出版社，2003.

[4] 许勇.黄樟油素植物资源的介绍[J].中国野生植物资源，1994（4）：21-23.

[5] 李端，周立刚，姜微波，等.伞形科植物抗菌成分的研究进展[J].西北农林科技大学学报（自然科学版），2005（33）：161-166.

[6] 刘布鸣，魏绣枝，赖茂祥.广西精油资源与药用[J].广西林业，2004（6）.

[7] 刘布鸣，赖茂祥.广西药用精油植物资源与应用[J].广西中医学院学报，2005，8（2）：83-87.

[8] 吕洪飞.药用芳香植物资源的开发和研究[J].中草药，2000，31（9）：711-715.

[9] 苗青，赵祥升，杨美华，等.芳香植物化学成分与有害物质研究进展[J].中草药，2013，44（8）：1062-1068.

[10] 田玉红，张祥民，黄泰松，等.植物精油资源及其在烟草工业中的应用

[J].广东化工，2007，34（10）：73-75.

[11] 刘冬恋，马松涛，曾仁勇，等.肉桂挥发油对小鼠的半数致死量测定[J].
西南国防医药，2010，20（5）：481-482.

[12] 马松涛，刘冬恋，兰小平，等.丁香挥发油对小鼠的半数致死量测测定
[J].辽宁中医药大学学报，2010.12（5）：67-68.

[13] 孙凌峰，汪洪武，汤敏燕.黄樟油素及其在合成香料中的利用[J].香料
香精化妆品，1998（2）：14-18.

[14] 田玉红，张祥民，黄泰松，等.植物精油资源及其在烟草工业中的应用
[J].广东化工，2007，34（10）：73-75.

[15] 李燕君，孔维军，李梦华，等.植物精油抑制真菌及真菌毒素的研究进
展[J].中草药，2011-2018，47（11）.

[16] 余伯良，罗惠波，周健，等.柠檬醛抗真菌及抑制黄曲霉产毒的试验报
告[J].食品科技，2002（4）：47-49.

[17] 程世法，朱亮锋，陆碧瑶，等.勒党果精油化学成分和抑菌活性的研究
[J].植物学报，1990，32（1）：49-53.

[18] 朱亮锋，陆碧瑶，余耀新.天然水果保新剂的研究[J].全国食品添加剂
通讯，1991（4）：28-29.

[19] 徐汉虹，赵善欢，朱亮锋，等.齿叶黄皮精油的杀虫作用与有效成分研
究[J].华南农业大学学报，1994，15（2）：36-40.

[20] 徐汉虹，赵善欢，朱亮锋.精油对储粮害虫种群的繁殖抑制作用研究[J].
中国粮油学报，1993，8（2）：290-296.

[21] 贾金莲.松节油工业现状与市场[J].林产化学与工业，2001，21（4）：60-
64.

[22] 刘志秋，陈进，徐勇.黄樟素植物资源的开发利用现状及前景[J].香料
香精化妆品，2001（4）：14-19.

[23] 中国医药公司上海化学试剂采购供应站.试剂手册[M].2版.上海：上海
科学技术出版社，1985：1074.

[24] 程必强，喻学俭，丁靖垲，等.中国樟属植物资源及其芳香成分[M].昆
明：云南科学技术出版社，1997.

[25] 程必强，许勇，曾仙仙.滇南樟属植物资源的开发利用[C]//.热带植物

研究论文报告集.昆明：云南大学出版社，1994: 38-45.

[26] 庞建光，张明霞，韩俊杰. 植物精油的研究及应用[J]. 邯郸农业高等专科学校学报，2003，20（1）: 26-29.

[27] 张有林，张闰光，钟玉. 百里香精油的化学成分、抑菌作用、抗氧化活性及毒理学特性[J]. 中国农业科学，2011，44（9）: 1888-1897.

[28] Bazana MT, Codevilla CF, de Menezes CR. Nanoencapsulation of bioactive compounds: Challenges and perspectives. Curr Opin Food Sci 26, 2019, 47–56.

[29] Luo M，Jiang L K，Zou G L. Acute and genetic toxicity of essential oil extracted from Litsea cubeba（Lour.）Pers [J]. J Food Prot，2005，68（3）: 581-588.

[30] Mishra AP, Devkota HP, Nigam M, Adetunji CO, Srivastava N, Saklani S, Shukla I, Azmi L, Shariati MA, Melo Coutinho HD, Mousavi Khaneghah A. Combination of essential oils in dairy products: A review of their functions and potential benefits. Lwt 133, 2020, 110116.

[31] Morcia C，Malnati M，Terzi V.In vitro antifungal activityof terpinen-4-ol，eugenol，carvone，1,8-cineole（eucalyptol）and thymol against mycotoxigenic plant pathogens [J]. Food Addit contam A，2012，29（3）: 415-422.

[32] Singh P，Kumar A，Dubey N K，et al. Essential oil of Aegle marmelos as a safe plant-based amtimicrobial against postharvest microbial infestations and aflatoxin contamination of food commoolities [J]. J Food sci，2009，74（6）: 302-307.

附录　汉英化合物名称对照

一画

1-乙基-2,3-二甲基苯	1-ethyl-2,3-dimethylbenzene
3-乙基-2,5-二甲基吡嗪	3-ethyl-2,5-dimethylpyrazine
4-乙基-2,6-二叔丁基苯酚	4-ethyl-2,6-di-t-butylphenol
乙基丁基醚	ethyl butyl ether
1-乙基丙基苯	1-ethylpropylbenzene
乙基丙基醚	ethyl propyl ether
4-乙基-2-甲氧基苯酚	4-ethyl-2-methoxyphenol
6-乙基四氢-2,2,6-三甲基-吡喃-3-醇	6-ethyltetrahydro-2,2,6-trimethyl-pyran-3-ol
2-乙基呋喃	2-ethylfuran
乙基吡嗪	ethylpyrazine
2-乙基环丁醇	2-ethylcyclobutanol
乙基苯	ethylbenzene
乙基苯甲醛	ethylbenzaldehyde
乙基苯酚	ethylphenol
乙基叔丁基醚	ethyl t-butyl ether
2-乙基庚酸	2-ethylheptanoic acid
乙烯氧基苯	ethenoxybenzene
2-乙烯基-2,5-二甲基-4-己烯醇	2-ethenyl-2,5-dimethyl-4-hexenol
6-乙基二氢-2,2,6-三甲基-2H-吡喃 -3 (4H)-酮	6-ethyldihydro-2,2,6-trimethyl-2H-pyran -3(4H)-one
3-乙基-1,2-二硫杂-4-环己烯	3-ethyl-1, 2-dithi-4-cycloherene
1-乙基-2-己烯基环丙烷	1-ethyl-2-hexenylcyclopropane
7-乙烯基-4a,5,6,7,8,8a-六氢 -1,4-二甲基-2 (1H)-萘酮	7-ethyl-4a,5,6,7,8,8a-hexahydro- 1,4-dimethyl-2(1H)-naphthalenone
1-乙烯基-4-甲氧基苯	1-ethenyl-4-methoxy benzene
2-(5-乙烯基四氢-5-甲基-2-呋喃基) -6-甲基-5-庚烯-3-酮	2-(5-ethenyltetrahydro-5-methyl 2-furanyl) -6-methyl-5-hepten-3-one

4-乙烯基环己基甲醇	4-ethenylcyclohexymethanol
乙烯基苯甲醛	ethenylbenzaldehyde
α-乙酰氧基苯乙酸甲酯	methyl α-acetoxyphenylacetate
2-乙酰基吡咯	2-acetylpyrrole
乙酸-1-乙氧基乙酯	1-ethoxyethyl acetate
乙酸乙酯	ethyl acetate
乙酸二氢葛缕酯	dihydrocarvyl acetate
乙酸-1-十二烯酯	1-dodecenyl acetate
乙酸-11-十四烯酯	11-tetradecenyl acetate
乙酸丁酯	butyl acetate
乙酸-2-己烯酯	2-hexenyl acetate
乙酸己酯	hexyl acetate
乙酸小茴香酯	fenchyl acetate
乙酸壬酯	nonyl acetate
乙酸月桂烯酯	myrcenyl acetate
乙酸戊酯	pentyl acetate
乙酸龙脑酯	bornyl acetate
乙酸-3-甲基-2-丁烯酯	3-methyl-2-butenyl acetate
乙酸-2-甲基丁酯	2-methylbutyl acetate
乙酸-2-甲基-2-丙烯酯	2-methyl-2-propenyl acetate
乙酸-2-甲苯酯	2-tolyl acetate
乙酸苯甲酯	benzyl acetate
乙酸达玛二烯酯	dammaradienyl acetate
乙酸芳樟酯	linalyl acetate
乙酸辛烯酯	octenyl acetate
乙酸辛酯	octyl acetate
乙酸苯乙酯	phenylethyl acetate
乙酸苯丙酯	phenylpropyl acetate
乙酸-1-乙酸松油酯-8-酯	1-p-menthen-8-yl acetate
乙酸松油酯	terpinyl acetate
乙酸松油醇-4-酯	terpin-4-yl acetate
乙酸β-侧柏酯	β-thujyl acetate

乙酸金合欢酯	farnesyl acetate
乙酸庚酯	heptyl acetate
乙酸柏木酯	cedryl acetate
乙酸香叶酯	geranyl acetate
乙酸香兰酯	vanillyl acetate
乙酸香芹酯	carvacryl acetate
乙酸香茅酯	citronellyl acetate
乙酸癸酯	decyl acetate
乙酸桂酯	cinnamyl acetate
乙酸桃金娘酯	myrtan acetate
乙酸桧酯	sabinyl acetate
乙酸菊酯	chrysanthenyl acetate
乙酸β-蛇麻-7-烯酯	β-humnlenyl-7 acetate
乙酸葛缕酯	carvyl acetate
乙酸薄荷酯	menthyl acetate
乙酸橙花酯	neryl acetate
乙酸麝香草酯	thymol acetate

二画

1,1-二乙氧基乙烷	1,1-diethoxyethane
1,1-二乙基丙基苯	1,1-diethylpropylbenzene
3,3-二乙基戊烷	3,3-diethylpentane
1,6-二乙酰氧基已烷	1,6-diacetoxyhexane
二丁酮	dibutanone
二戊烯	dipentene
1,4-二丙基苯	1,4-dipropylbenzene
1,4-二甲氧基-3,3,5,6-四甲基苯	1,4-dinethoxy-3,3,5,6-tetramethyl-benzene
1,2-二甲氧基苯	1,2-dimethoxybenzene
3,5-二甲氧基甲苯	3,5-dimethoxytoluene
3,4-二甲氧基苯甲酸	3,4-dimethoxybenzoic acid
3,4-二甲氧基苯甲醛	3,4-dimethoxyphenylaldehyde
1,2-二甲基-4-乙烯基苯	1,2-dimethyl-4-ethenyibenzene
3,5-二甲基二氢-2(3H)-呋喃酮	3,5-dimethyldihydr o-2(3H)-furanone
4,7-二甲基十一烷	4,7-dimethylundecane

6,10-二甲基-2-十一酮	6,10-dimethyl-2-undecanone
6,10-二甲基-5,9-十一碳二烯-2-酮	6,10-dimethyl-5,9-undecadien-2-one
2,8-二甲基十一酸甲酯	methyl 2,8-dimethylundecanoate
2,8-二甲基十三烷	2,8-dimethyltridecane
1,2-二甲基十三酸甲酯	methyl 1,2-dimethyltridecanoate
2,3-二甲基-2-丁醇	2,3-dimethyl-2-butanol
2,3-二甲基丁酸	2,3-dimethylbutyric acid
2,2-二甲基丁酸香叶酯	geranyl 2,2-dimethylbutyrate
2,4-二甲基己烷	2,4-dimethylhexane
3,4-二甲基-3-己烯-2-酮	3,4-dimethyl-3-hexene-2-one
2,2-二甲基己醛	2,2-dimethylhexanal
2,4-二甲基己酮	2,4-dimethylhexanone
2,5-二甲基壬烷	2,5-dimethylnonane
3-(4,8-二甲基-3,7-壬二烯基)呋喃	3-(4,8-dimethyl-3,7-nonadienly)furan
5,5-二甲基-2-丙基-1,3-环己二酮	5,5-dimethyl-2-propyl-1,3-cyclohexanedione
2,2-二甲基丙酸丁香酯	eugenol 2,2-dimethylpropionate
2,2-二甲基丙酸香叶酯	geranyl 2,2-dimethylpropionate
3,3-二甲基丙烯酸叔丁酯	t-butyl 3,3-dimethylacrylate
2,4-二甲基-2-戊烯	2,4-dimethyl-2-pentene
3,4-二甲基戊醇	3,4-dinethylpentanol
4,4-二甲基-2-戊醇	4,4-dimethyl-2-pentanol
2,4-二甲基-2-戊酮	2,4-dimethyl-2-pentanone
2,2-二甲基-3-(2-甲基-1-丙烯基)环丙羧酸乙酯	2,2-dimethyl-3-(2-methyl-1-propenyl)-cyclopropane carboxlic ethyl ester
2,3-二甲基-5-甲氧基苯酚	2,3-dimethyl-5-methoxyphenol
7,11-二甲基-3-亚甲基-1,6,10-十二碳三烯(Z)	7,11-dimethyl-3-methylene-1,6,10-dodecatriene(Z)
2,2-二甲基-1-亚甲基-3-丁烯-2-酮	2,2-dimethyl-l-methylene-3-butene-2-one
8,8-二甲基-4-亚甲基-1-氧杂螺[2.5]辛-5-烯	8,8-dimethy1-4-methylene-1-Oxaspiro[2.5]oct-5-ene
2-(3,3-二甲基亚环己基)乙醇	2-(3,3-dimethylcyclohexylidene)-ethanol
2,6-二甲基-6-(4-异己烯基)环[3.1.1]庚-2-烯	2,6-dimethyl-6-(4-isohexenyl)-cylo[3.1.1]hept-2-ene
4,10-二甲基-7-异丙基二环[4.4.0]癸-1,4-二烯	4,10-dimethyl-7-isopropyclo[4.4.0]deca-1,4-diene

1,5-二甲基-2-异丙基-2,5-桥氧环戊醇	1,5-dimethyl-2-isoproyl-2,5-eporycyclopentanol
1,6-二甲基-4-异丙基萘	1,6-dimethyl-4-isopropylnaphthalene
β-二甲基苏合香烯	β-dimethylstyrene
5,5-二甲基-2-呋喃酮	5,5-dimethyl-2-furanone
2,5-二甲基吡嗪	2,5-dimethylpyrazine
2,7-二甲基辛烷	2,7-dimethyloctane
2,6-甲基-2,4,6-辛三烯	2,6-dimethyl-2,4,6-octatriene
3,6-二甲基-1,6-辛二烯-3-醇	3,6-dimethyl-1,6-octadiene-3-ol
1,4甲基-3-环己烯醇	1,4-dinethyl-3-cyclohexenol
1,3-二甲基环戊烷	1,3-dimethylcyclopentane
2,6-二甲基苯乙烯	2,6-dimethylstyrene
2,4-二甲基苯乙酮	2,4-dimethylphenylethanone
2,6-二甲基庚烷	2,6-dimethylheptane
2,6-二甲基庚烯醛	2,6-dimethylheptenal
2,2-二甲基-2,4-庚二烯醛	2,2-dimethyl-2,4-heptadienal
2,4-二甲基-2-癸烯	2,4-dimethyl-2-decene
2,6-二甲基-5-癸醛	2,6-dimethyl-5-decanal
二环[6.1.0]壬-5,8-二烯-4-酮	bicyclo[6.1.0]non-5,8-diene-4-one
3,5-二叔丁基水杨醛	3,5-di-t-butyl-salicylaldehyde
2,6-二叔丁基对甲酚	2,6-di-t-butyl-p-cresol
2,6-二叔丁基对苯醌	2,6-di-t-butyl-1,4-benzoquinone
2,5-二氢-3,4-二甲基呋喃	2,5-dihydro-3,4-dimethylfuran
3,4-二氢-2,5-二甲基-2H-吡喃-2-甲醛	3,4-dihydro-2,5-dimethyl-2H-pyran-2-carboxaldehyde
2,3-二氢-3,5-二羟基-6-甲基吡喃-4-酮	2,3-dihydro-3,5-dihydroxy-6-methylpyran-4-one
3,4-二氢-4,6,8-三甲基-1(2H)-萘酮	3,4-dihydro-4,6,8-trimethyl-1(2H)-naphthalenone
二氢土木香内酯	dihydroalantolactone
二氢木香内酯	dihydrocostuslactone
2,3-二氢-4-甲基呋喃	2,3-dihydro-4-methylfuran
3a,7a-二氢-5-甲基茚-1,7(4H)二酮	3a,7a-dihydro-5-methylindene-1,7(4H) -dione
二氢芳樟醇	dihydrolinalool
二氢金合欢醇	dihydrofarnesol
2,3-二氢苯并呋喃	2,3-dihydrobenzofuran
二氢脱氢广木香内酯	dihydrodehydrocostuslactone, Mokko lactone
3,4-二氢-1(2H)-萘酮	3,4-dihydro-1(2H)-naphthalenone
二氢紫罗兰酮	dihydroionone

3,4-二氢-8-羟基-3-甲基-1*H*-2-苯并吡喃-1-酮	3,4-dihydydro-8-hydroxy-3-methyl-1*H*-2-benzopyran-l-one
二氢葛缕酮	dihydrocarvone
二氢猕猴桃(醇酸)内酯	dihydroactinidiolide
二烯丙基硫醚	diallyl sulfide
二烯丙基二硫醚	diallyl disulfide
二烯丙基三硫醚	diallyl trisulfide
二烯丙基四硫醚	diallyl tetrasulfide
二硫杂环戊烯	dithiacyclopentene
二聚戊二烯氧化物	dipentadiene oxide
十一烷	undecane
十一烯	undecene
1,4-十一碳二烯	1,4-undecadiene
十一醇	undecanol
10-十一烯醇	10-undecenol
十一醛	undecanal
2-十一烯醛	2-undecenal
2-十一酮	2-undecanone
十一酸	undecanoic acid
十一酸异戊酯	isopentyl undecanoate
十二烷	dodecane
1,11-十二碳二烯	1,ll-dodecadiene
十二醇	dodecanol
十二醛	dodecanal
2-十二烯醛	2-dodecenal
2-十二酮	2-dedecanone
十二酸乙酯	ethyl dodecanoate
十二酸甲酯	methyl dodecanoate
十七烷	heptadecane
1-十七烯	l-heptadecene
十七醛	heptadecanal
2-十七酮	2-heptadecanone
十七酸	heptadecanoic acid
十八烷	octadecane

十八烯	octadecene
十八醛	octadecanal
9-十八烯醛	9-octadecenal
9,12-十八碳二烯醛	9,12-octadecadienal
十八酸乙酯	ethyl octadecanoate
十八烯酸乙酯	ethyl octadecenoate
2,5-十八碳二烯酸甲酯	methyl 2,5-octadecadienoate
十八炔酸甲酯	methyl octadecynoate
十九烷	nonadecane
1-十九烯	1-nonadecene
十九酸	nonadecanoic acid
1-十三烯	1-tridecene
十三醛	tridecanal
2-十三酮	2-tridecanone
十三酸甲酯	methyl tridecanoate
十五烷	pentadecane
1-十五烯	1-pentadccene
2-十五烯醇	2-pentadecenol
十五醛	petadecanal
十五烯醛	petadecenal
十五酸	pentadecanoic acid
十五酸乙酯	ethyl pentadecanoate
7,10-十五碳二炔酸	7,10-pentadecadiynoic acid
十六烷	hexadecane
十六醇	hexadecanol
十六醛	hexadecanal
十六烯醛	hexadecenal
十六酸	hexadecanoic acid
十六酸甲酯	methyl hexadecanoate
9-十六烯酸	9-hexadecenoic acid
十六氢芘	hexadecahydropyrene
十四烷	tetradecane
十四醛	tetradecanal

十四酮	tetradecanone
十四酸	tetradecanoic acid
十四酸乙酯	ethyl tetradecanoate
十四酸丁酯	butyl tetradecanoate
十甲酸甲酯	methyl tetradecanoate
十氢1,5,5,8a-四甲基-1,2,4-亚甲基薁	decahydro-1,5,6,8a-tetramethyl-1,2,4-methyleneazulene
十氢-1,1,4,7-四甲基-4aH-环丙[e]薁-4a-醇	decahydro-1,1,4,7-tetramethyl-4aH-cyclopro[e]azulen-4a-ol
1,2,3,4,4a,7,8,8a-八氢-1,6-二甲基-4-(1-甲基乙基)萘醇	1,2,3,4,4a,7,8,8a-octahydro-l,6-dimethyl-4-(1-methylethyl) naphtha lenol
1,3-丁二醇	1,3-butanediol
丁二酮	butanedione
丁子香烯	clovene
丁香酚	eugenol
丁香酚甲酯	methyl eugenol
2-丁基四氢呋喃	2-butyltetrahydrofuran
3-丁基苯酞	3-butylphthalide
2-丁烯酸甲酯	methyl 2-butenoate
丁醇	butanol
丁酸	butyric acid
丁酸乙酯	ethyl butyrate
丁酸己酯	hexyl butyrate
丁酸 3-己烯酯	3-hexenyl butyrate
丁酸戊酯	pentyl butyrate
丁酸龙脑酯	bornyl butyrate
丁酸甲酯	methyl butyrate
丁酸 1-甲基丙酯	1-methylpropyl butyrate
丁酸 3-甲基丁酯	3-methylbutyl butyrate
丁酸异丁酯	isobutyl butyrate
丁酸苄酯,丁酸苯甲酯	benzyl butyrate

三画

1,2,4-三乙基苯	1,2,4-triethylbenzene
1,2,4-三甲苯	1,2,4-trimethylbenzene
三甲氧基苯丙烯	trimethoxyphenylpropene

3,4,5-三甲氧基苯甲醛	3,4,6-trimethoxyphenylaldehyde
三甲氧基烯丙基苯	trimethoxy-allylbenzene
2,6,6-三甲基-2-乙烯基-5-羟基吡喃	2,6,6-trimethyl-2-vinyl-5-hydroxy-pyran
1,1,5-三甲基-2-乙酰基-2,5-环己二烯-4-酮	1,1,5-trimethyl-2-acetyl-2,5-cyclohexadiene-4-one
3,6,6-甲基二环[3.1:1]庚-2-烯	3,6,6-trimethyl bicyclo[3.1.1]hept-2-ene
2,3,3-三甲基二环[2.2.1]庚-2-醇	2,2,3-trimethyl bicyclo[2.2.1]hept-2-ol
4,6,6-三甲基二环[3.1.1]庚-3-烯-2-酮	4,6,6-trimethyl bicyclo[3.1.1]hept-3-ene-2-one
6,10,14-三甲基十五烷	6,10,14-trimethyl pentadecane
6,10,14-三甲基-2-十五酮	6,10,14-trimethyl-2-pentadecanone
3,5,5-三甲基己醇	3,5,5-trimethylhexanol
3,7,11-三甲基-14-(1-甲基乙基)-1,3,6,10-环十四碳四烯	3,7,11-trimethyl-14-(1-methylethy1)-1,3,6,10-cyclotetradecatetraene
2,2,6-三甲基-1,4-环己二酮	2,2,6-trimethyl-1,4-cyclohexanedione
1-(2,6,6-三甲基-1,3-环己二烯基)-2-丁烯酮	1-(2,6,6-trimethyl-1,3-cyclohexadienyl)-2-utenone
4,4,6-三甲基环己二烯酮	4,4,6-timethylcyclohexadienone
3,3,5-三甲基环己烯	3,3,5-trimethy cyclohexene
4,4,5-三甲基-2-环己烯酮	4,4,6-trinethyl-2-cyclohexenone
3,5,5-三甲基-2-环己烯-1,4-二酮	3,6,5-trimethyl-2-cyclohexene-1,4-dione
2,6,6-三甲基-1-环己烯基甲醛	2,6,6-trimethyl-1-cyclobexenylcarbaldehyde
4-(2,6,6-三甲基-2-环己烯亚基)-2-丁酮	4-(2,6,6-trimethyl-2-cyclohexen-1-ylidene)-2-butanone
2,2,3-三甲基-3-环戊烯基乙醛	2,2,3-timethyl-3-cyclopentenylacetaldehyde
3,4,5-三甲基-2-环戊烯酮	3,4,5-trimethyl-2-cyclopentenone
2,4,6-三甲基苯甲醛	2,4,6-trimethylphenylaldehyde
2,5,5-三甲基-1,6-庚二烯	2,5,5-trimethyl-1,6-heptadiene
2,5,5-三甲基-2,6-庚二烯-4-酮	2,5,6-trimethyl-2,6-heptadien-4-0ne
2,5,5-三甲基-1,3,6-庚三烯	2,5,5-trimethyl-1,3,6-heptatriene
8,6,6-三甲基-2-降蒎醇	3,6,6-trimethyl-2-norpinaneol
1,3,3-三甲基-2-氧杂二环[2.2.2]辛烷	1,3,3-trimethyl-2-oxabicyclo[2.2.2]octane
3,3,5-三甲基-4-(3-羟基-1-丁烯基)-2-环己烯-1-酮	3,3,5-trimethyl-4-(3-hydroxy-1-butenyl)-2-cyclohexen-1-one
三环石竹烯	tricyclocaryophyllene

三环[3.2.1.01.5]辛烷	tricyclo[3.2.1.01.5]octane
三环烯,三环萜	tricyclene
土木香醇	alantol
土木香酸	alantic acid
土木香内酯	alantolactone
土青木香烯	aristolene
大茴香醚	anethole
大茴香醛	anisaldehyde
大根香叶酮	germacrone
万寿菊烯酮	tegatenone
广木香烯	costene
广藿香烷	patchoulane
广藿香烯	patchoulene
广藿香醇	patchoulol
广藿香吡啶	patchoulipyridine
广藿香奠醇	pogostol
2,4-己二烯醇	2,4-hexadienol
2,4-己二烯醛	2,4-hexadienal
2,4-己二烯酸甲酯	methyl 2,4-hexadienoate
1-己烯	1-hexene
3-己烯醇	3-hexenol
2-己烯醛	2-hexenal
5-己烯-2-酮	5-hexen-2-one
3-己烯酸	3-hexenoic acid
2-己烯酸乙酯	ethyl 2-hexenoate
2-己烯酸甲酯	methyl 2-hexenoate
己醇	hexanol
己醛	hexanal
己酸	hexanoic acid
己酸乙酯	methyl hexanoate
己酸甲酯	ethyl hexanoate
己酸-2-甲基丙酯	2-methypropyl hexanoate
己酸己酯	hexyl hexanoate
己酸-3-己烯酯	3-hexenyl hexanoate

己酸苄酯	benzyl hexanoate
5-己基二氢-2(3H)-呋喃酮	5-hexyldihydro-2(3H)-furanone
己基环己烷	hexylcyclohexane
2-己基塞吩	2-hexylthiophene
α-小茴香烯	α-fenchene
小茴香醇	fenchol
小茴香酮	fenchone
小茴香酸	fencholic acid
马齿苋醛	Portulal
马桑毒素	coriamyrtin
马兜铃烯-1,9-二烯	aristolene-1,9-diene
马兜铃烯	aristolene
9-马兜铃烯-1-醇	9-aristolene-l-ol
β-马榄烯	β-maaliene
马榄烯醇	maglianol
马鞭草烯醇	verbenol
马鞭草烯酮	verbenone
马鞭草酮	verbanone

四画

廿一烷	henicosane
1-廿二烯	1-docosene
廿三烷	tricosane
1-廿三烯	1-tricosene
廿五烷	pentacosane
1-廿五烯	1-penta cosene
1-廿烯	1-eicosene
廿碳二烯	eicosadiene
廿醇	eicosanol
廿醛	eicosanal
木香烯	costene
木香醇	costol
木香酸	costic acid
木香（烯）内酯	costunolide
比沙泊烯	bisabolene

瓦伦烯	valencene
(-)-16α-贝壳杉醇	(-)-kauran-16α-ol
壬烷	nonane
壬烯	nonene
1-壬烯-3-醇	1-nonen-3-ol
3-壬烯醛	3-nonenal
2,4-壬二烯醛	2,4-nonadienal
8-壬烯-2-酮	8-nonen-2-one
壬醇	nonanol
壬醛	nonanal
2-壬酮	2-nonanone
壬酸	nonanoic acid
2-壬烯酸	2-nonenoic acid
壬酸甲酯	methyl nonanoate
壬内酯	nonanolide
长叶烯	longifolene
长叶环烯	longicyclene
反式-对蓋-9-醇	trans-p-methan-9-ol
反式-2-辛烯醛	trans-2-octenal
反式-蓋二烯	trans-methadiene
反式-紫花前胡内酯	trans-marmelolactone
反式-4-羟基-3-甲基-6-异丙基-2-环己烯酮	trans-4-hydroxy-3-methyl-6-isopropyl-2-cyclohexenone
月桂烯	myrcene
月桂烯醇	myrcenol
月桂醇	laurye alcohol
月桂酸	lauric acid
月桂酸乙酯	ethyl laurate
风毛菊内酯	saussurea lactone
巴豆酸乙酯	ethyl crotonate
双花醇	(-)-cis-2,6,6-trimethyl-2-ethenyl-5-hydroxytetrahydropyran
水合-1,8-松油二醇	1,8-terpin hydrate
水合莰烯	camphene hydrate

水合桧烯	sabinene hydrate
水芹烯	phellandrene
水芹醇	phellandrol
水菖蒲烯	calarene
水杨酸	salicylic acid
水杨酸乙酯	ethyl salicylate
水杨酸丁酯	butyl salicylate
水杨酸丙酯	propyl salicylate
水杨酸甲酯	methyl salicylate
水杨酸苯甲酯	benzyl salicylate
1,2,3,4,4a,7-六氢-1,6-二甲基-4-(1-甲基乙基)萘	1,2,3,4,4a,7-hexahydro-1,6-dimethyl-4-(1-methylethyl)naphthalene
六氢金合欢基丙酮	hexahydrofarmesyl acetone
无环单萜	acyclic monoterpeneoid
无环倍半萜	acyclic sesquiterpenoid
双环单萜	bicyclicterpenoid

五画

正十七醇	n-heptadecanol
正廿醇	n-eicosanol
正戊基苯	n-pentylbenzene
6-去甲氧基胜红蓟素	6-demethyoxyageratochromen
去氢木香内酯	dehydrcostuslactone
去氢白菖烯	calamenene
β-去氢香薷酮	β-dehydroelsholtzione
去氢松香烷	dehydroabietane
去氢喇叭茶醇	dehydroledol
甘松醇	nartachinol
甘松酮	nardosinone
甘松奠醇	nardol
甘松素	jatamansin
古芸烯	gurjnene
6-丙基二环[3.2.0]庚-6-烯-2-酮	6-propylbicyclo[3.2.0]hept-6-en-2-one
6-丙基二环[1.2.0]庚-3,6-二烯-2-酮	6-propylbicyclo[3.2.0]hept-3,6-dien-2-one
丙基环丙烷	pronylcyelpropane

丙基苯甲醛	propylphenylaldehyde
3-丙基苯甲酸里酯	methyl 3-propylbenzoate
丙烯基焦儿茶酚	propenyl pyrocatechol
丙氧基苗香醚	propoxyanisole
丙酸乙酯	ethyl propionate
丙酸丁香酯	eugenyl propionate
丙酸2-甲基丁酯	2-methylbutyl propionate
丙酸龙脑酯	bornyl propionate
丙酸芳樟酯	linalyl propionate
丙酸β-苯乙酯	β-phenylethyl propionate
丙酸香叶酯	geranyl propionate
丙酸香茅酯	citronellyl propionate
丙酸顺式3-己烯酯	cis-3-hexenyl propionate
丙酸橙花酯	neryl propionate
石竹烯	caryophyllene
石竹烯醇	caryophyllenol
布黎醇	bulnesol
戊二烯基环戊烷	pentadienylcyclopentane
5-戊氧基-2-戊烯	5-pentoxy-2-pentene
2-戊基呋喃	2-pentylfuran
戊基环丙烷	pentylcyclopropane
3-戊醇	3-pentanol
戊醛	pentanal
3-戊酮	3-pentanone
戊酸	pentanoic acid
戊酸乙酯	ethyl pentanoate
戊酸丁香酯	eugenyl pentanoate
戊酸甲酯	methyl pentanoate
戊酸苯甲酯	benyzl pentanoate
戊酸麝香草酯	thymyl pentanoate
龙烯	bornene
龙脑	borneol

2,10-龙脑二醇	2,10-bornediol
龙脑烯	bornylene
α-龙脑烯醛	α-camphorenal,α-campholenic aldehyde
卢杷烯醇	luparenol
卢杷醇	luparol
1-甲氧基-3,7-二甲基-2,6-辛二烯	1-methoxy-3,7-dimethyl-2,6-octadiene
4-甲氧基二苯基乙炔	4-methoxydiphenylacetylene
4-甲氧基丁酸甲酯	methyl 4-methoxybutyrate
2-甲氧基-4-(1-丙烯基)苯酚	2-methoxy-4-(1-propenyl)phenol
2-甲氧基-4-(2-丙烯基)苯酚	methoxy-4-(2-propenyl)phenol
4-甲氧基甲苯	4-methoxytoluene
3-甲氧基-3-甲基-2-丁酮	3-methoxy-3-methyl2-butanone
2-甲氧基-4-异烯丙基苯酚	2-methoxy-4-isoallylphenol
1-(4-甲氧基苯基)乙酮	1-(4-methoxyphenyl)ethanone
2-甲氧基苯甲醛	2-methoxyphenylaldehyde
2-甲氧基苯甲酸甲酯	methyl 2-methoxybenzoate
2-甲氧基苯酚	2-methoxyphenol
4-甲氧基-1-叔丁氧基苯	4-methoxy-1-t-butoxybenzene
2-甲氧基桂醛	2-methoxycinnamic aldehyde
5-(1-甲基乙基)二环[3.1.0]己-2-酮	5-(1-methylethyl)bicyclo[3. 1. 0]hexan-2-one
4-甲基-4-乙基-2-环己烯酮	4-methyl-4-ethyl-2-cyclohexenone
1-甲基-2-乙基苯	1-methyl-2-ethylbenzene
4-甲基-6-乙酰基己醛	4-methyl-6-acetylhexanal
1-甲基-5,6-二乙基环己酮	1-methyl-5, 6-diethylcyclohexanone
1-甲基-1,2-二乙烯基-5-环己烯	1-methyl-l,2-diethyl-5-cyclohexene
1-甲基-1,2-二亚乙基-5-环己烯	I-methyl-l,2-diethylene-5-cyclohexene
6-甲基二环[3.2.0]庚-6-烯-2-酮	6-methylbicyclo[3.2.0]hept-6-en-2-one
4-甲基-2,6-二叔丁基苯酚	4-methyl-2,6-di-t-butylphenol
12-甲基二氢木香内酯	12-methyldihydrocostunolide
4-甲基-1,2-二硫杂-3-环戊烯	4-methyl-1,2-dithi-3-cyclopentene
甲基十二酮	methylododecanone
8-甲基十七烷	8-methylheptadecane
16-甲基十七酸甲酯	methyl 16-methylheptadecanoate

12-甲基十三酸甲酯	methyl 12-methyltridecanoate
14-甲基十五酸甲酯	methyl 14-methylpentadecancate
2-甲基十六烷	2-methylhexadecane
2-甲基十四烷	2-methyltetradecane
2-甲基十氢萘	2-methyl decahydronaphthalene
2-甲基丁烷	2-methylbutane
3-甲基丁醇	3-methylbutanol
2-甲基-2-丁烯醇	2-methyl-2-butenol
2-甲基-3-丁烯-2-酮	2-methyl-3-buten-2-ol
2-甲基丁醛	2-methylbutanal
2-甲基-2-丁烯醛	2-methyl-2-butenal
3-甲基-3-丁酮	3-methyl-3-butanone
甲基丁基醚	methyl butyl ether
2-甲基丁酸	2-methyl butanoic acid
2-甲基丁酸乙酯	ethyl 2-methylbutanoate
3-甲基丁酸丁酯	butyl 3-methylbutanoate
2-甲基丁酸丁香酯	eugenyl 2-methylbutanoate
2-甲基丁酸龙脑酯	bornyl 2-methylbutanoate
2-甲基丁酸甲酯	methyl 2-methylbutanoate
3-甲基丁酸2-甲基丁酯	2-methylbutyl 3-methylbutanoate
2-甲基丁酸2-甲基丙酯	2-methylpropyl 2-methylbutanoate
2-甲基丁酸苯甲酯	benzyl 3-methylbutanoate
3-甲基-2-丁烯酸乙酯	ethyl 3-methyl-2-butanoate
3-甲基-2-丁烯酸甲酯	methyl 3-methyl-2-butanoate
2-甲基己烷	2-methylhexane
5-甲基-2-己醇	5-methyl-2-hexanol
5-甲基己醛	5-methylhexanal
2-甲基己酸	2-methylhexanoic acid
甲基壬酮	methylnonanone
7-甲基壬酸甲酯	methyl 7-methylnonanoate
α-甲基-α-(2-丙烯基)-苯甲醇	α-methyl-α-(2-propenyl)-phenylmethanol
2-甲基-6-丙基十一烷	2-methyl-6-propylundecane
3-(2-甲基丙基)-2-环己烯酮	3-(2-methylpropyl)-2-cyclohexenone

1-甲基-2-(2-丙基)环戊烷	1-methyl-2-(2-propyl)cyelohexane
2-甲基-8-丙基癸烷	2-methyl-8-propyldecane
2-甲基丙酸乙酯	ethyl 2-methylpropionate
2-甲基丙酸丁香酯	eugenyl 2-methylpropionate
2-甲基丙酸丙酯	propyl 2-methylpropionate
2-甲基丙酸2-甲基丁酯	2-methylbutyl 2-methylpropionate
2-甲基丙酸2-甲基丙酯	2-methylpropyl 2-methylonate
2-甲基丙酸2-苯基乙酯	2-phenylethyl 2-methylpropionate
2-甲基丙酸香叶酯	geranyl 2-methylpropionate
2-甲基-2-丙烯酸2-甲基丙酯	2-methylpropyl 2-methyl-2-propenoate
3-甲基-2-(1,3-戊二烯基)-2-环戊烯酮	3-methyl-2-(l,8-pentadienyl)-2-cyclopentenone
3-甲基-2-戊基-2-环己烯酮	3-methyl-2-pentyl-2-cyclohexenone
1-甲基-2-戊基环丙烷	1-methyl-2-pentylcyclopropane
甲基戊基酮	methyl pentyl ketone
甲基戊烯	methylpentene
4-甲基-3-戊烯醛	4-methyl-3-pentenal
4-甲基-4-戊烯-2-酮	4-methyl-4-pentene-2-one
2-甲基-3-戊醇	2-methyl-3-pentanol
甲基戊酮	methylpentanone
2-甲基-5-(1-甲基乙基)-2,5-环己二烯-1,4-二酮	2-methyl-5-(1-methylethy)-2,5-cyclohexadiene-1,4-dione
1-甲基-4-(1-甲基乙基)环己醇	1-methyl-4-(1-methylethyl)cyclohexanol
2-甲基-3-(1-甲基乙基)环氧乙烷	2-methyl-3-(1-methylethyl)oxirane
2-甲基-5-(1-甲基乙烯基)环己酮	2-methyl-5-(1-methylethenyl)cyclohexanone
2-甲基甲酰胺	2-methylformamide
2-甲基-6-亚甲基-1,7-辛二烯-3-酮	2-methyl-6-methylene-1,7-octadiene-3 -one
1-甲基-4-异丙基-1,3-环己二烯	1-methyl-4-isopropyl-1,3-cyclohexadiene
5-甲基-2-异丙基-2-环己烯	5-methyl-2-isopropyl-2-cyclohexenene
1-甲基-2-异丙基苯	1-methyl-2-isopropylbenzene
2-(1-甲基-2-异丙烯基环丁基)乙醇	2-(1-methyl-2-isopropenylcyclobutyl)-ethanol
4-(5-甲基-2-呋喃基)-2-丁酮	4-(5-methyl-2-furanyl)-2-butanone
5-甲基-2-呋喃醛	5-methyl-2-furanaldehyde
3-甲基-1H-吡唑	3-methyl-1H-pyrazole
甲基吡嗪	methylpyrazine

3-甲基环戊烯	3-methylcyclopentene
2-甲基环戊烯醇	2-methylcyclopentenol
2-甲基环戊醇	2-methylcyclopentanol
甲基苯乙基醚	methyl phenylethyl ether
甲基苯乙醇	methyl phenethy alcohol
甲基苯甲醚	methyl benyl ether
2-甲基苯并呋喃	2-methyl benzofuran
1-甲基-3-苯基丙醇	1-methyl-3-phenylpropanol
2-甲基-2-苯基丙酸乙酯	ethyl 2-methyl-2-phenylpropanoate
4-甲基-5-苯基-3-羰基戊酸甲酯	methyl 4-methyl-5-phenyl-3-carbonylpentanoate
甲基庚二烯酮	methyl heptadienone
6-甲基-5-庚烯-2-醇	6-methyl-5-hepten-2-ol
甲基庚烯酮	methylheptenone
6-甲基庚醇	6-methylheptanol
4-甲基-2-庚酮	4-methyl-2-heptanone
7-甲基-4-癸烯	7-methyl-4-decene
1-甲基萘	1-methylnaphthalene
4-甲基-4-烃基-2-戊酮	4-methyl-4-hydroxy-2-pentanone
3-甲基-6-烃基苯乙酮	3-methyl-6-hydroxyacetophenone
甲基烯丙基三硫醚	methyl allyl trisulfide
甲基烯丙基四硫醚	methyl allyl tetrasulfide
甲基烯丙基五硫醚	methyl allyl pentasulfide
3-甲基硫丙酸乙酯	ethyl 3-methylthiopropionate
甲基硫烃乙酸甲酯	methyl methylthiolacetate
甲酚	cresol
2-甲酰氨基苯甲酸甲酯	methyl 2-formylaminobenzoate
甲酸乙酯	ethyl formate
甲酸1-己烯酯	1-hexenyl formate
甲酸己酯	hexyl formate
甲酸龙脑酯	bornyl formate
甲酸2-甲基丁酯	l-methylbutyl formate
甲酸环己烯酯	cyclohexenyl formate
甲酸苯甲酯	benzyl formate
甲酸香叶酯	geranyl formate

甲酸香茅酯	citronellyl formate
甲酸顺式-葛缕酯	cis-caraway formate
2,4,6,14-四甲基十五烷	2,4,6,14-tetramethyl pentadecane
3,7,11,15-四甲基-2-十六烯醇	3,7,11,15-tetramethyl-2-hexadecenol
1,2,9,10-四脱氢马兜铃烷	1,2,9,10-tetradehydroaristolane
四檀香醇	tetrasantalol
白术内酯A	butenolide A
白菖烯	calarene
白菖醇	calameneol
白柠檬亭	limettin
加州月桂酮	umbellulone
外-异莰烷酮	exo-isocamphenone
对乙基苯甲醛	*p*-ethylbenzaldehyde
对二甲苯	*p*-dimethylbenzene
对丙基苯酚	*p*-propylphenol
对甲氧基丙基苯	methoxy propyl bengene
对甲氧基桂酸乙酯	ethyl *p*-methoxycinnamate
对甲基乙基苯甲醛	*p*-methylethyl-benzaldehyde
对凹顶藻醇	oppositol
对异丁基甲苯	*p*-isobutyltoluene
对异丙基苯甲酸	*p*-isopropyl benzoic acid
对辛基甲氧基苯	*p*-octylmethoxybenzene
对苯二甲酸二乙酯	diethyl *p*-benzdioate
对苯二酚	*p*-benzenediol
对叔丁基苯甲醇	*p*-t-butylpheny methanol
2,4-对蓋二烯	2,4-*p*-menthadiene
1,4-对蓋二烯醇	1,4-*p*-menthadienol
1-对蓋烯-8-醇	1-*p*-menthen-8-ol
1-对蓋烯-9-醛	1-*p*-menthen-9-al
1-对薄荷烯-9-醇	1-*p*-menthen-9-ol
对檀香醇	teresantalol
母菊萜	chamanzulene
印防己苦内酯	picrotin
印防己毒内酯	picrotoxinin

六画

吉玛烯-D	germacrene-D
吉玛酮	germacrone
亚丁基-2-苯并[C]呋喃酮	butylidenephthalide
3-亚甲基-2,2-二甲基二环[2.2.1]庚烷	3-methylene-2,2-dimethylbicyclo[2.2.l] heptane
2-亚甲基丁酸甲酯	methyl 2-methylenebutyrate
1-亚甲基-2-甲基-4-异丙基环己烷	I-methylene-2-methyl-4-isopropylcyclohexane
2亚甲基-6-甲基-5,7-辛二烯	2-methylene-6-methyl-5,7-octadiene
1,1-(1,4-亚苯基)双乙酮	1,1-(1,4-phenylene)biethanone
亚油酸	linoleic acid
亚油酸乙酯	ethyl linoleate
亚油酸甲酯	methyl linoleate
亚麻酸乙酯	ethyl linolenate
亚麻酸甲酯	methyl linolenate
当归素	angelicin
当归酸正丁酯	butyl angetate
肉豆蔻醚	myristicin
肉豆蔻酸	myristic acid
肉豆蔻酸乙酯	ethyl myristate
肉豆蔻酸甲酯	methyl myristate
优藏茴香酮	eucarvone
伪柠檬烯	pseudo-limonene
伞花烃	cymene
伞花醇	cymol
异丁酸丁酯	butyl isobutyrate
异丁酸正己酯	n-hexyl isobutyrate
异丁酸辛酯	octyl isobutyrate
异丁酸苯乙酯	phenylethyl isobutyrate
异丁酸庚酯	heptyl isobutyrate
异丁酸香叶酯	geranyl isobutyrate
异丁酸麝香草酯	thymyl isobutyrate

异己酸甲酯	methyl isohexanoate
异水菖蒲二醇	iso-calamendiol
2-异丙基-5-甲基茴香醚	2-isopropyl-5-methylanisole
4-异丙基-3-环己基甲醇	4-isopropyl-3-cyclohexylmethanol
异丙基-2-环己酮	isopropyl-2-cyclohexanone
4-异丙基草酚酮	4-isopropyltropolone
异丙酸甲酯	methyl isopropionate
异戊二烯	isopenoid
异戊醇	isopentanol
异戊酸十六酯	hexadecyl isovalerate
异戊酸龙脑酯	bornyl isovalerate
异戊酸3-甲基丁酯	3-methylbutyl isovalerate
异戊酸苯酯	phenyl isovalerate
异戊酸β-苯乙酯	β-phenylethyl isovalerate
异戊酸顺式-3-己烯酯	cis-3-hexenyl isovalerate
异白菖二醇	isocalamendid
异辛醇	isooctanol
异辛醛	isooctanal
异松油烯	terpinolene
异胡椒烯酮	isopiperitenone
异胡薄荷醇	isopulegol
异菖蒲二醇	iso-calamendiol
异蒎樟脑酮	isopinocamphorone
4-异硫氯基-1-丁烯	4-isothiocyanate-1-butene
3-异硫氰基-1-丙烯	3-isothiocyanate-1-propene
异硫氰酸丙酯	propyl isothiocyanate
异硫氰酸苯甲酯	benzyl isothiocyanate

七画

玛索依内酯	messoialactone
麦由酮	mayurone
芮木烯	rimuene
花柏烯	cuparene
花柏酮	cuparenone

芹子-3,7(11)-二烯	selina-3,7(11)-diene
芹子烯	selinene
芹子醇	selinenol
芹菜脑	apiole
苍术素	atractylocin
苍术酮	atractylone
苄基丙酮	benzyl acetone
芳樟醇	linalool
苏合香烯	styrene
杜松烯	cadinene
杜松醇	cadinol
杜松烯醇	cadinenol
杜基醛	duryl aldehyde
杜鹃次烯	neocurzerene
杜鹃烯	neofuranodinene
2-呋喃甲醛,2-呋喃醛	2-furaldehyde
1H-吡咯-2-甲醛	1H-pyrrole-2-formaldehyde
1-(1H-吡咯-2-基)乙酮	1-(1H-pyrrole-2-ly)ethanone
吡啶	pyridine
2-吡啶腈	2-pyridinecarbonitrile
4-吡啶羧酸酰肼	4-pyridinecarboxylic acid hydrazide
吲哚	indole
别二氢香芹酮	carvotanacetone
别罗勒烯	allo-ocimene
别香树烯	alloaromadendrene
佛手烯	bergamotene
邻二甲苯	o-dimethylbenzene
邻二甲氧基苯	o-dimethoxybenzene
邻丙烯基甲苯	o-propenyltoluene
邻甲基苯乙酮	o-methylacetophenone
邻甲基茴香醚	o-methylanethole
邻苯二甲酸二乙酯	diethyl phthalate
邻苯二甲酸二丁酯	dibutyl phthalate

邻苯二甲酸二辛酯	dioctyl phthalate
邻苯二甲酸丁基异丁基酯	n-butylisobutyl phthalate
邻癸基羟胺	o-hydroxylamine
邻氨基苯甲酸甲酯	methyl anthranilate
邻羟基苯甲酸苯酯	phenyl o-hydroxybenzoate
9,12-辛二烯醛	9,12-octadialdehyde
γ-辛内酯	γ-caprylolactone
辛烷	octane
2-辛烯	2-octene
辛烯醇	octenol
2-辛烯酸甲酯	methyl 2-octenoate
辛基环丙烷	octylcyclopropane
辛醇	octanol
辛醛	octanal
2-辛酮	2-octanone
辛酸	octanoic acid
辛酸乙酯	ethyl octanoate
辛酸甲酯	methyl octanoate
驱蛔素	scaridole

八画

5-环己烯基乙酮	5-cyclohexenylethanone
1-(1-环己烯基)-2-丙酮	1-(1-cyclohexenyl)-2-propanone
3-环己烯基甲醇	3-cyclohexenyimethano
环己醇	cyclohexanol
环己酮	cyclohexanone
环小茴香烯	cyclofenchene
环丙烷壬酸甲酯	methyl cyclopropanenonanoate
1,3,5,7-环辛四烯	1,3,5,7-cyclooctatetraene
1,3,5-环庚三烯	1,3,5-cycloheptatriene
环庚烷	cycloheptane
环癸醇	cyclodecanol
环氧石竹烯（石竹素）	epoxycaryophyllene
环氧蛇麻烯	epoxyhumulene

玫瑰醚	rose oxide
表樟脑	epicamphor
表愈创吡啶	epiguaipyridine
茉莉内酯	jasmine lactone
茉莉酮	jasmone
茉莉酮酸甲酯	methyl jasmonate
苯	benzene
1,4-苯二甲醛	1,4-phthalic aldehyde
苯乙腈	benzyl cyanide
苯乙醇	phenylethanol
苯乙醛	phenylacetaldehyde
苯乙酮	acetophenone
苯乙酸异丁酯	isobutyl phenylacetate
苯丙醇	phenylpropanol
苯丙醛	phenylpropanal
苯丙酸乙酯	ethyl phenylpropionate
5-苯甲氧基戊醇	5-phenylmethoxypentanol
苯甲腈	cyanobenzene
苯甲醇	benzyl alcohol
苯甲醛	benzaldehyde
苯甲醚	phenyl methyl ether
苯甲酸乙酯	ethyl benzoate
苯甲酸-3-己烯酯	3-hexenyl benzoate
苯甲酸丙烯酯	propenyl benzoate
苯甲酸甲酯	methyl benzoate
苯甲酸异戊酯	isopentyl benzoate
苯甲酸环己酯	cyclohexyl benzoate
苯甲酸-2-苯乙酯	2-phenylethy benzoate
苯甲酸苯酯	phenyl benzoate
苯甲酸苄酯,苯甲酸苯甲酯	benzyl benzoate
苯并呋喃	benzofuran
苯并噻唑	benzothiazole
2-苯基乙醇	2-phenylenthanol

2-苯基乙酸3-甲基丁酯	3-methylbutyl 2-phenydlacetate
4-苯基-2-丁酮	4-phenyl-2-butanone
2-苯基-2-丙醇	2-phenyl-2-propanol
3-苯基-2-丙烯醇	3-phenyl-2-propenol
3-苯基-2-丙烯醛	3-phenyl-2-propenal
苯基-2-丙醇	phenyl-2-propanone
3-苯基-2-丙烯酸乙酯	ethyl 3-phenyl-2-propenoate
3-苯基-2-丙烯酸甲酯	methyl 3-phenyl-2-propenoate
N-苯基甲酰胺	N-phenylformamide
N-苯基苯胺	N-phenylanilline
N-苯基-1-萘胺	N-phenyl-1-naphthalenamine
苯酚	phenol
茅术醇	hinesol
松油烯	terpinene
松油醇	terpinol
松樟酮	pinocamphone
枞三烯	abietatriene
枞油烯	sylevestrene
刺柏脑	juniper camphor
欧芹酚甲醚	osthole
鸢尾酮	irone
叔丁基间甲酚	t-butyl-m-cresol
岩兰草烯	vetivene
岩兰草醇	vetiverol
岩兰草酮	vetiver ketone
岩兰草酸	vetivenic acid
罗勒烯	ocimene
罗汉松烯	podocarprene
罗汉柏烯	thujopsene, widdrene
d-[2(12)]罗汉柏烯-3-α-醇	d-[2(12)]thujopsene-3-α-ol
侧柏烯	thujene
侧柏烯醇	thujenol
侧柏醇	thujanol

侧柏酮	thujone
依兰油烯	ylangoilene
依兰油醇	ylangoilalcohol
依兰烯	ylangene
斧柏烯	thujopsene
金合欢烯	farnesene
金合欢醇	farneseol
金合欢醛	farneseal
金钟柏醇	occidentalol
γ-庚内酯	γ-onantholactone
庚烯醇	heptenol
2-庚烯醛	2-heptenal
2-庚烯酸	2-heptenoic acid
庚醇	heptanol
庚醛	heptanal
2-庚酮	2-heptanone
庚酸	heptanoic acid
庚酸甲酯	methyl heptanoate
单紫杉烯	aplotaxene
油酸乙酯	ethyl oleate
油酸甲酯	methyl oleate
β-波旁烯	β-bourbonene
降拉帕醇	norlapachol
细辛醛	asarylaldehyde
细辛酮	asatone
细辛醚	asaricin
细辛脑	asarone
细胞原生体	protoplasma
单环单萜	monocyclic monoterpene
单环倍半萜	monocyclic sesquiterpene
甲羟戊酸	mevalonate(MVA)
青蒿素	artemisinin

九画

珂玐烯	copaene
珂玐烯醇	copaenol
荜澄茄烯	cubibene
茵陈炔	capillene
茵陈炔酮	capillone
茴香醚	anisole
胡芦巴碱	trigonelline
胡萝卜醇	carotol
胡椒醇	piperitol
胡椒醛	piperonal
胡椒酮	piperitone
胡椒烯酮	piperitenone
胡椒酚	chavicol
胡椒酚甲醚	piperfenol methylether
胡薄荷酮	pulegone
枯茗醇	cumic alcohol, (cuminol)
枯茗醛	cumaldehyde, cuminic aldehyde
柏木烯	cedrene
柏木烯醇	cedrenol
柏木脑	cedrol
柏木酮	cedrone
柠檬烯	limonene
柠檬醛	citral
牻牛儿酮	germacrone
香叶基丙酮	geranyl acetone
香叶烯	geranene
香叶醇	geraniol
香叶醛	geranial
香叶酸甲酯	methyl geranate
香兰素	vanillin
香豆素	coumarin
香豆烷	coumarane

香芹二烯酮	carvendione
香芹酚,香荆芥酚	carvacrol
香附烯	cyperene
香附酮	cyperone
香茅烯	citronellene
香茅醇	citronellol
香茅醛	citronellal
香茅酸甲酯	methyl citronellate
香树烯	aromadendrene
香桦烯	betulene
香桦烯醇	betulenol
香榧醇	torreyol
香薷酮	elsholzione
顺式-1,3-二甲基-8-异丙基-3-癸烯	cis-1,3-dimethyl-8-isopropyl-3-decene
顺式-4,11,11-三甲基-8-亚甲基二环 [7.2.0]十一碳-4-烯	cis-4,11,11-trimethyl-8-methylenebicyclo [7.2.0] undec-4-ene
顺式-水合萜二醇	cis-terpin hydrate
顺式-对蓋-9-醇	cis-p-menthan-9-ol
顺式-4-癸烯酸乙酯	ethyl cis-4-decenoate
顺式-8-蓋烯	cis-8-menthene
胜红蓟素	ageratochromene
β-恰米烯	β-chamigrene
姜黄烯	curcumene
姜黄烯醇	curcumenol
姜黄酮	turmerone
姜烯	zingiberene
姜黄烯	α-curcumene
前异白菖二醇	preisocalamendiol
洋茉莉醛	piperonal
癸二烯醛	decadienal
γ-癸内酯	γ-decalactone

癸烷	decane
1-癸烯	1-decene
2-癸烯醛	2-decenal
9-癸烯酸	9-decenoic acid
4-癸烯酸甲酯	methyl 4-decenoate
癸醇	decanol
癸醛	decanal
2-癸酮	2-decanone
癸酸	decanoic acid
癸酸乙酯	ethyl decanoate
癸酸甲酯	methyl decanoate
癸酸异丙酯	isopropyl decanoate
独角金醇	strigol

十画

莰烯	camphene
莳萝醛	dillapiolal
莳萝脑	dillapiole
莪术酮	curzerenone
桂醇	cinnamic alcohol
桂醛	cinnamic aldehyde
桂酸	cinnamic acid
桂酸乙酯	ethyl cinnamate
桂酸丙酯	propyl cinnamate
桂酸甲酯	methyl cinnamate
桃金娘醇	myrtanol
桃金娘醛	myrtanal
桃金娘烯醇	myrtenol
桃金娘烯醛	myrtenal
桧烯	sabinene
桧醇	sabinol
桧脑	juniper camphor
1,8-桉叶油素	1,8-cineole
β-桉叶醇	β-eudesmol

氧化石竹烯	caryophyllene oxide
氧化玫瑰(吡喃型)	robe oxide (pyran type)
氧化芳樟醇(呋喃型)	linalool oxide(furan type)
氧化芳樟醇(吡喃型)	linalool oxide (pyran type)
氧化柠檬烯	limonene oxide
氧化蛇麻烯	humulene oxide
氧化辣薄荷酮	piperitone oxide
9-氧化橙花叔醇	9-oxonerolidol
8-氧杂二环[5.1.0]辛烷	8-oxabicyclo[5.1.0]octane
9-氧杂二环[6.1.0]壬烷	9-oxabicyclo[6.1.0]nonane
氨基甲酰苯甲酸酯	carbamyl benzoate
β-倍半水芹烯	β-sesquiphellandrene
倍半萜香茅烯	sesquicitronellene
爱草脑	estragole
海松-8,15-二烯	pimara-8, 15-diene
诺卡烯	nootkatene
诺卡酮	nootkatone
莽草毒素	anisatin

十一画

黄柏烯	nootkatene
黄柏酮	nootkatone
黄葵内酯	ambrettolide
黄樟油素	safrole
萘	naphthalene
菖蒲二烯	acoradiene
菊醇	chrysanthenol
1,8-孟二烯	1,8-menthadiene
盖烯	menthene
菠烯	bornene
樫木烯	dysoxylonene
雪松烯	himachanlene
紫苏烯	perillene
紫苏醛	perillaldehyde

紫罗兰醇	ionol
紫罗兰醛	ionic aldehyde
紫罗兰酮	ionone
蛇麻酞内酯	cnidilide
蛇麻二烯酮	humuladienone
蛇麻烯	humulene
甜没药烯	bisabolene
α-甜没药醇	α-bisabolol
α-甜没药醇氧化物β	α-bisabolol oxide β
甜旗烯	calacorene
α-甜橙醛	α-sinensal
喇叭茶醇	ledol
脱氢木香内酯	dehydrocostunolide
脱落酸	abscisic acid
3-基丁酸乙酯	ethyl 3-hydroxyobutyrate
3-羟基-2-丁酮	3-hydroxy-2-butanone
3-羟基己酸乙酯	ethyl 3-hydroxyhexanoate
羟基马桑毒素	tutin
α-羟基-马兜铃烯-9	α-hydroxy-aristolene-9
2-羟基-5-甲氧基苯乙酮	2-hydroxy-5-methoxyacetophenone
4-羟基-3-甲氧基苯甲醛	4-hydroxy-3-methoxyphenylaldehyde
1-（2-羟基-4-甲氧基苯基）乙酮	1-(2-hydroxy-4-methoxyphenyl)ethanone
2-羟基-3-甲基戊酸甲酯	methyl 2-hydroxy-3-methylpentanoate
4-羟基-5-甲基苯乙酮	4-hydroxyy-5-methylacetophenone
4-羟基-1,1,4,7-四甲基十氢环丙基薁	4-hydroxy-1,1,4,7-tetramethyldecahydro-cyclopropylazulene
2-羟基肉豆蔻酸	2-hydroxymyristic acid
羟基苯乙腈	4-hydroxyphenylacetonitrile
羟基苯乙酮	hydroxyacetophenone
2-羟基苯甲醛	2-hydroxyphenylaldehyde
2-羟基苯甲酸乙酯	ethyl 2-hydroxybenzoate
2-羟基苯甲酸甲酯	methyl 2-hydroxybenzoate
2-羟基苯甲酸苯甲酯	benzyl 2-hydroxybenzoate

2-羟基苯甲酰肼	2-hydroxybenzhydrazide
5-羟基-顺式-7-癸烯酸乙酯	ethyl 5-hydroxy-cis-7-decenoate
烯丙基环己烷	allylcyclohexane
烯丙基甲基二硫醚	allyl methyl disulfide
豚草素	ambrosin
银胶菊素	parthenin
萜	terpene

十二画

斯潘连醇	spathulenol
联苯	phenylbenzene
葑烯	fenchene
葑醇	fenchol
葑酮	fenchone
(+)-蒈烷-顺-4-醇	(+)-caran-cis-4-ol
蒈烯	carene
葛缕二烯酮	carvendione
葛缕醇	carveol
葛缕醇甲基醚	carveol methyl ether
葛缕酮	carvone
蒌叶酚	chavibetol
蒎烯	pinene
蒎葛缕醇	Pinocarveol
植物醇,植醇	phytol
植物酮	phytol ketone
植烷	phytane
棕榈油酸	palmitoleic acid
棕榈酸	palmitic acid
棕榈酸乙酯	ethyl palmitate
棕榈酸甲酯	methyl palmitate
硬脂酸乙酯	ethyl stearate
硬脂酸甲酯	methyl stearate
雅槛蓝二烯酮	eremophiladienone
雅槛蓝烯	eremophilene

雅槛蓝酮	eremophilone
黑椒酚	chavicol
黑椒酚甲醚	methyl chavicol
惕各酸己酯	hexyl tiglate
惕各酸甲酯	methyl tiglate
惕各酸苯甲酯	benzyl tiglate
惕各酸顺式-3-己烯酯	cis-3-hexenyl tiglate
蛔蒿素	somtonin

十三画

瑟丹烯内酯	sedanenolide
蒽	anthracene
蒿素	artemisin
蒿酮	artemisia ketone
榄香脂素	elemicin
榄香烯	elemene
榄香醇	elemol
蜂花醇	myricyl alcobol
愈创木烯	guaiene
愈创木醇	guaiol
愈创木薁	guaiazulene
新薄荷醇,新薄荷脑	neomenthol
新蛇床内酯	neocnidilide

十四画

雌诱素	Sirenin
4-酮基-β-紫罗兰酮	4-ketone-β-ionone, 4-oxo-β-ionone
薰衣草醇	lavandulol
碳酸二新戊酯	dineopentyl carbonate
辣薄荷醇	piperitol
辣薄荷酮	piperitone

十五画

薁	azulene
樟脑	camphor
α-樟脑烯醛	α-camphorene aldehyde

缬草烯	valerene
缬草酮	valeranone
缬草酸	valeri acid

十六画

薄荷烯	menthene
薄荷醇,薄荷脑	memthol
薄荷酮	menthone
橙花醇	nerol
橙花醛	neral
橙花叔醇	nerolidol

十七画

藏红花醛	safranal
藁本内酯	ligustilide, ligusticum lactone
檀香烯	santalene
檀香醇	santalol
檀烯	santene
螺[4.5]癸-1-烯	spiro[4.5]dec-1-ene
糠醇	frufuryl alcohol
糠醛	furfural, 2-furaldehyde
糠酸乙酯	ethyl furoate
糠酸甲酯	methyl furoate
磷酸甲基赤藓糖	methylerythritol phosphate (MEP)

廿一画

麝香草氢醌二甲醚	thymyl hydroquinone dimethylether
麝香草酚	thymol
麝香草酚甲醚	thymol methyl ether

Essential Oil of Plants

Liang-feng Zhu

Han-xiang Li

Winnie Yeung

Ze-xian Li

Chief Editor

Preface

Plant essential oils are a natural product that has been utilized by human ancestors for a long time. As early as ancient Egypt, Mesopotamian civilization and China's Shang Zhou era, there are records of the rhizomes, branches, leaves and fruits of various plants being directly or simply extracted for aromatherapy, cleansing, decontamination, refreshing, and curing diseases. After tens of thousands of years of human evolution, this ancient natural product has become an important and continuously renewable natural resource for humans today. Whether it is food, medicine, or daily products, it is indispensable. The Chinese version of the book "Plant Essential Oils", has been translated into English by many practitioners, in order to make plant essential oils more widely understood and accepted by humans.

I would like to express my gratitude to Ms. Liu Chi, Mr. Lu Hantao and Mr. Xu Yujian for their help and support in the publication of the English version of this book. The principle of organizing the overview of aromatic plants and the plant essential oil resources developed in China in the first chapter is to classify them into different categories according to the practical production needs and the importance of strategic applications. The first and second chapters of the English version of this book have been updated and enhanced in accordance with the development of the industry based on the Chinese version.

Chapter 1
Overview of Plant Essential Oils Resources in China

Section 1
Overview of Aromatic Plants

From a purely botanical perspective, it can be argued that plant essential oil resources should be considered aromatic plant resources, since essential oils are obtained exclusively from aromatic plants. In essence, a part of the aromatic plant is subjected to steam distillation or other methods to obtain a volatile, non-water-soluble liquid with a strong scent, which is the essential oil. Essential oils are widely distributed in various parts (organs) of the aromatic plants.

China has a vast territory, located in temperate, subtropical, and tropical regions, with diverse climates, soils, and vegetation, providing extremely favorable natural conditions for the growth of various aromatic plants. Therefore, aromatic plants in China are not only abundant but also diverse. According to incomplete statistics, aromatic plants include more than 1,000 introduced species, belonging to 77 families and 192 genera. China's development and utilization of aromatic plants date back a long time, as early as the Qin and Han Dynasties more than 2,000 years ago, using plants from the Asteraceae family like Artemisia spp. for bathing and fumigation to clean the body and prevent and treat diseases, a method that is still in use today. During the Tang and Song Dynasties, there were records of using Jasmine (*Jasminum sambac*) to make scented tea, which is still a popular tea today. Nowadays, plant essential oils are widely used in the chemical, light industry, food, cosmetics, pharmaceutical, medical, and other industries. They are also a traditional export commodity in China, including turpentine oil, peppermint oil, Chinese cedar oil, ginger oil, various camphor oils, and some floral pastes, such as jasmine paste, osmanthus paste, and bitter water rose oil. However, in response to the dynamic shifts in the global market, the export of these products is subject to constant adjustment.

A considerable portion of China's land is in the South Asian tropical region, and the Yangtze River Basin and areas south of the Yangtze River are the best places for growing aromatic plants. Thus, this vast area is home to many resources. Among them, the aromatic plants that have been developed and are waiting to be developed include *Pinus massoniana* and *P. elliottii* from the Pinaceae family, Juniperus formosana from the Cupressaceae family, *Cupressus funebris* from the Cypress family, *Cinnamomum aromaticum* and *C. burmannii* from the Lauraceae family, *C. camphora* and *C. parthenoxylon* from the Camphor family, and various Eucalyptus trees such as *Eucalyptus citriodora* and *E. globulus* from the Myrtaceae family. Among them, the Camphor genus of the Lauraceae family has great potential for the development of essential oils. There is now encouragement to change the destructive production methods of logging and root digging to using branches and leaves as raw materials for essential oil production. In terms of cultivation, superior plants are selected as 'mother plants' for asexual reproduction, establishing germplasm resources and expanding into planting bases.

1.1 Essential oil resources of *Cinnamomum* in the Lauraceae family

Studies have discovered some important plant essential oil resources within the genus based on chemical types, which are detailed in the second section. Here we provide an overview of the essential oil resources of *Cinnamomum.*

Lauraceae family

1.*Cinnamomum appelianum*

Branches and leaves contain 1.05% essential oil. The chemical constituents of the essential oil are eucalyptol 37.6%, linalyl acetate 16.2%. It is an endemic species in China, distributed in Hunan, Jiangxi, Guangdong, Guangxi, Guizhou, Sichuan, and Yunnan.

2.*C. austro-yunnanense*

Branches and leaves contain essential oil. The chemical constituent of the essential oil is γ-terpinene 20.8%. It is distributed in Yunnan.

3.*C. bejolghota*

Branches and leaves contain 0.3%~0.4% essential oil. The chemical constituents of the essential oil are camphor 84.0%, linalool 68.4%. It is distributed in Yunnan and Guizhou.

4.*C. bodinieri*

Branches and leaves contain 0.3%~0.4% essential oil. The chemical constituents of the essential oil are camphor 84.0%, linalool 68.4%. It is distributed in Yunnan and Guizhou.

5.*C. bodinieri* var. *hupehanum*

Leaves contain 1.4%~1.5% essential oil. The chemical constituents of the essential oil are camphor 88.5%, citral 95.0%. It is distributed in Hubei and Sichuan.

6.*C. burmannii*

Branches and leaves contain essential oil. The chemical constituents of the essential

Cinnamomum burmannii

oil are linalool 57.0%, citral 76.8%. It is produced in Yunnan. Mume tree, branches and leaves contain 0.3%~0.4% essential oil. The chemical constituents of the essential oil are d-Borneol 51.3%, linalyl acetate 7%. It is distributed in Guangdong, Fujian, Yunnan, and Guangxi. Ji tree, branches and leaves contain 0.3%~0.4% essential oil. The chemical constituent of the essential oil is eucalyptol 65.5%. It is distributed in Guangdong. Youji tree, branches and leaves contain 0.25%~0.3% essential oil. The chemical constituents of the essential oil are sabinene 27.5%, eucalyptol 27.6%. It is ditributed in Guangdong.

7.C. burmannii f. heyneanum

Branches and leaves contain 0.54% essential oil, barks contain 0.81%. The chemical constituent of the essential oil is camphor 97.5%. It is distributed in Hubei, Sichuan, Guizhou, Guangxi, and Yunnan.

8.C. camphora

Camphor type: Leaves and twigs contain 1.0% essential oil, camphor 83.9%. It is distributed in Jiangxi. Linalool type: Leaves and twigs contain 0.3%~0.8% essential oil, linalool 90.6%. It is distributed in Jiangxi and Guangdong. Cineole type: Leaves and twigs contain 0.75% essential oil, cineole 50.0%. Borneol type: Leaves and twigs contain 0.8% essential oil, borneol 81.8%.It is distributed in Jiangxi and Hunan. Floral type: Leaves and twigs contain 0.4% essential oil, linalool oxide 57.7%. Fresh leaves contain 1.6%~2.0% essential oil, citral 69.9%. It is distributed in Xishuangbanna, Yunnan.

Cinnamomum camphora

9.C. camphora var. *linaloolifera*

Fresh leaves contain 1.95% essential oil, the chemical constituent of the essential oil is linalool 92.7%. Leaves and twigs contain 1.3% essential oil, the chemical constituent of the essential oil is linalool 90.1%. It is distributed in Xishuangbanna, Yunnan and Pucheng, Fujian.

10. *C. aromaticum*

Leaves and twigs contain 0.35% essential oil, the chemical constituent of the essential oil is cinnamaldehyde 74.1%. Dried barks contain 0.8%~2.2% essential oil, the chemical constituent of the essential oil is trans-cinnamaldehyde 97.1%. It is distributed in Zhaoqing, Guangdong and Yunnan.

11. *C. caudiferum*

Leaves and twigs contain essential oil. The chemical constituent of the essential oil are camphor 60.4%, cineole 53.9%. It is distributed in Xichou, Yunnan.

12. *C. contractum*

Leaves and twigs contain 0.47% essential oil. The chemical constituents of the essential oil is trans-cinnamaldehyde. It is distributed in Yunnan.

13. *C. glanduliferum*

Leaves and twigs contain essential oil. The chemical constituents of the essential oil are cineole 45.8%, citral 54.4%, alpha-terpinene 65.7%. Safrole 90% in fruit skin. It is distributed in Yunnan.

14. *C. japonicum*

Roots and woods contain essential oil. The chemical constituents of the leaf essential oil are eugenol 15.3%, cineole 13.2%; bark: eugenol 29.5%, cineole 16.1%, linalool 11.4%, camphor 30.3%, safrole 45.9%. It is distributed in the East China region.

Cinnamomum aromaticum

Cinnamomum japonicum

Cinnamomum micranthum

15. *C. ilicioides*

Branches and leaves contain 0.83% essential oil. The chemical constituent of the essential oil is yellow camphor oil 82.1%. Dry lateral roots contain 0.21% essential oil. The chemical constituent of the essential oil is yellow camphor oil 82.7%. It is distributed in Yunnan and Guangxi.

16. *C. longepaniculatum*

Branches and leaves contain 1.2% essential oil. The chemical constituent of the essential oil are 1,8-cineole 58.6%, α-terpineol 15.4%, β-eucalyptol 41.1%, Branches and leaves contain 2.3% essential oil. The chemical constituent of the essential oil is camphor 90.5%. It is distributed in Sichuan Province and Xishuangbanna, Yunnan Province.

17. *C. micranthum*

Fresh flowers contain 0.12% essential oil. The chemical constituents of the essential oil are decanal 13.8%, decanoic acid 15.9%, linalool 14.3%. Tree trunks contain 0.09% essential oil. The chemical constituent of the essential oil is yellow camphor oil 61.3%. Roots contain 1.5% essential oil. The chemical constituent of the essential oil is yellow camphor oil 97.7%. It is distributed in Fujian, Guangxi, Guangdong, Hunan, and Jiangxi.

18.*C. osmophloeum*

Leaves contain 0.24%~1.45% essential oil. The chemical constituents of the essential oil are trans-cinnamaldehyde 79.5%, benzyl alcohol 83.3%. Branches and leaves contain essential oil. The chemical constituents of the essential oil are safrole 43.9%, benzaldehyde 35.7%, leavesbenzyl alcohol 40.4%, trans-cinnamaldehyde 32.9%. It is distributed in Taiwan.

19.*C. pauciflorum*

Branch leaves contain 2.4% essential oil.The chemical constituent of the essential oil is yellow camphor oil 69.7%. It is distributed in Jiuzhaigou, Sichuan. Branch leaves contain 2.5% essential oil. The chemical constituent of the essential oil is yellow camphor oil 99.3%. Wood contains 0.44% essential oil. The chemical constituent of the essential oil is yellow camphor oil 99.4%. Fresh tree barks contain 1.6% essential oil. The chemical constituent of the essential oil is yellow camphor oil 99.0%. It is distributed in Xishuangbanna, Yunnan.

20.*C. pingbienense*

Dry leaf contains 0.28% essential oil. The chemical constituents of the essential oil are benzyl alcohol 38.4%, 1,8-cineole 18.5%. It is distributed in Pingbian, Yunnan.

21.*C. platyphyllum*

Fresh leaf contains 0.57% essential oil. The chemical constituents of the essential oil are methyl isoeugenol 94.1%. It is distributed in Shaanxi, Gansu, Sichuan, and Hubei.

22.*C. parthenoxylon*

Twigs and leaves contain 0.6%~0.8% essential oil. The chemical constituents of the essential oil are α-phellandrene 22.4%, sabinene 12.7%, borneol 21.2%. It is distributed in Guangxi, Guangdong, Fujian, Jiangxi, Hunan, Guizhou, Yunnan. 'Da Ye You Zhang' type: twigs and leaves contain 0.2% essential oil, the chemical constituents of the essential oil is 1,8-cineole 62.4%; 'Da Ye Fang Zhang'type: twigs and leaves contain 1.1%~1.4% essential oil, the chemical constituents of the essential oil is d-linalool 82.8%; 'Da Ye Nao Zhang'type: twigs and leaves contain 0.8%~1.0% essential oil, the chemical constituents of the essential oil is d-camphor 86.7%; ginger camphor type: twigs and leaves

Cinnamomum parthenoxylon

contain 0.5%~0.8% essential oil, the chemical constituents of the essential oil is citral 72.1%; sesquiterpene type: twigs and leaves contain 0.3%~0.8% essential oil, the chemical constituents of the essential oil are 9-epi-rose oxide 24.2%, rose oxide 54.8%.

23. *C. rigidissimum*

Roots contain 1.4% essential oil, twigs and leaves contain 0.04%. The chemical constituents of the essential oil are eugenol methyl ether 28.6%, borneol-4 18.7%. It is distributed in Jianfengling, Hainan and Guangdong, Guangxi, Taiwan. Leaves contain 1.03% essential oil, roots contain 1.44%. The chemical constituents of the essential oil are benzyl benzoate 82.8%, safrole 61.7%. It is distributed in Xishuangbanna.

24. *C. saxatile*

Leaves and twigs contain 0.19% essential oil. The chemical constituents of the essential oil are rose oxide 30.7%, benzyl salicylate 12.1%, benzyl alcohol 7.6%. Dry root contains 1.46% essential oil. The chemical constituent of the essential oil is Safrole 94.4%. It is distributed in Malipo, Yunnan.

25. *C. septentrionale*

Leaves and twigs contain 1.10% essential oil. The chemical constituents of the essential oil are t-methyl eugenol methyl ether 85.7%. It is distributed in Lichuan, Hubei and Sichuan, Shaanxi, Gansu, Guangxi.

26. *C. subavenium*

Leaves and twigs contain 0.24% essential oil. The chemical constituents of the essential oil are 1,8-cineole 76.1%, eugenol 26.4%. It is distributed in Xishuangbanna. Leaves and twigs contain 0.1%~0.4% essential oil. The chemical constituents of the essential oil are borneol-4 21.4%, 1,8-cineole 15.3%. Bark contains 0.24% essential oil. The chemical constituent of the essential oil is eugenol 67.4%. It is distributed in Yunnan, Sichuan, Hubei, Guangdong, Guangxi, Anhui, Zhejiang, Jiangxi, Fujian.

27. *C. temala*

Leaves and twigs contain 1.1% essential oil. The chemical constituents of the essential oil are d-citronellal 69.1%. Leaves and twigs contain 0.47% essential oil. The chemical constituent of the essential oil is safrole 44.3%. Dry barks contain 5.42% essential oil. The chemical constituent of the essential oil is safrole 98.8%. It is distributed in southern Yunnan.

28. *C. tenuipilum*

A cultivated species. Branches and leaves contain essential oil. The chemical constituents of the essential oil are l-fenchone 97.51%, camphor 1.4%, and it is the camphor type. The essential oil of the branches and leaves of the cultivated variety contains eugenol 92.5%, camphor 1.55%~2.04%, and it is the eugenol type. The essential oil of the branches and leaves of the cultivated variety of *C. tenuipilum* contains methyl eugenol 70.03%, and it is the Methyl Eugenol type. The essential oil of the branches and leaves of the cultivated variety contains methyl eugenol 84.3%, and it is the methyl eugenol type. The essential oil of the branches and leaves of the cultivated variety contains camphor 85.7%, and it is the camphor type. The essential oil of the branches and leaves of the cultivated variety of *C. tenuipilum* contains 1,8-cineole 50%~57%, and it is the oil camphor type. The essential oil of the branches and leaves of the cultivated variety of *C. tenuipilum* contains 45% bornyl acetate. The essential oil of the branches and leaves of the cultivated variety of *C. tenuipilum* contains 59%~75% citral, and it is the borneol type. The essential oil of the branches and leaves of the cultivated variety of *C. tenuipilum* contains fatty substance 84%, linalool 1.0%~1.5%, and it is the sesquiterpene type. The essential oil of the root of the cultivated variety contains 33.1% elemicin, and 0.44% belongs to the sesquiterpene type.

29. *C. tonkinense*

Branches and leaves contain essential oil. The chemical constituents of the essential oil are eugenol 37.3%, l-fenchone 21.6%, and methyl eugenol 57.5%, and belongs to the eugenol type. The chemical constituents of the bark of the essential oil are safrole 36.9%, methyl eugenol 73.4%, and belongs to the camphor type. The chemical constituents of the branches and leaves of the essential oil are citral 39.6%, l-fenchone 13.3%, and belongs to the fresh leaf type. The chemical constituent of the bark of the essential oil is methyl eugenol 67.7%. It is distributed in Malipo, Yunnan.

30. C. wilsonii
Branches and leaves contain essential oil. The chemical constituents of the essential oil are l-fenchone 24.1%, 1,8-cineole 18.2%, and bornyl acetate 26.1%, and belongs to the camphor type. It is distributed in Shaanxi, Sichuan, Hubei, Hunan, Jiangxi, Guangdong, and Guangxi.

31. C. validinerve
Leaves contain essential oil. The chemical constituents of the essential oil are camphor 43.8%, linalool 17.4%. It is distributed in Yunnan.

32. C. villosulum
Fresh leaves contain 0.25%~0.30% essential oil. The chemical constituent of the essential oil is camphor 59.5%. It is distributed in Guangxi.

33. C. zeylanicum
Fresh leaves contain 1.5%~2.13% essential oil. The chemical constituents of the essential oil are eugenol 81.3%, benzyl and benzoate 8.10%. It is distributed in Xishuangbanna, Yunnan.

Cinnamomum zeylanicum

1.2 Chemical types within the *Cinnamomum* genus of the Lauraceae family

Table 1-1 Overview of Chemical Types within the *Cinnamomum* Genus of the
Lauraceae Family (Branch and Leaf Essential Oil)

Name	Chemical Constituents	Chemical Type
1. *Cinnamomum bejolghota*	Cinnamaldehyde 82.6%	Cinnamaldehyde type Mixed type
2. *C. bodinieri*	Linalool oxides 68.4%	Orange blossom camphor type
	Safrole 68.0%	Yellow camphor type
	d-Borneol 51.3%	Mei tree (borneol type)
3. *C. burmannii*	1,8-Cineole 65.5%	Oil calculating tree (cinnamomum type)
	1,8-Cineole 27.6% Paracymene 27.5%	Mixed transitional type
	Camphor 88.9%	Orange blossom camphor type
	1-Cinnamyl alcohol 90.6%	Camphor tree (camphor type)
4. *C. camphora*	1,8-Cineole 50.0%	Oil calculating tree (cinnamomum type)
	d-Borneol 81.8%	Methyl eugenol tree (borneol type)
	Safrole 57.7%	Orange blossom camphor type
5. *C. caudiferum*	1,8-Cineole 53.9%	Oil calculating tree (cinnamomum type)
	Camphor 60.4%	Camphor tree (camphor type)
	Citral 54.4%	Ginger camphor type
6. *C. glanduliferum*	Safrole 90.0%	Yellow camphor type
	1,8-Cineole 45.8%	Camphor tree (camphor type)
	α-Terpineol 65.7%	Terpene type

(continued)

Name	Chemical Constituents	Chemical Type
7. *C. burmannii* f. *heyneanum*	Safrole 98.5%	Yellow camphor type
	Camphor 48.7%	Methyl eugenol tree (borneol type)
	1,8-Cineole 33.6% Camphor 35.0%	Mixed transitional type
8. *C. longepaniculatum*	1,8-Cineole 58.6%	Camphor tree (camphor type)
	Camphor 90.5%	Oil calculating tree (cinnamomum type)
	β-Phellandrene 41.1%	Sesquiterpene alcohol type
9. *C. longipetiolatum*	Citral 71.7%	Ginger camphor type
	1,8-Cineole, Safrole	Mixed type
10. *C. osmophloeum*	t-Cinnamaldehyde 79.5%	Cinnamaldehyde type
	Linalool 83.3%	Linalool type
	Linalool 43.9% Benzaldehyde 35.7%	Mixed transitional type
11. *C. pauciflorum*	Yellow camphor oil 97.5%	Yellow camphor type
	Lemon aldehyde α-pinene 1,8-cineole	Mixed type
	1,8-cineole 62.4%	Oil camphor type
12. *C. parthenoxylon*	d-linalool 94.3%	Linalool type
	d-camphor 86.7%	Camphor type
	α, β-citral 72.1%	Ginger camphor type
	Eugenol methyl ether 71.5%	Eugenol type
13. *C. rigidissimum*	Benzyl benzoate 82.8%	Aromatic type
	Eugenol methyl ether 28.6% 4-pinus alcohol 18.7%	Mixed type

(continued)

Name	Chemical Constituents	Chemical Type
14. *C. saxatile*	Camphor 40.5%	Camphor type
	t-isobutyl methoxyphenol 85.7%	Eugenol type
15. *C. septentrionale*	t-eugenol methyl ether 85.7%	Eugenol type
	1,8-cineole 76.8%	Oil camphor type
	Camphor 85.0%	Camphor type
16. *C. subavenium*	1,8-cineole 76.1%	Oil camphor type
	Eugenol methyl ether 84%~89%	Eugenol type
17. *C. tamala*	Citronellal 69.1%	Citronellal type
	Safrole 44.3%	Yellow camphor type
	Acacia alcohol 64%~70%	Terpenoid type
18. *C. tenuipilum*	Linalool 92.5%	Linalool type
	γ-terpinene 56.4%	Terpene type
	Benzene	Sesquiterpene type
	α, β-citral 53.2%	Ginger camphor type
19. *C. tonkinense*	Linalool 37.0% geraniol 22.1%	Flower fragrance type
	Eugenol 81.3%	Eugenol type
20. *C. zeylanicum*	Benzyl benzoate 8.1%	Aromatic type
	t-cinnamaldehyde 31.2% Linalool 34.1%	Mixed type

1.3 Essential oil resources of trees and shrubs

In addition to the Pinaceae, Cupressaceae, and Lauraceae families mentioned above, China has a very rich resource of essential oil from woody plants, most of which are obtained from branches, leaves, fruits, and seeds. This not only plays an important role in resource regeneration and the ecological environment, but also ensures the sustainable production of essential oil resources.

Taxaceae family

Pseudotaxus chienii

Leaves contain essential oil. The chemical constituents of the essential oil are α-pinene, sabinene, β-pinene, δ-3-carene, lauryl acetate, limonene, β-caryophyllene, t-muurolol, para-mentha- 8-ol, t-cadinol, α-eudesmol, β-caryophyllene oxide, t-muurolol, and nerolidol. It grows in evergreen broad-leaved forests and deciduous broad-leaved forests in provinces and regions such as Jiangxi, Zhejiang, Hunan, Guangdong, and Guangxi. It is also used as an ornamental tree species and its wood is suitable for carving.

Pseudotaxus chienii

Torreya grandis

Seeds contain essential oil. The chemical constituents of the essential oil are α-pinene, lauryl acetate, 2-carene, bornyl acetate, α-terpineol, linalool, α-cubebene, β-cubebene, β-eudesmol, β-cubebene isomer, β-caryophyllene, and δ-cadinene. It is distributed in provinces such as Zhejiang, Jiangsu, Jiangxi, Fujian, Hunan, and Guizhou. Its seeds are a well-known dried fruit used for food, and its essential oil is used in the daily chemical industry, such as toothpaste and soap. The seeds are also used medicinally as insect repellents.

Podocarpaceae family

Podocarpus nagi

Branches and leaves contain essential oil. The chemical constituents of the essential oil are 3-hexen-1-ol, α-pinene, 7-octen-4-ol, 1,8-cineole, α-terpineol, α-terpinene, eudesma- 4(15),7-dien-1-ol, β-caryophyllene, γ-cadinene, γ-cadinene isomer, δ-cadinene, and nerolidol. It grows in low-altitude evergreen broad-leaved forests and is distributed in provinces and regions, such as Taiwan, Fujian, Zhejiang, Jiangxi, Hunan, Guangdong, and Sichuan. It is a high-quality timber species and is used as an ornamental tree. Its seeds are rich in fatty oil, and are used for industrial and food purposes.

Podocarpus nagi

Pinaceae family

Abies ernestii

Branches and leaves contain 0.15%~0.28% essential oil. The chemical constituents of the essential oil are 3-ethanol, limonene, 2,3,3-trimethyl-2-cyclohexen-1-ol, longifolene, pinocarveol-4, alpha-terpineol, piperitone, pulegone, dragon ester, cyclohexane-2-hexanoic acid, delta-cadinene, and beta-eucalyptol. It is primarily used for construction and its essential oil is used as natural fragrance materials. It is native to western and northern Sichuan and eastern Tibet in China.

Rutaceae family

Murraya kwangsiensis

Leaves contain 0.27% essential oil. The chemical constituents of the essential oil are alpha-phellandrene, beta-pinene, linalool, limonene, p-cymene, gamma-terpinene, benzyl alcohol, citronellal, isomenthone, menthone, alpha-terpineol, 2-(3,3-dimethyl)cyclohexyl ethanol, beta-citral, nerol, alpha-citral, geraniol, and ethyl salicylate. Its fruit can be used to relieve pain, its roots can be used to improve digestion, and its leaves can be used to treat shingles and fractures. Its branches and leaves can be used to dispel wind, relieve pain, promote Qi flow, and stop coughing. It is native to Guangxi, China, and grows in mountainous forests at an altitude of 800 meters.

Pinaceae family

Abies fargesii

Leaves contain 0.2%~0.3% essential oil. The chemical constituents of the essential oil are 3-carene, limonene, 2,3,3-trimethyl-2-cyclohexen-1-ol, bornyl acetate, pinene, α-terpineol, cyclohexane-2-ylidene ethyl acetate, δ-cadinene, and β-eudesmol, among other compounds. It grows in provinces such as Gansu, Shaanxi, Henan, Hubei, and Sichuan, forming pure or mixed forests, at altitudes of 2500~3700 meters above sea level. The essential oil from the leaves can be used as a fragrance for soap, while the wood, being lightweight and soft, can be used as a construction material.

A. faxoniana

Leaves contain 0.4%~0.7% essential oil. The chemical constituents of the essential oil are santene, tricyclene, α-phellandrene, sabinene, β-phellandrene, lauric aldehyde, limonene, methoxy-eugenol, α-ionone, linalool, nerol, and other compounds. It grows in provinces such as Gansu and Sichuan, in cold and humid climates, in well-drained acidic brown and gray soil and meadows, at altitudes of 2700~3900 meters above sea level. The essential oil from the leaves can be used as a solvent for coatings.

A. holophylla

Branches, leaves, and resin contain essential oil. The essential oil extracted from the branches and leaves. The chemical constituents of the essential oil are tricyclene, α-pinene, sabinene, β-pinene, lauryl, phellandrene-3, limonene, camphor, bornyl acetate, terpinolene, and β-selinene. It is native to the northeastern region of China, including the Mudanjiang River basin, Changbai Mountain area, and the eastern part of the Liao River. It also grows in North Korea. Liaodong spruce thrives in cold and humid grey-brown forest soil at an altitude of 500~1200 meters. The essential oil is used in traditional medicine to prevent colds.

A. nephrolepis

Fresh leaves contain 2.2% essential oil. The chemical constituents of the essential oil are pinene, tricyclene, α-pinene, sabinene, β-pinene, lauryl, cuminene, limonene, camphor, bornyl acetate, terpinyl acetate, α-terpineol, α-copaene, β-selinene, nerolidol, and α-cedrene. It is native to the southern slope of the Lesser Khingan Mountains, Changbai Mountains, and also grows in Shanxi and Hebei provinces of China. It prefers cold and humid environments with well-drained slopes at an altitude of 300~2100 meters.

Cedrus deodara

Branches and leaves contain 0.75% essential oil. The chemical constituents of the essential oil are α-pinene, camphene, β-pinene, β-myrcene, limonene, 2-carene,

Cedrus deodara

Cedrus deodara

p-cymene, bornyl acetate, camphor, borneol, pinocarveol, thujopsene, and delta-cadinene. It is commonly cultivated in gardens and is a popular ornamental tree in famous gardens worldwide. It is distributed in southwestern Tibet and Beijing.

Picea jezoensis var. *microsperma*

Reins contain essential oil. The chemical constituents of the essential oil are α-pinene, camphene, lauryl alcohol, delta-3-carene, α-limonene, α-phellandrene, camphor, borneol, lauryl alcohol, terpinolene, and caryophyllene. It is distributed in Heilongjiang, Jilin, and Liaoning, growing on gentle slopes and hillsides at an altitude of 300~800 meters above sea level. The essential oil of the resin is used in folk medicine to prevent and treat colds.

P. meyeri

Branches and leaves contain 0.35% essential oil. The chemical constituents of the essential oil are tricyclene, α-pinene, camphene, β-myrcene, delta-3-carene, limonene, p-cymene, linalool, camphor, borneol, 2,3,3-trimethyl-2-norbornanol, pinocarveol, beta- caryophyllene, and delta-cadinene. It grows in cool, wet, and brown soil forests. The lightweight wood is a major afforestation species in northern China, and the essential oil from its branches and leaves can be used as a solvent in coatings.

P. wilsonii

Leaves contain 0.24% essential oil. The chemical constituents of the essential oil are tricyclene, α-pinene, camphene, bornyl acetate, β-pinene, β-linalool, limonene, 2-carene, camphor, longifolene, pine oil alcohol-4, α-terpineol, citronellol, dragon brain acetate, δ-juniperene, β-acaciaene, etc. It is distributed in Gansu, Hebei, Hubei, Nei Mongol, Qinghai, Shaanxi, Shanxi, Sichuan. It is well adapted to dry and cold climates, growing at elevation of 700~2800 meters. The timber is used for construction, furniture, and wood pulp. The species is also cultivated for afforestation and as an ornamental. The essential oils from the branches and leaves can be used as solvents for general coatings.

Pinus armandii

Leaves contain 0.60% essential oil. The chemical constituents of the essential oil are Triene, α-pinene, camphene, bornyl acetate, β-pinene, β-linalool, β-selinene, 2-carene, α-terpineol, camphor, longifolene, pine oil alcohol-4, dragon brain acetate, β-caryophyllene, α-caryophyllene, γ-juniperene, δ-juniperene, δ-juniperol, and its isomer, etc. It grows in provinces such as Shanxi, Shaanxi, Gansu, Henan, Hubei, Sichuan, Yunnan, Guizhou, and Tibet at an altitude of 1600~3600 meters on slopes and valleys of pure forests and mixed forests. Pine resin can be extracted to obtain turpentine, the wood is an excellent raw material for papermaking, the bark can be used to extract oak gum, and the seeds can be used to extract edible oil.

P. bungeana

Leaves contain 0.90% essential oil. The chemical constituents of the essential oil are Triene, α-pinene, camphene, bornyl acetate, β-pinene, β-linalool, limonene, 3,6,6-trimethyl-2-cyclohexene-1,4-dione, camphor, α-terpineol, β-caryophyllene, α-caryophyllene, γ-juniperene, β-cedrene, δ-juniperene, etc. It is distributed in provinces such as Shanxi, Henan, Shaanxi, Gansu, Sichuan, and Hubei, etc. It is suitable for dry and cold climates and can withstand low temperatures of -30℃. It grows in acidic soil and limestone. It is used as a timber and ornamental tree species, and its seeds are edible. The essential oil can be used for turpentine.

P. caribaea

Pine reins contain 10%~12% essential oil. The chemical constituents of the essential oil are tricyclene, alpha-pinene, alpha-terpineol, camphene, beta-pinene, lauricene, phellandrene, beta-selinene, beta-cymene, elemol, artemisia ketone, etc. Originally from the Caribbean Sea area, it is introduced to Guangdong. In recent years, superior varieties have been introduced, and pine resin can be used to produce turpentine and rosin. Also known as Slash Pine.

Pinus caribaea

P. densiflora

Leaves contain 0.6%~0.8% essential oil. The chemical constituents of the essential oil are tricyclene, alpha-terpineol, camphene, beta-pinene, lauricene, beta- cymene, phytol, tert-butyltoluene, ethyl acetate, beta-caryophyllene, alpha-caryophyllene, beta-bisabolene, gamma-muurolene, delta-cadinene, etc. It grows in the eastern part of Heilongjiang, Changbai Mountains, central Liaoning, the Liaodong Peninsula, coastal areas of northern Jiangsu and Shandong. It is a high-quality tree species and pine needle essential oil can be used as turpentine.

P. elliottii

Reins contain 19%~28.6% essential oil. The chemical constituents of the essential oil are terpenes, alpha-terpineol, camphene, beta-pinene, limonene, alpha-terpinene, pinocarvone, linalool, paeonol, califonolide, rhodinol, sabinene, etc. Originally from the wet and warm regions of the eastern United States, it is introduced to southern and eastern China. It is a high-quality fast-growing tree species, with high resin yield and high content of turpentine and pinene.

Pinus elliottii

P. griffithii

Leaves contain 0.7%~0.8% essential oil. The chemical constituents of the essential oil are tricyclene, alpha-terpineol, camphene, beta-pinene, lauricene, alpha-cymene, limonene, phytol, beta-caryophyllene, gamma-elemene, delta-cadinene isomer, etc. It grows in Yunnan and Tibet, as well as Afghanistan, Myanmar, and Nepal, on mountain slopes and valleys at altitudes of 1 200~3 000 meters. It is a timber species, and both pine resin and the essential oil of the branches and leaves can be used as turpentine.

P. kesiya var. *langbianensis*

Reins contain essential oil. The chemical constituents of the essential oil are α-pinene, camphene, β-pinene, myrcene,phellandrene-3, β-phellandrene, bornyl acetate, α-terpineol, longifolene, and α-cadinene. It is economically valuable and has a higher value than Pinus massoniana due to its higher content of β-phellandrene in its resin. It is a pine species found in the southern Yunnan province of China, including areas such as Simao, Malipo, Puer, Jingdong, and Xishuangbanna.

P. koraiensis

Leaves contain 0.5% essential oil, with barks contain essential oil. The chemical constituents of the leaf essential oil are α-pinene, camphene, myrcene, phellandrene, limonene, α-terpineol, camphor, borneol, pinene-4, para-cymene-α-ol, α-terpinolene, and cis-caryophyllene. Its bark is also used for medicinal purposes. The chemical constituents of the bark essential oil are α-pinene, camphene, myrcene, phellandrene, limonene, α-terpineol, camphor, borneal, pinene-4, α-terpineol, cis-ocimene, cubenol, longifolene, cis-caryophyllene, and β-selinene. It is a pine species found in the northeastern regions of China, including the Changbai Mountains, Jilin, and the Xiaoxing'an Mountains. It grows in humid and cold temperate forests at altitudes ranging from 150 to 1800 meters.

P. kwangtungensis

Pine reins contain turpentine oil. The chemical constituents of the essential oil are α-pinene, camphene, lauryl alcohol, β-pinene, p-cymene, α-terpinolene, limonene, camphor, borneol, 4-terpineol, lauryl alcohol, t-sabinene, [α]-β-agathic acid, sabinene. It is an endemic species in China, distributed in provinces such as Guizhou, Guangxi, Guangdong, and Hunan. It thrives in warm and humid climates with deep acidic soil, rocky slopes, and is used for construction, mining, furniture, and the production of turpentine oil from its resin.

P. massoniana

Pine reins contain essential oil, with leaves contain 0.2% essential oil. The chemical constituents of the pine resin essential oil are α-pinene, camphene, lauryl alcohol, β-pinene, longipinene, limonene, β-caryophyllene, α-selinene. It is distributed in the Yangtze River basin in Anhui, Shaanxi, and southern Henan, as well as Guangdong and Guangxi. It is drought-resistant and can grow in poor red soil. The resin is a major raw material for producing turpentine oil, and pine needle and pinecone essential oil can be used in cosmetics and fragrances. The chemical constituents of the leaf essential oil are α-pinene, camphene, β-pinene, lauryl alcohol, limonene, p-cymene, lauryl alcohol, borneol, α-terpineol, ethyl laurate, β-linalool, β-caryophyllene, α-selinene, β-selinene, γ-terpinene, and more.

P. pumila

Leaves contain 1.0% essential oil, barks contain 4.1% essential oil. The chemical constituents of the leaf essential oil are α-pinene, α-phellandrene, sabinene, bornyl acetate, lauryl alcohol, β-caryophyllene, terpinolene, γ-muurolene, caryophyllene oxide, limonene, camphene, α-terpineol, myrcene, and β-phellandrene. The chemical constituents of the bark essential oil are α-pinene, sabinene, β-pinene, myrcene, limonene, terpinolene, camphene, β-caryophyllene, fenchyl alcohol, and

Pinus massoniana

caryophyllene oxide. It is distributed in Daxing'anling Baila Mountain, Xiaoxing'anling, over 1000 meters above sea level. Both pine needle essential oil and bark essential oil can be used to produce turpentine.

P. sibirica

Leaves contain 2.8% essential oil, while the bark of the tree contains 4.16%. The chemical constituents of the leaf essential oil are α-pinene, camphene, β-pinene, myrcene, phellandrene-3, limonene, β-sylvestrene, 1,8-cineole, sabinene, pinene, bornyl acetate, ε-cadinene, and γ-cadinene, among others. The essential oil from the barks contain α-pinene, camphene, β-pinene, myrcene, phellandrene-3, pinosylvin, ε-cadinene, γ-cadinene, and bornyl acetate, among others. It is a species of pine tree native to the northwestern part of the Altai Mountains in Xinjiang, China, along the Kanasi River and Homu River valleys. It grows at altitudes of 1600~2300 meters and is commonly found mixed with larch. It is a high-quality timber tree species, and its needles are rich in essential oil, which can be used to produce pine needle oil.

P. sylvestris var. mongolica

Leaves contain 0.67% essential oil. The chemical constituents of the essential oil are tricyclene, α-pinene, sabinene, β-pinene, myrcene, phellandrene, α-terpineol, artemisia ketone, bornyl acetate, limonene, t-cadinol, β-caryophyllene, γ-elemene, α-elemene, γ-terpinene, α-terpinene, γ-terpinene, dehydro-β-agarofuran, and labda-8(17),12-diene-15,16-dial. It is distributed in Heilongjiang, and grows on mountains and sand dunes at an altitude of 400 to 800 meters. The tree can produce resin, which is used to make rosin and turpentine. The wood has strong resistance to decay and can be used for construction purposes.

P. sylvestris var. sylvestriformis

Leaves contain 0.78% essential oil. The chemical constituents of the essential oil are tricyclene, α-pinene, sabinene, β-pinene, myrcene, para-cymene, γ-terpinene, α-terpinene, cubebene, β-caryophyllene, cis- caryophyllene, bisabolene, β-bisabolene, β-bisabolol, δ-cadinene, and dehydro-β- agarofuran. It grows in in Jilin Province on mountains at an altitude of 800~1600 meters. The tree can produce resin, which is used for the same purposes as the Scotch Pine.

P. tabuliformis

Leaves contain 0.50% essential oil. The chemical constituents of the essential oil are tricyclene, α-pinene, sabinene, longifolene, β-pinene, β-myrcene, β-laurene, phellandrene-2, bornyl acetate, α-cedrene, β-cedrene, γ-terpinene, δ-cadinene, γ-elemene, γ-terpinene, and δ-cadinol isomers. It is distributed in provinces and regions such as Jilin, Liaoning, Hebei, Henan, Shandong, Shanxi, Shaanxi, Gansu, Ningxia, Qinghai, and Sichuan. It grows on deep and well-drained soil at an altitude of 100~2600 meters. It is the main tree species used for lumber in China. The tree can produce resin and the bark can be used to extract tannin.

P. taeda

Also known as pitch pine, it is a species of pine native to North America. Leaves contain essential oil. The chemical constituents of the essential oil are α-pinene, limonene, β-pinene, myrcene, and others. It has been introduced and grown successfully in various provinces and regions of China, particularly in the eastern and southern parts of the country, and is considered a recent and superior pine species.

P. thunbergii

Leaves contain 1.0% essential oil. The chemical constituents of the essential oil are α-pinene, limonene, β-pinene, β-linalool, p-cymene, α-terpineol, and others. Xingkai Lake pine (Pinus takahasii) is a pine species native to Heilongjiang, China. It grows in sandy dunes near the lakeside and on the top of mountains. The tree is used for construction, furniture, and vehicle materials due to its lightness and corrosion resistance. The essential oil of the tree contains terpenes such as α-pinene, limonene, β-pinene, myrcene, β-linalool, β-caryophyllene, and others. It is native to Japan and North Korea and has been planted in Liaodong Peninsula, Shandong, Jiangsu, Zhejiang, Fujian, and Taiwan. It prefers a temperate oceanic climate and can produce resin in its trunk.

Cupressaceae family

Cunninghamia lanceolata

Stems contain 0.9%~1.1% essential oil. The chemical constituents of the

Cunninghamia lanceolata

essential oil are tricyclene, alpha-pinene, camphene, beta-pinene, beta- myrcene, limonene, para-cymene, alpha-terpineol, bornyl acetate, hinokitiol, and acetate ester of cis-3-hexenol. It is widely planted in the Yangtze River Basin and the Qinling area in southern China. Its roots are used as lifebuoys and bottle stoppers, and it is also used as a tree species for water network areas and garden landscaping.

Glyptostrobus pensilis

Stems contain contains 0.4%~0.6% essential oil. The chemical constituents of the essential oil are tricyclene, alpha-pinene, camphene, beta-pinene, beta-myrcene, limonene, para-cymene, alpha-terpineol, bornyl acetate, hinokitiol, and acetate ester of cis-3-hexenol. It is a unique tree species in China, distributed in Guangdong, Guangxi, and Yunnan. It grows in low-altitude wetlands and is resistant to water but not to cold.

Glyptostrobus pensilis

Cryptomeria fortunei

Leaves contain contains 0.62% essential oil. The chemical constituents of the essential oil are hinokitiol, 1,8-cineole, gamma-terpinene, camphene, alpha-terpineol, bornyl acetate, delta-selinene, beta-eudesmol, and longifolene. It is distributed in in Guangdong, Guangxi, Yunnan, Guizhou, Sichuan, Jiangsu, Anhui, Shandong, and Henan provinces. It is commonly used for garden landscaping and ornamental purposes.

Taxodium ascendens

Branches and leaves contain 0.3%~0.5% essential oil. The chemical constituents of the essential oil are 2-hexenal, 3-hexenol, alpha-pinene, beta-pinene, camphene, limonene, and others. It is primarily grown in provinces such as Jiangsu, Zhejiang, Henan, and Hubei in China, and is resistant to strong winds. Pond cypress is used as a tree for landscaping on the banks of water systems and is also a source of timber. Its essential oil can be used in the production of turpentine.

Cupressus funebris

Stems and roots contain 3%~5% essential oil. The chemical constituents of the

Cryptomeria fortunei

Taxodium ascendens

Cupressus funebris

Fokienia hodginsii

Thuja orientalis

essential oil are alpha-pinene, beta-pinene, camphene, limonene, and others. It is primarily grown in provinces such as Zhejiang, Fujian, Jiangxi, Hunan, Guangdong, Guizhou, Yunnan, and Sichuan in China, and also in Vietnam. Its wood is of high quality and resistant to decay, and its essential oil is primarily used as a natural fragrance.

Fokienia hodginsii

Leaves contain 0.24% essential oil. The chemical constituents of the essential oil are alpha-pinene, beta-pinene, camphene, limonene, and others. It is primarily grown in provinces such as Zhejiang, Fujian, Guangdong, Guizhou, Yunnan, and Sichuan in China. It is a unique species in China, and its essential oil is used in the production of fragrances and cosmetics due to its special aroma.

Thuja orientalis

Leaves contain 0.8% essential oil, while the barks contain 0.31%, the dried fruit husk contains 0.4%, and the dried woods contain 1.1%. The chemical constituents of the leaf essential oil are α-thujene, α-pinene, sabinene, β-pinene, myrcene, α-terpinene, p-cymene, bornyl acetate, terpinolene, 2,4-terpinadienol, and δ-3-carene. It is distributed throughout various regions of China and has medicinal properties that can help treat cough and relieve phlegm.

The essential oil has a refreshing scent and can be used in cosmetics, soap, fragrance, disinfectants, and insecticides. The chemical constituents of the bark essential oil are α-thujene, α-pinene, carvone, β-terpinene, myrcene, p-cymene, umbellulone, limonene, terpinolene, t-cadinol, linalool, and β-caryophyllene. The fruit husk essential oil contains α-thujene, α-pinene, sabinene, β-pinene, myrcene, p-cymene, umbellulone, terpinolene, 2,4(8)-terpinadienol, 4-terpineol, α-terpineol, β-pinene, and β-caryophyllene. The wood essential oil contains α-thujene, β-caryophyllene, β-funebrene, cuparene, α-terpineol, thujopsene, huoxiangenol, speusic acid, d-[2(12)]-rohan-8α-ol-3-one, and kusunokinin. The wood essential oil has a long-lasting, pleasant fragrance.

Juniperus chinensis

Leaves contain essential oil. The chemical constituents of the essential oil are tricyclene, alpha-pinene, camphene, beta-pinene, myrcene, beta-phellandrene, alpha-terpinene, limonene, elemene, and more. It is found in Inner Mongolia, Hebei, Shanxi, Shandong, Jiangsu, Zhejiang, Fujian, Anhui, Jiangxi, Henan, Shaanxi, Gansu, Sichuan, Yunnan, Guizhou, Guangdong, Guangxi, and Tibet. Grows in neutral or slightly acidic soil. The essential oil from its leaves can be

Juniperus chinensis

used as a folk medicine for ailments such as colds and coughs. The wood also contains essential oil and is resistant to decay.

J. przewalskii

Leaves(dried)contains 6% essential oil. The chemical constituents of the essential oil are alpha-pinene, beta- pinene, alpha-terpineol, limonene, phellandrene, thujone, terpinen-4-ol, alpha-bisabolene, gamma-terpinene, cubebene, t-muurolol, cedrol, delta-cadinene, and more. It is found in Qinghai, Gansu, and Sichuan. It grows on south-facing slopes at altitudes of 2600~4000 meters. The leaves can be used for medical purposes and can stop bleeding and coughing, and treat conditions such as hemoptysis, hematemesis, and hematuria.

Magnoliaceae family

Magnolia coco

Leaves contain 0.24% essential oil. The chemical constituents of the essential oil are α-phellandrene, β-pinene, β-ocimene, camphene-4, 1,8-cineole, cyclohexene, caryophyllene-3, linalool, pinen-4-ol, α-terpineol, α-bergam otene, β-eudesmol, dehydroaromadendrene, β-caryophyllene, α-curcumene, and others. It is widely planted in southern China and Southeast Asia, night blooming jasmine is a common ornamental tree with fragrant flowers that can be used to scent tea.

Magnolia coco

M. denudata

Leaves contain 0.04%~0.05% essential oil. The chemical constituents of the essential oil are α-phellandrene, sabinene, β-ocimene, limonene, 1,8-cineole, para-cymene, cis- 3-hexenol, linalool, β-bourbonene, α-terpineol, bornyl acetate, β-caryophyllene, α-amorphene, γ-elemene, terpinen-4-ol, ε-elemol, β-eudesmol, and others.

M. sprengeri

Leaves and flower buds contains 0.20%essential oil. The chemical constituents of the essential oil are α-phellandrene, kessane, β-phellandrene, 1,8-cineole, para-cymene, camphor, linalool, bornyl acetate, terpinolene, amorphene, γ-terpinene, citral, curcumene, fokienol, γ-eudesmol, β-caryophyllene, α-caryophyllene, and others. It grows in mountain mixed forests, the flower buds and leaves of Wudang magnolia can be developed into natural fragrances.

Illiciaceae family

Illicium tashiroi

Fruits contain 0.83% essential oil. The chemical constituents of the essential oil are α-phellandrene, limonene, β-selinene, anethole, camphene, safrole, methyl eugenol, eucalyptol, t-patchoulene, methoxydienol, δ-cadinene, etc. It grows in Sichuan, Hunan, Guangdong, Guangxi, Taiwan, and other provinces and regions. It is also found found in Japan. It grows in mountain gullies, water sides, or sunny shrubs. It is a resource for developing aniseed flavors.

I. dunnianum

Fruit peels contain 0.42% essential oil. The chemical constituents of the essential oil are α-phellandrene, β-phellandrene, lauryl alcohol, limonene, β-selinene, α-terpinene, 1,8-cineole, asarone, linalool, bornyl acetate, cubenol, α-terpineol, undecanone, camphor, t-muurolol, t-β-farnesene, δ-cadinene, nerolidol, β-eudesmol, and cypressene. It grows in Fujian, Hunan, Guangdong, Guangxi, Guizhou, and other provinces and regions. It grows in valleys and streams at an altitude of 500~700 meters and in dense forests along rivers and mountains. Fruit peel essential oil is a natural flavoring resource waiting to be developed.

Illicium dunnianum

I. henryi

Fruits contain 2.25% essential oil. The chemical constituents of the essential oil are 3-carene, limonene, β-selinene, β-phellandrene, sabinene, elemol, nerol, cis-muurola-4(14),5-diene-3-one, paeonol, 1,8-cineole, α-terpineol, eucalyptol, methyl eugenol, myristicin, ethyl eugenol, methyl salicylate, hinokitiol, cubenol, and celery ketone. It grows in Shaanxi, Anhui, Jiangsu, Jiangxi, Fujian, Henan, Hubei, Sichuan, Guizhou, Yunnan, and other provinces. It grows in the undergrowth or shrubs on slopes and gullies at an altitude of 750~1500 meters. Both the fruit and leaves are used in medicine for their effects in relaxing tendons and activating blood circulation, stopping bleeding and pain, and promoting Qi flow.

I. lanceolatum

Fruits contain essential oil. The fruit essential oil is a natural fragrance with potential for development. It is distributed in provinces and regions such as Jiangsu, Anhui, Zhejiang, Jiangxi, Fujian, Guangdong, Guangxi, and Yunnan. It grows in damp ditches or on both sides of mixed forests.

I. majus

Fruits contain 1.70% essential oil. The fruit essential oil is also a natural fragrance with potential for development. The chemical constituents of the essential oil are dimethyl benzene ethylene, paracymene, methyl phenyl ketone, 2-pinene, linalool, δ-cadinene-3, limonene, α-phellandrene, β-phellandrene, khellactone, anisaldehyde, perillaldehyde, camphor, piperonal, heliotropin, t-geosmin, p-geosmin, turpentine alcohol-4, eugenol methyl ether, acetic acid borneol ester, acetic acid turpentine ester, δ-juniperene, t-caryophyllene, elemicin, etc. It is distributed in provinces and regions such as Sichuan, Guizhou, Hunan, and Guangxi, and grows in mixed forests at altitudes ranging from 200 to 2000 meters.

I. minwanense

Fruits contain essential oil. The chemical constituents of the essential oil are α-pinene, β-pinene, sabinene, myrcene, limonene, β-phellandrene, α-terpineol, α-phellandrene, 1,8-cineole, eucalyptol, terpinen-4-ol, α-terpinolene, acetate ofterpinolene, acetate of linalyl, γ-terpinene, β-eudesmol, etc. It is found in Fujian. It grows at altitudes of 1000~1850 meters in ravines and forest edges. The fruit essential oil is a natural flavoring that has yet to be fully developed.

I. simonsii

Fruits contain 1% essential oil. The chemical constituents of the essential oil are α-pinene, β-pinene, sabinene, myrcene, limonene, linalool, camphor, bornyl acetate, linalyl acetate, β-citronellene, etc. Found in Guizhou. Used as a spicy condiment instead of star anise by locals.

I. ternstroemioides

Fruits contain essential oil. The chemical constituents of the essential oil are α-pinene, β-pinene, limonene, β-phellandrene, α-terpinolene, 1,8-cineole, eucalyptol, terpinen-4-ol, α-terpinol, cis- chrysanthemum ene, linalool, cubebene, β-olive oil, elemene, α-yirane, γ-cadinene, linalyl acetate, γ-terpinene, etc. Found in Guangdong. Grows in high-altitude dense forests or near streams. The essential oil of the fruit is a natural flavoring resource that has not yet been fully developed.

I. verum

Fruits contain 8%~12% essential oil, while its fresh leaves contain 0.3~0.4% essential oil. It is widely cultivated in Guangxi, Guangdong, Fujian, Yunnan, Guizhou, and other provinces in China, and grows in humid and warm valleys. The fruit and fresh leaves, which are rich in anethole, are the best raw materials for extracting anethole and are primarily used as a seasoning ingredient. They are also used in medicine to stimulate the appetite, warm the stomach, dispel cold, and relieve pain. They are the main raw material for synthesizing the estrogen hormone estrone. The chemical constituents of the essential

Illicium verum

oil are α-pinene, myrcene, limonene, linalool, terpineol-4, eucalyptol, trans-anethole, β-citronellene, and eugenol methoxyacetate. The essential oil of fresh leaves contain contains limonene, linalool, piperonyl methyl ether, cis-anethole, para- anisaldehyde, β-caryophyllene, trans-anethole, and α-terpineol.

Annonaceae family

Annona muricata

Fruits contain essential oil. It is native to tropical America and is now cultivated in Guangdong, Guangxi, Fujian, Yunnan, Taiwan, and other provinces in China. The chemical constituents of the essential oil are 3-methyl-3-buten-1-ol, 3-methyl-2-butenol, ethyl butyrate, methyl 3-methylbutyrate, ethyl 3-methyl-2-butenoate, α-pinene, β-pinene, methyl 2-hydroxy-4-methylpentanoate, β-caryophyllene, hexyl acetate, 7-methyl-4-decene, 1,8-cineole, limonene, methyl octanoate, terpineol-4, α-terpineol, and ethyl octanoate. The fruit is edible and the wood is used as a shipbuilding material.

Uvaria macrophylla

Branches and leaves contain essential oil. The chemical constituents of the essential oil are 1,8-cineole, 3,6,6-trimethyl-2-cyclohexene-1-carboxaldehyde,

Uvaria macrophylla

Uvaria macrophylla

α-terpineol, cubenol, β-ocimene, β-caryophyllene, α-caryophyllene, β-selinene, γ-cadinene, δ-juniperene, etc. It is found in southern China and also distributed in Vietnam. It grows in low-altitude shrubs and is used as an ornamental plant in gardens.

Myristicaceae family

Myristica fragrans

Dried flowers contain essential oil. The chemical constituents of the essential oil are α-pinene, α-phellandrene, sabinene, β-phellandrene, β-ocimene, β-myrcene, terpinene-3, α-terpinene, γ-terpinene, terpinen-2, 4-terpineol, safrole, isoeugenol, myristicin, etc. Originally from Malaysia, it is now widely grown in Guangdong, Yunnan, and Taiwan. The seeds are

Myristica fragrans

used in medicine, and the essential oil is used as a food flavoring.

Carica papaya

Fruits contain essential oil. The chemical constituents of the essential oil are Methylcyclohexane, 2-methylbutane, methyl acetate, toluene, methyl hexanoate, cis-β-ocimene, cis-linalool oxide (pyran type), trans-linalool oxide (pyran type), linalool, methyl decanoate, benzaldehyde, benzyl acetate, phenyl hydrogen sulfide ester, etc. It is originally from tropical and subtropical regions of Asia and America. It is a common fruit in China.

Actinidia chinensis

Fruits contain essential oil. The chemical constituents of the essential oil are ethyl acetate, propyl acetate, methyl butyrate, ethyl butyrate, t-2-hexenal, t-2-

Baeckea frutescens

hexenol, methyl caproate, ethyl caproate, methyl benzoate, linalool, ethyl benzoate, and β-damascenone. It is mainly grown in Shaanxi, Hubei, Henan, Jiangsu, Anhui, Zhejiang, Jiangxi, Fujian, Guangxi, and Guangdong, in forests or shrubs. The fruit is rich in vitamins and can be used as a fruit or for medicinal purposes.

Myrtaceae family

Baeckea frutescens

Branches and leaves contain 1.4% essential oil. The chemical constituents of the essential oil are 1,8-cineole, γ-terpinene, linalool,Baeckea frutescens, terpineol-4, α-terpineol, eugenol, β-caryophyllene, α-pinene, and δ-selinene. It is mainly grown in Jiangxi, Fujian, Guangdong, Guangxi, and other provinces and regions on low hills, barren mountains, and grassy

Baeckea frutescens

slopes. It is used in folk medicine to treat intestinal inflammation, diarrhea, athlete's foot, skin itching, and mosquito bites.

Eucalyptus calophylla

Branches and leaves contain 0.4%~0.5% essential oil. The chemical constituents of the essential oil are α-pinene, camphene, β-pinene, p-cymene, 1,8 -cineole, γ-terpinene, carvacrol, terpinolene, terpineol-4, ethyl myrtoliate, apiol, methyl chavicol, β-caryophyllene, caryophyllene oxide, γ-muurolene, and elemicin. It is native to Australia, and is grown in Guangdong for medicinal and industrial purposes, and as a source of honey.

E. camaldulensis

Branches and leaves contain 0.14%~0.28% essential oil. The chemical constituents of the essential oil are α-pinene, β-pinene, α-terpinene, limonene, 1,8 -cineole, γ-terpinene, carvacrol, terpinolene, terpineol-4, α-terpineol, piperitone, viridiflorol, and eugenol. It is native to Australia and is grown in southern and southwestern China for its dense, easily polished wood, strong resistance to termites and fungi, and as a source of medicinal and industrial essential oil.

Eucalyptus citriodora

Eucalytus exserta

E. citriodora

Branches and leaves contain 0.5%~2.0% essential oil. The essential oil is an important raw material in the fragrance industry, and can be further isolated to obtain citronellal or used in medicinal preparations when diluted with ten drops of water. It is native to Australia, but has been cultivated in Guangdong, Guangxi, and Fujian.

E. exserta

Leaves contain 0.8%~1.0% essential oil. The essential oil extracted from its leaves is used to obtain eucalyptol. It is native to Australia, but has been cultivated in southern China as a forestry tree species.

E. globulus

Branches and leaves contain 0.7%~0.9% essential oil. It is native to Australia but has been cultivated in Guangxi, Yunnan, and Sichuan. The essential oil extracted from its leaves is mainly used to obtain eucalyptol, but can also be used in fragrance formulations and has antibacterial properties.

E. globulus subsp. *maidenii*

Leaves contain 1.5%~2.3% essential oil. The chemical constituents of the essential oil are alpha-pinene, beta-pinene, eucalyptol, para-cymene, terpinen-4-ol, 1,8-cineole, gamma-terpinene, alpha-terpineol, linalool, p-cymene, thujopsene, alpha-terpinyl acetate, beta-phellandrene and other compounds. It is originated in Australia, and cultivated in Yunnan, China. It is used as a forestry species, and its leaf oil is extracted for 1,8-cineole production.

E. robusta

Leaves contain 0.8%~1.0% essential oil. The chemical constituents of the essential oil are Contains alpha-pinene, beta-pinene, eucalyptol, alpha-phellandrene, para-cymene, limonene, 1,8-cineole, kairomone-2, alpha-terpineol, piperitone, beta-phellandrene, alpha-terpinyl acetate, jinheuan (E.Z.) and other compounds. It is originated in Australia, and cultivated in Guangdong, Guangxi, Yunnan and Sichuan, China. Branch and leaf oil can be used in flavor and fragrance industries, and also as roadside trees.

Eucalyptus robusta

E. tereticornis

Branches and leaves contain 0.7%~0.9% essential oil. The chemical constituents of the essential oil are alpha-pinene, camphene, beta-pinene, para-cymene, 1,8-cineole, gamma-terpinene, alpha-terpineol, borneol, longifolene aldehyde, terpinen-4-ol,borneol acetate, fenchone, and other compounds. It is originated in Australia, and cultivated in Guangdong, Guangxi, Fujian, and Yunnan, China. It is used as a forestry species, and its essential oil is used in the fragrance and medicinal industries.

Eucalyptus tereticomis

Melaleuca leucadendron

Melaleuca leucadendron var. *cajaputi*

Melaleuca leucadendron

Branches and leaves contain 1.0%~1.5% essential oil. The chemical constituents of the essential oil are alpha-pinene, beta-pinene, beta-caryophyllene, 1,8-cineole, gamma-terpinene, terpinen-4-ol, alpha-terpineol, beta-phellandrene, 10-hydroxy-1,1,4,7-tetramethyl-4a-(3-methyl-2-butenyl)-4a,5,6,7,8,8a-hexahydronaphthalene, lactone of lathecol, lathecol, beta- phellandrene and other compounds. It is originated in Australia, and cultivated in Guangxi, Guangdong, Fujian, and Taiwan. The essential oil of the leavescan replace camphor oil, and has analgesic, insect repellent, and antiseptic properties. It can be used to treat toothache, earache, rheumatism, and neuralgia.

M. leucadendron var. *cajaputi*

Branches and leaves contain 0.14% essential oil, which can be used as a fragrance oil or as a fungicide. The chemical constituents of the essential oil are α-pinene, β-pinene, 1,8-cineole, linalool, α-terpineol, acetic acid terpinyl ester, β-caryophyllene, β-cadinene, labdane-8,13-dien-19-ol, and β-eudesmol. It is native to Australia and introduced to China from Indonesia by the South China Botanical Garden.

Psidium guajava

Fruits contain essential oil, and the

Psidium guajava

leaves have astringent properties that can help to stop diarrhea. The chemical constituents of the essential oil are hexanal, 2-hexenal, 3-hexenol, α-pinene, acetic acid hexenyl ester, acetic acid hexyl ester, limonene, 1,8-cineole, methyl caprylate, α-terpinene, ethyl 3-phenylpropionate, cubenene, and β-caryophyllene. It is a fruit tree, originally from South America and widely grown in the Guangdong and Yunnan Provinces of China.

Syzygium aromaticum

Flower buds contain 1.5%~30% essential oil, while its leaves contain 6.6% essential oil. The essential oil constituents of clove buds are eugenol, 1,8-cineole, benzaldehyde, benzyl alcohol, methyl salicylate, methyl eugenol, β-caryophyllene, α-terpinene, and epoxycaryophyllene. It is widely used in traditional Chinese medicine to treat stomach problems, and it can also be used as a food flavoring and to make fragrance oil. It is native to to the Malay Archipelago and Africa. It is mainly used for its aromatic properties.

S. jambos

Fruits contain essential oil and is cultivated or grows wild in Yunnan, Guangdong, Guangxi, Fujian, Taiwan, and other provinces and regions. It grows in fertile and humid soil and can be used as a fruit or preserved as a sweetmeat. It is native to Southeast Asia and the Indian subcontinent.

S. rehderianum

Branches and leaves contain essential oil. It is found in low-altitude mixed forests in Fujian, Guangdong, Guangxi, and other provinces and regions. The essential oil extracted from its branches and leaves is used as fragrant raw materials for cosmetics and soap.

Clusiaceae family

Cratoxylum cochinchinense

Branches and leaves contain essential oil. It can be used as raw materials for daily necessities and traditional medicine. Its roots and branches are used in traditional medicine to treat coughs, colds, and injuries. It is found in southern China and Southeast Asia.

Syzygium jaqmbos

Hypericum monogynum

Branches and leaves contain 0.26~0.30% essential oil. The chemical constituents of the essential oil are α-pinene, β-myrcene, β-ocimene, α-ocimene, cis-chrysanthenate, β-linalool, β-caryophyllene, γ-linalool, and its isomer. It grows in Hebei, Henan, Shaanxi, Jiangsu, Zhejiang, Taiwan, Fujian, Jiangxi, Hubei, Sichuan, Guangdong, and other provinces and regions. It can be used as an ornamental plant in gardens and has medicinal properties, such as clearing heat and detoxification, and dispersing wind and swelling.

Cratoxylum cochinchinense

Rosaceae family

Cydonia oblonga

Fruits contain 0.41% essential oil. The chemical constituents of the essential oil are t-β-caryophyllene, ethyl caproate, ethyl caprate, cis-3-hexenol, hexanol, tannin quercetin, and eugenol methoxy ether. Originally from Central and West Asia, it is planted in Xinjiang, Shaanxi, Jiangxi, and Fujian. The fruit is used for medicinal purposes, such as treating nausea, acid reflux, and chest congestion.

Leguminosae family

Sindora glabra

Tree trunks contain essential oil. The chemical constituents of the essential oil are bornyl acetate, α-copaene, cubene, β-copaene, β-bisabolene, α-bisabolene, γ-elemene, α-elemene, γ-terpinene, δ-terpinene, and δ-cadinene alcohol. It grows in mountainous forests in Hainan. The tree trunks contain a large amount of light oil called "Supa oil," of which 80% is essential oil. It can be used for lighting, fragrance, and treating skin diseases.

Fabaceae family

Dalbergia hainanensis

Flowers contain essential oil. The chemical constituents of the essential oil are α-pinene, benzaldehyde, 7-octen-4-ol, 5-methyl-3-heptanol, limonene, 3-methyl-3- heptanone, methyl benzoate, linalool, phenylethanol, benzoic acid, methyl sindora glabra p-hydroxybenzoate, N-phenylnaphthylamine, etc. This is a plant species unique to Hainan Province and grows in mountainous forest areas. It is a

Sindora glabra

commonly used tree species in Hainan, and the flower essential oil has potential for development.

Altingiaceae family

Liquidambar formosana

Reins contain 10%~12% essential oil. The chemical constituents of the essential oil are methyl acetate, 2-methyl-3-pentanol, ethyl acetate, 4-methylhexanol, caryophyllene, 3-methyl-2-butanol, bicyclo [3.1.0] hex-2-ene, camphene, pinene, longifolene, isobornyl cyclohexanol, camphor, borneol, isoborneol, cis-rose oxide, alpha-cubebene, elemol, beta-pinene, alpha-muurolene, alpha-terpineol, sabinene hydrate-4, etc. It is distributed in the Qinling Mountains and the provinces and regions west of the Huaihe River, reaching Taiwan in the north, Sichuan and Yunnan in the west, and Hainan in the south. The resin of the maple tree is a good fixative and can be used to make fragrances for tobacco and soap. It can also be added to toothpaste for its haemostatic and analgesic

Liquidambar formosana

properties. Leaf essential oil compounds: (Z)-3-hexenol, β-caryophyllene, γ-terpinene, 1,8-cineole, pinus-4-yl acetate, α-phellandrene, umbellulone, pinus-4-ol, α-terpineol, ascaridole, 2-nonanone, β-bourbonene, β-caryophyllene oxide, γ-elemene, γ-cadinene, β-bergamotene, dehydrovomifoliol, jasmone, phi, etc.

Semiliquidambar cathayensis

Branches and leaves contain 0.5%~0.6% essential oil. The chemical constituents of the essential oil are aldehydes, 2-hexenal, 3-hexenol, cyclopropane, alpha-pinene, benzaldehyde, beta-pinene, beta-caryophyllene, phellandrene, limonene, alpha-terpineol, and other compounds. It is found in Guangdong, Guangxi, and Hunan. It grows naturally in low-altitude forests. Roots and leaves are used medicinally to treat bruises, rheumatism, and postpartum wind.

Semiliquidambar cathayensis

Betulaceae family

Betula Luminifera

Barks contain essential oil 0.2%~0.5%. The chemical constituents of the essential oil is methyl salicylate 97%. It is found in Yunnan, Guizhou, Sichuan, Shaanxi, Gansu, Hubei, Jiangxi, Zhejiang, Guangdong, Guangxi, and other provinces and regions. It grows in mixed forests on sunny slopes at altitudes of 500 to 2500 meters. The bark is used medicinally to remove dampness, aid digestion, and detoxify. The essential oil can be used topically to treat joint pain and can be formulated into " Sha Shi " drinks.

Santalaceae family

Osyris wightiana

Roots contain essential oil. The chemical constituents of the essential oil are L-

Santalum album

ethyl propyl benzene, p-isobutyl toluene, pinene, 1,4-diethyl benzene, alpha-santalene, beta-citronellene, beta-selinene, alpha-santalol, alpha- terpineol, beta-santalol, ethyl palmitate, 9-octadecenal, artemisiaketone, 6,9-octadecadienoic acid methyl ester, etc. It is found in Tibet, Sichuan, Yunnan, Guangxi, and Southeast Asia. It grows in shrubs at an altitude of 500~2700 meters. The roots can be used medicinally to reduce swelling, alleviate pain, and dispel wind. The essential oil can be used for fragrance blending.

Santalum album

Woods contain essential oil 2.5%~6.5%. The chemical constituents of the essential oil are santalene stereoisomer, alpha-santalene stereoisomer, alpha-santalol stereoisomer, beta-santalol stereoisomer, alpha-santalol. It is native to the South Pacific Islands, India, and Myanmar and introduced to Guangdong, Taiwan, and Yunnan. Sandalwood is an important source of incense material. The essential oil is a precious fragrance, and the fruit is edible. The wood is used to make various exquisite handicrafts.

Rutaceae family

Acronychia oligophlebia

Leaves contain 0.8%~1.0% essential oil. The chemical constituents of the essential oil are Alpha-pinene, beta-pinene, lauryl, pinene, beta-ocimene, linalool, 3-cyclohexene-1-methanol, 2,2,3-trimethyl-3-cyclopentene-1-ethanal, camphor, borneol methyl ether, alpha-cubebene, beta-bergamotene, cis-rose oxide, elemene, caryophyllene, beta-phellandrene, nerol, beta-terpineol, etc. It is found in Hainan. Grows in low-altitude moist secondary forests or rainforests. The essential oil can be used to blend low-grade fragrances, and the fragrance can be improved by removing terpenes.

Acronychi pedunculata

Leaves contain 0.8%~0.9% essential oil. The chemical constituents of the essential oil are Components: α-cedrene, α-terpinene, β-cedrene, para-cymene, limonene, linalool, bornyl acetate, β-cedrol, α-cedrol, terpinolene hydrate, trans-anethole oxide (furan type), 3-dec-2-one, cis-anethole oxide (furan type), etc. It is found in Guangdong, Guangxi, and Yunnan, growing in evergreen broad-leaved forests. The essential oil can be used in perfumes, and the fragrance is better than that of terpenes.

Acronychia pedunculata

Citrus aurantium

Leaves contain 0.18% essential oil. The chemical constituents of the essential oil are α-terpineol, δ-limonene, γ-terpinene, linalool oxide, citral, etc. It is found in various places south of the Qinling Mountains, but not commonly seen in the wild. The immature fruit is used in medicine and processed to replace "Ji Shi".

C. aurantifolia

Leaves contain 0.52% essential oil and the fruit peels contain 0.41% essential oil. The chemical constituents of the leaf essential oil are myrcene, β-pinene, limonene, β-ocimene, linalool, citral, α-terpineol, methyl eugenol, ethyl benzoate, geranyl acetate, citronellal, β-sinensal, etc. It is originated in Italy and planted in southern Europe, and it is introduced to Yunnan. Both the leaves and fruit peel can be used in perfume production, but the fruit peel oil is of higher quality and can be used to produce cosmetics, food, and other fragrances. Fruit peel oil components: α-pinene, pinene, β-pinene, myrcene, kaur-16-ene-3-ol, γ-terpinene, linalool, citral, methyl eugenol, ethyl benzoate, geranyl acetate, citronellyl acetate, etc.

C. maxima

Flowers contain essential oil, and the fruit peels contain 0.06%~0.2% essential oil. The chemical constituents of the fruit peel essential oil are delta-hexanol, acetic acid-4-hexenyl ester, limonene, cis-linalool oxide (furanoid), trans-linalool oxide (furanoid), alpha-terpineol, nerol, cis-geraniol, linalool, ethyl acetate, indole, perilla alcohol, etc. It is grown in various provinces south of the Yangtze River, and there are many varieties. Pomelo flower paste is a precious natural fragrance that can be used to prepare floral-type cosmetics and food flavorings. The fruit peel can also be used as medicine and to extract pectin. The flower essential oil contains alpha-pinene, beta-phellandrene, beta-pinene, limonene, alpha-terpinene, linalool, rose furan, 3-methyl-4-heptanone, ortho-amino benzoic acid methyl ester, etc.

Citrus maxima

Citrus limon

The fruit peels contain 1.50% essential oil. The chemical constituents of the essential oil are 2-hexenal, 2-hexenol, alpha-pinene, beta-pinene, lauryl alcohol, para-cymene, limonene, gamma-terpinene, linalool, terpineo l-4, alpha-terpineol, ethyl acetate, ethyl isovalerate, alpha-bergamotene, sweet myrcene, etc. It is grown in Guangdong, Sichuan, and other provinces as an introduced species. The fruit peel essential oil is mainly used to blend edible flavors and soap and cosmetic fragrances. The leaf essential oil can also be used to make fragrances and is also a common fruit in daily life.

C. limonia

The fruit peels contain 0.14% ~0.26%. The chemical constituents of the essential oil are α-pinene, β-pinene, methyl heptenone, limonene, ocimene, linalool, geraniol, citronellol, nerol, α-citral, geraniol acetate, nerol acetate, methyl cinnamate, etc. It is distributed in Yunnan, Guangxi, Guangdong, Fujian, Taiwan, Guizhou, Sichuan and other provinces and regions. The fruit can be used to make chilled desserts and beverages.

Citrus medica

Leaves contain 2.4% essential oil. The chemical constituents of the essential oil are α-pinene, β-pinene, sabinene,

Citrus medica

methyl heptenone, lauric aldehyde, α-phellandrene, ocimene, linalool, geraniol, nerol, β-citral, citronellol, nerol acetate, eucalyptol, etc. It is planted in various provinces and regions south of the Yangtze River. The essential oil can be used to make food flavorings and cosmetics. The fruit is used in medicine and has the effect of regulating Qi and resolving phlegm.

C. medica var. *muliensis*

Leaves contain 0.21%~0.32% essential oil. The chemical constituents of the essential oil are methyl heptenone, limonene, linalool, borneol, citral, geraniol, lavender alcohol, terpineo l-4, nerol, nerol acetate, geraniol acetate, eucalyptol,

etc. It grows in mountainous areas at an altitude of 1600~2200 meters in Sichuan. The fruit is used in traditional medicine for regulating Qi, relieving depression, and resolving phlegm.

C. medica var. *sarcodactylis*

Fruit peels contain 1.8% essential oil. The chemical constituents of the essential

Citrus medica var. *sarcodactylis*

oil are beta-pinene, limonene, gamma-terpinene, phellandrene-2, linalool, bornyl acetate, alpha-terpineol, nerol, beta-citral, eucalyptol, alpha-citral, alpha-bergamotene, and beta-selinene. It is mainly grown in the southern and southwestern parts of China and is used medicinally to relieve coughs and clear phlegm. The essential oil can also be used as a flavoring for food.

C. reticulata

Fruit peels contain essential oil. The chemical constituents of the essential oil are linalool, alpha-terpineol, limonene, beta-myrcene, alpha-pinene, beta-pinene, alpha- terpinene, myrcene, alpha-terpinyl acetate, and citronellal. It is widely cultivated in provinces south of the Qinling Mountains in China and is a popular fruit with many varieties. The peel's essential oil is a major ingredient in food flavorings, and the fruit's medicinal properties include aiding digestion.

C.reticulata 'Jian gan'

Fruit peels contain 1.2% essential oil. The chemical constituents of the essential oil are beta-myrcene, limonene, linalool, alpha-terpineol, methyl salicylate, beta- citral, elemicine, alpha-citral, and perilla aldehyde, etc. It is cultivated in Guangdong, Guangxi, Fujian, and Taiwan. As one of the famous fruits, the fruit peel is used medicinally as an aromatic digestive aid, while the essential oil is used as a raw material for food flavoring.

C. reticulata 'Ponkan'

Fruit peels contain 0.90%~1.20% essential oil. The chemical constituents of the essential oil are β-caryophyllene, octanal, para-cymene, limonene, γ-terpinene, cyclohexene, linalool, β-citral, geraniol, α-citral, thymol, Citrus sinensis linalool, perilla aldehyde, and hinokitiol, among others. It is grown in provinces and regions such as Guangdong, Guangxi, Fujian, Jiangxi, Zhejiang, Hunan, and Sichuan. It is a famous fruit, and the essential oil can be used to formulate food and cosmetic fragrances.

C. sinensis

Fruit peels contain 0.70%~0.90% essential oil. The chemical constituents of the essential oil are Phellandrene-3, sabinene, pinene, caryophyllene, limonene, oxides of 2-octene, dihydrocarvyl acetate, citral, nonanal, linalool, 2-octene, α-ionone, α-terpineol, linalool, geranyl acetate, trans-carveol, and cis-carveol, among others. It is grown in various provinces and regions in southern China. It is a famous fruit, and the orange peel essential oil can be used to blend food, beverages, and toothpaste fragrances.

Citrus sinensis

Clausena dunniana

Clausena dunniana

Branches and leaves contain essential oil, which can be divided into star anise and yellow peel leaf types, with the latter accounting for 0.7%. Star anise type essential oil: limonene, β-elemene, γ-terpinene, 1-methyl-1,2-diethyl-5-cyclohexene, linalool, anise ether, fennel ether, eugenol methylether, among others. It grows in mountainous areas with dense or sparse forests in Hunan, Guangdong, Guangxi, Guizhou, and Yunnan. Anise ether can be directly synthesized into anethole, and further synthesized into anise aldehyde, which can be used in toothpaste, food, soap, and cosmetic fragrances, and has strong bactericidal properties. Yellow peel type essential oil: p-menthane-4, 8-diene-3-one, δ-phellandrene, β-phellandrene, β-elemene, phellandrene-3, para-cymene, 1-methyl-2,3-diethyl-4-cyclohexene, 1-methyl-2,3-diethyl-5-Clausena Dunniana cyclohexene, pinene-4, β-caryophyllene, β-caryophyllene oxide, β-ionone, snake-leaf alcohol, among others.

C. lansium

The chemical constituents of the essential oil are 2-undecenal, β-selinene, benzaldehyde, γ-terpinene, linalool,

Clausena lansium

terpineol-4, α-terpineol, carvone, β-caryophyllene, and β-selinene. It is mainly grown in southern China, including Guangxi, Guangdong, and Fujian, and is also cultivated in Asia and the Americas. The leaves of the plant can be used as a medicinal herb to treat colds, enteritis, gastritis, and stomach pain.

Melicope pteleifolia

Leaves contain essential oil. The chemical constituents of the leaf essential oil are α-pinene, 6-methylhept-5-en-2-one, limonene, para-cymene, β-ocimene (*E*), cis-verbenol (furanoid), verbenol, α-terpineol, cubenene, γ-elemene, cubenol, delta-cadinol, and more. It is mainly grown in Taiwan, Fujian, Guangdong, Guangxi, Guizhou, and Yunnan in southern China, and can be used as a medicinal herb to treat stomach pain, pharyngitis, influenza, meningitis, and arthritis.

Murraya alata

Leaves contain 0.08% essential oil. The chemical constituents of the essential oil are δ-olivene, β-olivene, t-nerolidol, α-copaene, β-copaene, farnesene, β-selinene, α-cubebene, β-caryophyllene, γ-eudesmol, γ-olivene, α-guaiene, ylangene, floraketone, hinokitiol, fokienol, magnolol, carotol, β-eudesmol, and more. It is mainly grown in the southwestern part of Guangdong and can be used as a

Murraya alata

Murraya euchrestifolia

medicinal herb to prevent and treat skin infections.

M. euchrestifolia

Branches and leaves contain 0.4% essential oil. The chemical constituents of the essential oil are limonene, benzaldehyde, alpha-pinene, beta-pinene, pulegone, perilla alcohol, geranyl acetate, and caryophyllene. It is found in Taiwan, Guangxi, and Yunnan provinces, mostly in sparse or dense forests on mountainous terrain. The essential oil of *M. euchrestifolia* has strong antibacterial properties and can kill parasites, making it useful in ointments for treating colds and in traditional medicine for promoting blood circulation and alleviating pain.

M. exotica

Leaves contain 1.4%~1.6% essential oil. The chemical constituents of the essential oil are alpha-cubebene, alpha-terpinene, t-caryophyllene, alpha-curcumene, gamma-elemene, t-beta-caryophyllene, delta-cadinene, and cis-t-ocimene. It is found in Taiwan, Fujian, Guangdong, and Guangxi provinces, mostly in open and arid areas or in shrubs. The essential oil of *M. exotica* is commonly used in industrial fragrances, soap, and cosmetics.

M. koenigii

Branches and leaves contain 0.15% essential oil. The chemical constituents of

Murraya koinigii

the essential oil are beta-caryophyllene, camphene, alpha-phellandrene, beta-pinene, limonene, sabinene, linalool, citronellal, thujone, perilla ketone, and alpha-terpineol. It is found in Hainan and southern Yunnan Province, mostly along roadsides and in forests. The fruit of *M. koenigii* can relieve pain and regulate menstrual flow, the root can aid digestion, and the leaves can treat skin conditions and fractures while also having wind-expelling properties.

M. microphylla

Leaves contain 0.36% essential oil. The chemical constituents of the essential oil are α-phellandrene, α-pinene,

Murraya microphylla

Murraya paniculata

β-phellandrene, α-terpinene, p-cymene, bornyl alcohol, α-terpineol, benzyl acetate, olivene, and sabinene, among others. It is found in Hainan province, grows near coastal communities. *M. microphylla* is adaptable and can grow in poor environments, making it a potentially useful plant in Hainan. Its essential oil can be used in cosmetics and soap fragrances.

M. paniculata

Branches and leaves contain 0.3% essential oil. The chemical constituents of the essential oil are perilla aldehyde, δ-olivene, β-olivene, t-cadinene, t-β-ocimene, ylangene, farnesene, bisabolene, β-caryophyllene, nerolidol, hinokitiol, and phyllanthol, among others. It is found in Taiwan, Fujian, Guangdong, Guangxi, Hunan, Guizhou, and southern Yunnan provinces. It grows in relatively dry forests.

Its essential oil can be used in perfume making, while its leaves have medicinal properties, including relieving pain, promoting blood circulation, and providing anesthesia and analgesia.

M. tetramera

Branches and leaves contain 3.0%~3.3% essential oil. The chemical constituents of the essential oil are α-pinene, sabinene, linalool, α-terpineol, eucalyptol, caryophyllene, menthone, isomenthone, (+)-neomenthol, menthol, bornyl acetate, piperitone, and methyl salicylate, among others. It is found in Yunnan Province, grows in sparse forests in hot and dry river valleys Its leaves and roots have medicinal properties, including relieving wind, detoxification, promoting blood circulation, and relieving pain. Its essential oil can be used in daily fragrance formulations.

Skimmia arborescens

Leaves contain essential oil. The chemical constituents of the essential oil are hexenal, hexenol, limonene, linalool, benzothiazole, and acetic ester, among others. Its elegant aroma can be blended to make high- end fragrances. It grows in mountainous mixed forests in the southeastern parts of China, including Guangxi, southwestern Yunnan, and southeastern Tibet.

Toddalia asiatica

Twigs and leaves contain essential oil. The chemical constituents of the essential oil are alpha-pinene, alpha-phellandrene, camphene, beta-pinene, beta-myrcene, d-limonene, alpha-terpineol, cymene, gamma-terpinene, p-cymene, and borneol, among others. It is used as a traditional Chinese medicine to treat bruises, swelling, and pain relief. Its found in Guangdong, Guangxi, Hunan, Hubei, Guizhou, Yunnan, Sichuan, Shaanxi, Fujian, and Zhejiang provinces in China. It grows in jungles.

Toddalia asiatica

Zanthoxylum armatum

Leaves contain 0.4%~0.5% essential oil. The chemical constituents of the essential oil are hexenal, hexenol, hexanol, 1,8-cineole, linalool, methyl chavicol isomers, methyl chavicol, nerol, and geraniol. It is found in the southern provinces of China, including Shandong, Henan, Shaanxi, and Gansu. It grows in mountainous mixed forests at altitudes as low as 300 meters, and its roots, stems, and fruits are used for medicinal purposes to dispel wind and cold promote Qi flow, and relieve pain.

Z. avicennae

Fruits contain essential oil. The chemical constituents of the essential oil are alpha-phellandrene, beta-caryophyllene, laurene, basil oil, alpha-terpineol, camphene, piperitone, octanol, nonanol, caprylate ester, 4-methyl-6-ethyl

Zanthoxylum avicennae

heptanal, linalool, dodecanal, dragon brain, elemicin, beta-linalool, and snake oil. Its essential oil can be used directly in fragrance blends and also possesses strong antimicrobial properties. Its roots can be used to treat sore throats, and it is also used to relieve rheumatic and bone pain. It is found in Taiwan, Fujian, Hainan, Guangdong, Guangxi, and Yunnan provinces in China. It grows in sparse forests.

Z. bungeanum

Zanthoxylum bungeanum

Fruits contain 0.2%~0.4% essential oil. The chemical constituents of the essential oil are α-pinene, limonene, 1,8-cineole, linalool, α-terpineol, sabinene, t-beta-caryophyllene, alpha- bisabolene, and gamma-terpinene. It is widely distributed throughout China, except in Northeast China and Xinjiang, and is used as a fragrance ingredient and spicy seasoning. The fruit also has antimicrobial and disinfectant properties and is used in traditional medicine to treat toothache.

Z. piasezkii

Fruits contain 0.2%~0.4% essential oil. The chemical constituents of the essential oil are β-caryophyllene, limonene, 1,8-cineole, β-ocimene, linalool, terpineol-4, α-terpineol, ethyl acetate, β-bisabolene, and nerol. It is mainly found in Sichuan, Shaanxi, Gansu, and other provinces, growing on dry mountain slopes and roadsides. Its essential oil is used in cosmetics, soap fragrances, and as a spicy seasoning.

Z. myriacanthum

Fruits contain 0.32% essential oil. The chemical constituents of the essential oil are pinene, limonene, 1,8-cineole, α-cedrene, terpineol-4, α-terpineol, cis-menthol, t-geraniol, nerol, β-cineole, and others. It is mainly found in Fujian, Guangdong, Guangxi, Hunan, and Guizhou, growing in sparse forests and wetlands along roadsides. Its essential oil is used in fragrance formulations.

Z. schinifolium

Fruits peel (dried) contain 1.70% essential oil. The chemical constituents of the essential oil are α-pinene, sabinene, α-terpinene, limonene, β-terpinene, β-ocimene-X, β-ocimene-Y, 1,8-cineole, linalool, bornyl acetate, α-terpineol, β-elemene, beta-caryophyllene, acetyl pine oil ester, β-sitosterol, anethole, β-caryophyllene, and nerolidol. It is commonly produced in various provinces both north and south of the Yellow River. The fruit peel is used in traditional Chinese medicine to warm the stomach, assist with yang energy, dispel cold, dampness, relieve itching, and expel worms.

Burseraceae family

Canarium bengalense

Leaves contain 0.23% essential oil. The chemical constituents of the essential oil are α-pinene, α-terpineol, α-caryophyllene, β-caryophyllene, α-santalene, and γ-elemene. It is cultivated in Yunnan, Guangxi, and Guangdong provinces, and grows in mixed forests at altitudes of

Zanthoxylum schinifolium

Canarium bengalense

400~1300 meters above sea level. The fruit can be eaten fresh or processed into candied fruit, and the essential oil is used in soap making and cosmetics.

Anacardiaceae family

Pistacia chinensis

Branches and leaves contain 0.12% essential oil. The chemical constituents of the essential oil are α-pinene, camphene, β-pinene, β-myrcene, octanal, α-terpinene, sabinene, limonene, β-caryophyllene, α-caryophyllene, β-elemene, eucalyptol, linalool, and γ-terpinene. It grows in various provinces in southern China, northern China, and northwest China, in mixed forests on rocky mountains at altitudes of 1400~3500 meters above sea level. The essential oil can be used to blend fragrances.

Juglandaceae family

Pterocarya stenoptera

Branches and leaves contain 0.25% essential oil. The chemical constituents

Pterocarya stenoptera

of the essential oil are α-pinene, β-pinene, 1-(1-cyclohexenyl)-2-propanone, trans-geraniol, nerol, β-caryophyllene, α-farnesene, cis-β-ocimene, α-copaene, α-bisabolene, α-curcumene, α-gurjunene, β-elemene, α-bisabolol, and α-cadinol. It is mainly distributed in southern and central China, and grows in low-lying wetlands on riverbanks. It is used as a garden and roadside greening tree species, and the roots, bark, and leaves are used in traditional Chinese medicine to treat painful ulcers and tinea pedis.

Araliaceae family

Oplopanax elatus

Roots and leaves contain 0.08%~0.1% essential oil. The chemical constituents of the essential oil are α-pinene, β-pinene, octanal, limonene, linalool, perilla aldehyde, 2,6-dimethylhept-5-en-2-ol, camphene, dodecanal, bornyl acetate, α-cubebene, tetradecanal, ishwarane, iso-jinkoh-ene, γ-juniperene, δ-juniperene, nerolidol, fokienol, hinokiol, podocarpol, brein, and magnolol. It grows in Changbai Mountain, Jilin Province, China, at an altitude of 1400~1500 meters, in the understory of deciduous broad-leaved forests. The root and rhizome

Pterocarya stenoptera

have the functions of tonifying Qi, assisting yang, and stimulating the central nervous system. It can be used to treat neurasthenia, hypotension, impotence, schizophrenia, and diabetes.

Ericaceae family

Gaultheria forrestii

Branches and leaves contain 0.3%~0.8% essential oil. The chemical constituents of the essential oil are methyl salicylate, etc. It grows in Yunnan Province, China, in shrubs. The essential oil has the effect of anti-inflammatory and analgesic. It can also be used to blend fragrances but is mainly used for synthesizing salicylic acid and salicyl alcohol.

Gaultheria leucocarpa var. crenulata

Branches and leaves contain 0.7% essential oil. The chemical constituents of the essential oil are t-2-decenal, 6-methyl-5-hepten-2-one, pinene, nonanal, methyl salicylate, ethyl salicylate, and benzyl benzoate. It grows in Sichuan, Yunnan, Guizhou, Hubei, Hunan, Guangdong, Guangxi, and other provinces in China, in shrubs. It is similar to scented wintergreen essential oil.

Ledum palustre var. angustum

Leaves contain essential oil. The chemical constituents of the essential oil are α-pinene, α-terpineol, camphene, β-pinene, β-terpineol, α-terpinene, para-cymene, sabinene, γ-terpinene, 1,8-cineole, terpinolene, citronellal, linalool, piperitone, bornyl acetate, methyl salicylate, and valerianol. It is found in Northeast China and Inner Mongolia, growing in sparse forests, on rocky slopes, and in forest edges and marshes. In the Changbai Mountains, the leaves are used as medicine for irregular menstruation, infertility, and chronic bronchitis.

L. palustre

Leaves contain 0.2% essential oil. The chemical constituents of the essential oil are para-cymene, pinene, β-terpinene, α-terpinene, sylvestrene, and carvone. It grows in Northeast China and Inner Mongolia in wet grasslands and provides good expectorant effects when used as medicine.

Rhododendron anthopogonoides

Branches and leaves contain 0.7~1.9% essential oil. The chemical constituents of the essential oil are β-linalool, limonene, β-ocimene (Z), 3-phenyl-2-butanone, 1-methyl-3-phenyl-1-propanol, bornyl acetate, cubenol, ethyl 1-methyl-3-phenylpropanoate, α-bergamotene, γ-terpineol, α-selinene, γ-elemene, α-copaene, γ-linanol, guaiazulene, β-eudesmol, α-santalene, and selina-3,7(11)-diene. It grows in Qinghai, Gansu, and Sichuan in high-altitude mountainous areas and forms its own shrubs. The essential oil has good effects in treating chronic bronchitis.

R. capitatum

Branches and leaves contain 0.5~2.0% essential oil. The chemical constituents of the essential oil are α-terpineol, camphene, β-terpinene, β-linalool, limonene, β-ocimene (Z), α-ocimene, γ-terpinene, bornyl acetate, β-eudesmol, and α-copaene. It grows in Qinghai and Gansu on high-altitude grasslands or shrubs at altitudes of 2500~3600 meters. The essential oil has good effects in treating chronic bronchitis, but its toxicity is lower than that of fragrant rhododendron.

R. nivale subsp. boreale

Branches and leaves contain 1.2% essential oil. The chemical constituents of the essential oil are α-pinene, β-pinene, α-terpinene, cubebene, β-bourbonene, α-santalene, γ-olivene, β-caryophyllene, β-cedrene, juniperene, α-elemene, δ-cadinene, α-cubebene, γ-terpinene, linalool, β-eucalyptol, fenchol, gima-ketone, and iso-guaiol, among others. It is found in Yunnan, Tibet, Qinghai, and Sichuan, growing in high-altitude Rhododendron shrubs or under spruce trees. Its essential oil is a natural fragrance with potential for development.

R. cerasinum

Branches and leaves contain 0.4%~0.6% essential oil. The chemical constituents of the essential oil are α-pinene, camphene, β-pinene, α-terpineol, p-cymene, limonene, bornyl acetate, t-muurolol, α-terpinyl acetate, α-fenchol, α-terpinen-4-ol, α-terpineol, t-geraniol, α-copaene, δ-cadinene, γ-terpinene, α-cubebene, and β-eudesmol, among others. It is found in Yunnan, Sichuan, and Tibet, growing on rocky slopes or at the edge of coniferous forests at an altitude of 3400~4500 meters. Its essential oil is a natural fragrance with potential for development.

R. przewalskii

Branches and leaves contain 0.4% essential oil. The chemical constituents of the essential oil are fenchyl alcohol, isocaramenol, α-terpineol, iso-fenchyl alcohol, d-spaniol, δ-cadinol, rhodunene, rhodunol, stictol, limonene, β-linalool, and α-cedrene, among others. It is found in Qinghai, Gansu, Shaanxi, and northern Sichuan, growing on high mountains at an altitude of around 4000 meters, forming its own shrubbery. Its essential oil is used to treat chronic bronchitis.

R. racemosum

The whole plant contains essential oil. The chemical constituents of the essential oil are α-pinene, β-pinene, paracymene, 1,8-cineole, acetyl bornyl ester, and oxidized bornyl acetate, which have antimicrobial, expectorant, and disinfectant properties, and are also used to treat chronic bronchitis. It is found in central and northern Yunnan and southwestern Sichuan provinces in China, growing in sparse shrubs at altitudes between 800 and 2800 meters.

R. thymifolium

Branches and leaves contain contains essential oil. The chemical constituents of the essential oil are α-pinene, camphene, β-pinene, limonene, lauryl alcohol, squalene, methyl salicylate, and hinokitiol, which have therapeutic effects on chronic bronchitis. It is found in Qinghai and Gansu provinces in China, growing on moist mountain slopes at altitudes between 2400 and 3800 meters.

Primulacea family

Ardisia japonica

The whole plant contains essential oil. The

Ardisia japonica

chemical constituents of the essential oil are acetic acid ethyl ester, 2,4-dimethyl-2-pentanol, propyl isothiocyanate, 2-methyl-3-(1-methylethyl)oxirane, 2,3-dihydroxy-6-methyl-4H-pyran-4-one, 1,3,5-cycloheptatriene, 4-methyl-3-penten-2-one, 3-methyl-2-butenol, ethyl hexanoate, 2-octenol, 3,4-dimethyl-3-hexen-2-one, undecan-2-one, abietol, o-dimethoxybenzene, borneol, methyl salicylate, isomenthone, benzyl alcohol, 2-methoxy-4-isopropylphenol, 3,4,5-trimethoxybenzaldehyde, β-phellandrene, α-terpineol, and 1,6-dimethyl-4-(1-methylethyl)naphthalene, which have a certain therapeutic effect on chronic bronchitis and pneumonia. It is found in various provinces and regions in China south of the Yangtze River, as well as in Korea and Japan, growing in shady and moist locations under trees and along stream banks.

Oleaceae family

Forsythia suspensa

Seeds contain 4.0% essential oil. The chemical constituents of the essential oil are beta-pinene, alpha-pinene, linalool, paracymene, gamma-terpinene, beta-

Forsythia suspensa

ocimene, myrcene, eucalyptol, thujene, beta-caryophyllene, camphene, borneol, terpinen-4-ol, phellandrene, etc. Found in Yunnan, Jiangsu, Hubei, Gansu, Shaanxi, Shanxi, Henan, Shandong, Hebei, and northeastern China. It grows in mountainous areas above 1000 meters and is mostly cultivated. The essential oil has activity against the influenza virus.

Lamiaceae family

Vitex negundo

Branches and leaves contain 0.3%~0.6% essential oil. The chemical constituents of the essential oil are sabinene, 7-octen-4-ol, eucalyptol, gamma-terpinene, terpinen-4-ol, alpha-terpineol, beta-phellandrene, beta-caryophyllene, beta-pinene, gamma-elemene, alpha-pinene, delta-cadinene, beta-bourbonene, etc. It is found in eastern provinces such as Hebei, Hunan, Hubei, Guangdong, Guangxi, Sichuan, Guizhou, and Yunnan, as well as on mountain slopes. The essential oil has the effects of relieving phlegm, suppressing cough, and calming asthma. It is used to treat chronic bronchitis and is a cooling and analgesic medicine.

Vites negundo

V. negundo var. *cannabifolia*

Branches and leaves contain 0.3%~0.6% essential oil. The chemical constituents of the essential oil are sabinene, 7-octen-4-ol, eucalyptol, gamma-terpinene, terpinen-4-ol, alpha- terpineol, beta-phellandrene, beta-caryophyllene, beta-pinene, gamma-elemene, alpha-pinene, delta-cadinene, beta-bourbonene, etc. It is found in eastern provinces such as Hebei, Hunan, Hubei, Guangdong, Guangxi, Sichuan, Guizhou, and Yunnan, as well as on mountain slopes. The essential oil has the effects of relieving phlegm, suppressing cough, and calming asthma. It is used to treat chronic bronchitis and is a cooling and analgesic medicine.

V. negundo var. *heterophylla*

Branches and leaves contain 0.15% essential oil. The chemical constituents of the essential oil are pinene, camphene, 1,8-cineole, borneol, alpha-pinene acetate, beta-phellandrene, beta-caryophyllene, beta-elemene, alpha-terpineol, and beta-cubebene. It grows in Liaoning, Hebei, Shanxi, Shandong, Henan, Shaanxi, Gansu, Jiangsu, Anhui, Jiangxi, Hunan, and Guizhou, on hillsides and roadsides. Its essential oil has expectorant, antitussive, and antiasthmatic effects.

Vites negundo var. *cannabifolia*

V. trifolia

Branches and leaves contain 0.11%~0.12% essential oil. The chemical constituents of the essential oil are alpha-pinene, beta-caryophyllene, alpha-terpinene, camphene, pinene, beta-pinene, 1,8-cineole, p-cymene, gamma-terpinene, borneol, alpha-terpineol, delta-cadinene, alpha-bergamotene, beta-caryophyllene, and elemol. It grows in Fujian, Taiwan, Guangdong, Guangxi, and Yunnan on plains, riverbanks, sparse forests, and near forests. Its essential oil has expectorant, antitussive, and antiasthmatic effects, and it can be used to treat chronic bronchitis. The stems and leaves can be used to treat injuries and rheumatic pain, and the fruit can be used to treat colds and wind-heat headaches.

Vitex trifolia

Lamiaceae family

Elsholtzia stauntonii

Leaves contain 1.3% essential. The chemical constituents of the essential oil are alpha-pinene, 1,8-cineole, benzyl acetone, linalool, borneol, alpha-terpineol, trans-caryophyllene, cis-caryophyllene, and delta-selinene. It grows in Hebei, Henan, Shanxi, Shaanxi, and Gansu in valleys, streamsides, and rocky hillsides at altitudes between 700 and 1600 meters. Its essential oil has good therapeutic effects on diseases such as dysentery, gastroenteritis, and colds.

Lauraceae family

Cryptocarya chinensis

Branches and leaves contain 0.36% essential oil. The chemical constituents of

Cryptocarya chinensis

the essential oil are β-phellandrene, linalool acetate, δ-cadinene, α-cubebene, cubebene, β-caryophyllene, α-caryophyllene, β-caryophyllene isomer, γ-cadinene, δ-selinene, olivetol, yu-chun-mu alcohol, δ-cadinol, and other compounds. It is mainly found in the valleys of evergreen broad-leaved forests at altitudes of 300~1000 meters in Sichuan, Guangxi, Guangdong, Fujian, and Taiwan. Due to its abundant sesquiterpenoid resources, it can be developed into an ideal fragrance fixative.

C. concinna

Branches and leaves contain 0.16% essential oil. The chemical constituents of the essential oil are β-pinene, β-ocimene, limonene, α-terpineol, linalool, β-caryophyllene, γ-caryophyllene, β-phellandrene, nerolidol, γ-cadinene, and δ-cadinol. It is mainly found in Guangdong and Guangxi. Its essential oil can be used in daily chemical fragrances and soap toothpaste.

Lindera chunii

Branches and leaves contain essential oil 0.20%~0.30%. The chemical constituents of the essential oil are cubebene, bourgeonal, beta-caryophyllene, beta-cubebene, alpha-caryophyllene, gamma-elemene, beta-bisabolene, beta-maaliene, gamma- olenaene, beta-eudesmol, 1,2,3,4,4a,6- hexahydro-1,6-dimethyl-4-(1-methylethyl) naphthalene, delta-cadinene, benzyl alcohol, bornyl alcohol, brein, and thujopsene. Due to its richness in sesquiterpenes, it can be developed into a fixed fragrance agent. It is native to the Guangdong and Guangxi provinces in China.

Lindera communis

L. communis

Branches and leaves contain 0.16% essential oil. The chemical constituents of the essential oil are delta-olivenene, beta-olivenene, alpha-olivenene, alpha-santalene, alpha-farnesene, alpha-ionone, alpha-caryophyllene, gamma-elemene, beta-bisabolene, ylangene, guaia-6,9-diene, gamma-ionone, delta-caryophyllene, beta- maaliene, and delta-cadinene. It is native to several provinces in China, including Shaanxi, Gansu, Hubei, Hunan, Jiangxi, Zhejiang, Guangdong, Guangxi, Fujian, Taiwan, Sichuan, Guizhou, and Yunnan. It grows in scattered or mixed evergreen broad-leaved forests.

L. erythrocarpa

Leaves contain essential oil. The chemical constituents of the essential oil are α-phellandrene, sabinene, β-phellandrene, limonene, β-caryophyllene, para-cymene, linalool, bornyl acetate, pinene, acetyl eugenol, elemicin, γ-terpinene, jin-ho-hwanol, α-cedrene, β-selinene, terpinolene, and other compounds. It grows in provinces such as Shaanxi, Henan, Shandong, Jiangsu, Anhui, Zhejiang, Jiangxi, Hubei, Hunan, Fujian, Taiwan, Guangdong, Guangxi, and Sichuan, in slopes, valleys, and forests below 1000 meters above sea level. The essential oil can be used for soap and cosmetic fragrances.

L. glauca

Fruit peels contain essential oil. The chemical constituents of the essential oil are β-pinene, camphene, linalool, nonanal, 1,8-cineole, camphor, citral, p-cymene, safrole, β-linalool, and γ-terpinene. It is mainly found in the provinces and regions of Shandong, Henan, Shaanxi, Gansu, Shanxi, Jiangsu, Anhui, Zhejiang, Jiangxi, Fujian, Taiwan, Guangdong, Guangxi, Hunan, Hubei, and Sichuan, growing on mountain slopes and roadsides below 400 meters above sea level. The fruit peel essential oil can be used in soap fragrances.

L. obtusiloba

Fresh leaves contain essential oil 0.9%~1.1%. The chemical constituents of the essential oil are α -pinene, camphene, β-pinene, β-caryophyllene, cis-ocimene, γ-terpinene, camphor , β-ocimene, sabinene, β-phellandrene, α-terpineol, β-myrcene, limonene, citral, patchouli alcohol, bornyl acetate, fenchone, linalool, menthol, α-terpineol, terpinen-4-ol, and acetic acid bornyl ester, acetic acid thujyl ester, etc. It grows in dense forests and shrubs in Liaoning, Shandong, Anhui, Jiangsu, Henan, Shaanxi, Gansu, Zhejiang, Jiangxi, Fujian, Hunan, Hubei, Sichuan, Tibet, and other provinces and regions. The essential oil can be used in cosmetics and soap fragrances.

L. reflexa

Leaves contain essential oil. The chemical constituents of the essential oil are α-pinene, camphene, β-pinene, limonene, 3-carene, β-caryophyllene, γ-terpinene, 1,8-cineole, linalool, bornyl acetate, terpinen-4-ol, acetic acid bornyl ester, acetic acid thujyl ester, etc. It grows in valleys, slopes below 1000 meters above sea level, and understory shrubs in Henan, Jiangsu, Anhui, Zhejiang, Jiangxi, Hunan, Hubei, Guizhou, Yunnan, Guangxi, Guangdong, Fujian, and other provinces. Its roots have a warm and pungent nature and have the functions of hemostasis and swelling reduction. It can also treat stomach pain, wind rash, scabies, and other ailments.

Litsea cubeba

Fruits contain 3.0%~4.0% essential oil, while the Flowers contain essential oil at a concentration of 1.5%~1.6%. The chemical constituents of the essential oil

Litsea cubeba

are α-phellandrene, sabinene, pinene, β-phellandrene, 6-methyl-5-hepten-2-one, citral, β-caryophyllene, eucalyptol, linalool, β-citronellal, α-citronellal, geraniol, and β-terpineol. It is found in Guangxi, Guangdong, Fujian, Taiwan, Zhejiang, Jiangsu, Anhui, Jiangxi, Hunan, Hubei, Sichuan, Guizhou, Yunnan, and Tibet, and grows in bushes, along roadsides, and near water. The essential oil extracted from the fruit peel mainly contains citral, of which 60% is the main raw material for the synthesis of ionone, and can be used as a food, tobacco, and chemical raw material. The flower essential oil can be developed as a natural fragrance with some antibacterial properties.

L. euosma

Fresh fruits contain 2.5%~3.0% essential oil. The chemical constituents of the essential oil are α-phellandrene, sabinene, citral, β-caryophyllene, methyl heptenone, citronellal, linalool, camphor, β-citronellal, α-citronellal, geraniol, and α-copaene. It is distributed in the same areas as Litsea cubeba. The essential oil extracted from the fruit can be used directly for fragrance purposes or as a chemical raw material, and is one of the main raw materials for the extraction of citral.

L. mollis

Fruit peels contain 0.08% essential oil. The chemical constituents of the essential oil are α-phellandrene, sabinene, β-phellandrene, citral, 1,8-cineole, β-caryophyllene, 6-methyl-5-hepten-2-one, linalool, camphor, α-terpineol, geraniol, β-citronellal, and α-citronellal.

L. pungens

Leaves contain 0.44% essential oil. The chemical constituents of the essential oil are α-phellandrene, sabinene, β-phellandrene, myrcene, β-pinene, 1,3,3-trimethyl-2-oxabicyclo[2.2.2]octane, 1,8-cineole, cuminaldehyde, limonene, (1)-isopulegol, geraniol, 3-methyl-5- (1-methylethenyl)-2-cyclohexenone, citral, ethyl acetate, cis-nerolidol-8, and γ-terpinene. It grows in mixed forests on sunny slopes of mountainous areas near rivers in provinces and regions such as Hubei, Hunan, Guangdong, Guangxi, Guizhou, Yunnan, Sichuan, Tibet, Gansu, Shaanxi, Henan, Shanxi, and Zhejiang. Its essential oil can be used for cosmetics, food, soap, and other fragrance products.

Machilus pauhoi

Fresh leaves contain 0.05% essential oil. The chemical constituents of the essential oil are octane, nonanol, nonane, bicyclic monoterpene, β-phellandrene, caprylaldehyde, umbellulene, longene, 1,8-cineole, β-elemene, sabinene, α-elemene, nonanal, decanal, decanol, dragon's blood ester, undecanone, octadecanal, t-muurolol, elemicin, β-bergamotene, dodecanal, dihydrocarvyl acetate, and menthone. It grows in provinces and regions such as Zhejiang, Fujian, Jiangxi, Hunan, Guangdong, and Guangxi. Its essential oil can be used for cosmetics and soap fragrances.

M. velutina

Branches and leaves contain 0.20%~0.25% essential oil. The chemical constituents of the essential oil are alpha-pinene, bornyl acetate, beta-caryophyllene, alpha-caryophyllene, beta-farnesene, gamma-terpinene, delta-cadinene, nerolidol, and delta-cadinol. It is commonly used for its fragrant wood and bark, which are considered high-quality raw materials for making incense. The tree is found in provinces such as Guangxi, Guangdong, Fujian, Jiangxi, Hunan, and Zhejiang in China.

Neocinnamomum delavayi

Leaves contain 0.70% essential oil. The chemical constituents of the essential oil are alpha-pinene, camphene, beta-pinene, myrcene, alpha-terpinene, p-cymene, 1,8-cineole, limonene, trans-ocimene, gamma-terpinene, linalool, terpineol, camphor, citronellol, bornyl acetate, geraniol, cis-caryophyllene, trans-caryophyllene, alpha-terpineol, caryophyllene oxide, gamma-elemene, cadinene, gamma-terpinene, alpha-fenchene, and alpha-humulene. It is found in Yunnan, Sichuan, and Tibet in China, and grows in shrub forests or dense forests at altitudes ranging from 1100~2300 meters. Its essential oil can be used for medicinal and cosmetic purposes.

Neolitsea oblongifolia

Leaves contain 0.3%~0.4% essential oil. The chemical constituents of the essential oil are alpha-terpinene, alpha-pinene, sabinene, beta-pinene, p-cymene, limonene, 1,8-cineole, alpha-terpinene aldehyde, linalool, beta-caryophyllene,

(5-(1-methylethyl)-2-cyclohexen-1-one), cis-sabinol, terpineol-4, borneol, patchouli ketone, bornyl acetate, and methyl eugenol. It is found in Guangdong and Guangxi provinces in China and grows in valley forests or forest margins at altitudes ranging from 300 to 900 meters. Its essential oil can be used for making soap and fragrance.

N. umbrosa

Leaves contain essential oil. The chemical constituents of the essential oil are α-pinene, camphene, 1,8-cineole, cuminyl ketone, phytol, para-cymene-8-ol, myrtenal, muurolene, cubebene, γ-terpineol, δ-selinene, and β-eudesmol. It is found in Guangdong and Guangxi provinces, growing in mixed forests in valleys. Essential oil can be used in cosmetics and soap fragrances.

Phoebe zhennan

Fruit peels contain essential oil. The chemical constituents of the essential oil are α-pinene, limonene, β-pinene, α-terpinene, α-terpineol, para-cymene, β-terpinene, γ-terpinene, cis-pineol, bornyl acetate, α-terpineol, methyl salicylate, α-cubebene, δ-selinene, linalool, olivetol, patchouli alcohol, fokienol, β-eudesmol, and 4-(5-methyl-2-furyl)-2- butanone. It is found in Hubei, Hunan, and Sichuan provinces, growing in broad-leaved forests below 1500 meters above sea level. It is a valuable tree species for its solid wood, which is used in construction, furniture, and shipbuilding. The essential oil can be used as a cosmetic raw material.

1.4 Herbal (including liana) essential oil resources

China has more abundant herbal essential oil resources. Herbal essential oil resources include annual and perennial herbs, which have short growth cycles, are more flexible in production, and are more adaptable to market demand. Herbaceous plants also have several advantages: strong regenerative potential, such as Pelargonium graveolens and Iris pallida; can be used as food seasonings, such as Piper nigrum and Foeniculum vulgare; can be used as light industrial raw materials, such as *Cymbopogon citratus*; can be used as medicinal herbs, such as Panax ginseng and Angelica sinensis. Here we provide an overview of the essential oil resources of herbal (including liana).

Aristolochiaceae family

Asarum caudigerum

The whole plant contains 0.08% essential oil. The chemical constituents of the essential oil are 1,8-cineole, isoeugenol methyl ether (*Z*), isoeugenol methyl ether, linalool, and iso-linalool (*Z*), 3,4,5-trimethyl enzaldehyde, and trimethyl-2-propenylbenzene (II) are the main components. It is found in Zhejiang, Jiangxi, Fujian, Taiwan, Hunan, Guangdong, Guangxi, etc. It grows in moist forests under the shade of streams. The whole plant is used as medicine, and as a substitute for Asarum.

A. delavayi

The whole plant contains 1.4% essential oil. The chemical constituents of the essential oil are α-pinene, camphene, β-pinene, 1,8-cineole, para-cymene, bornyl acetate, camphor, (4S)-p-mentha-1,8-dien-4-ol, α-terpineol, isobornyl acetate, 3,5-dimethoxytoluene, t-muurolol, β-caryophyllene, eugenol methyl ether, 2,3,5-trimethoxytoluene, 3,4,5-trimethoxytoluene, nerolidol, asarone, elemicin, and linalool are the main components. It is found in Sichuan, northern Yunnan. It grows on shady rocky slopes at an altitude of 800~1600 meters. The whole plant is used as medicine in Emei, Sichuan.

A. insigne

The whole plant contains 1.05% essential oil. The chemical constituents of the essential oil are α-pinene, β-myrcene, β-pinene, camphor, bornyl acetate, camphene, eugenol methyl ether, t-muurolol, isobornyl acetate, nerolidol, β-caryophyllene, α-copaene, asarone, elemicin, and linalool are the main components. It is distributed in Guangxi. It grows in the deep mountain streams and

Asarum caudigerum

Asarum insigne

shady areas. The local people also use it as a substitute for Asarum in medicine.

A. himalaicum

The whole plant contains 0.4% essential oil. The chemical constituents of the essential oil are camphor, linalool, bornyl acetate, naphthalene, isopropyl myristate, t-geijerene, acetic acid bornyl ester, cubene, 3,5-dimethoxytoluene, acetic acid pine oil ester, β-caryophyllene, eugenol methyl ether, dehydrovomifoliol, and elemol. It is found in Hubei, Sichuan, Yunnan, Guizhou, Tibet, Gansu, and Shaanxi provinces of China, and grows in damp areas under forests along streams at an altitude of 1300~3100 meters. It is used as a medicinal herb, also known as ginseng.

Piperaceae family

Piper arboricola

Stems and leaves contain 0.22%~0.28% essential oil. The chemical constituents of the essential oil are beta-caryophyllene, limonene, linalool, pine oil alcohol, decanal, decanol, pine oil alcohol-4, 2-undecanone, beta-olive oil, dodecanal, beta-fenchyl alcohol, alpha-fenchyl alcohol, gamma-terpinene, 2-tridecanone, beta-olive oil isomer, olive oil alcohol, and nerol. It is distributed from sourtheastern to sourthwestern China, growing in forests and climbing on rocks and tree trunks. It is a traditional aromatic herb used to treat migraines.

P. austrosinense

Branches and leaves contain essential oil at 0.4%~0.6%. The chemical constituents of the essential oil are alpha-pinene, sabinene, beta-pinene, beta-caryophyllene, alpha-terpinene, phytol-3, alpha-terpineol, abietatriene, gamma-terpineol, linalool, pine oil alcohol-4, cubenol, beta-caryophyllene, alpha-fenchyl alcohol, beta-cedrene, alpha-cedrene, beta-himachalene, beta-caryophyllene oxide, and gamma-olive oil. It is distributed in Guangdong and Guangxi, growing in forests and climbing on tree trunks and rocks. It is a traditional local aromatic herb.

P. betle

Branches and leaves contain essential oil. The chemical constituents of the essential oil are hydroxychavicol, eugenol, propenyl pyrocatechol, thymol, eugenyl acetate, para-cymene, 1,8-cineole, methyl eugenol, beta-caryophyllene, and alpha-pinene. It is distributed in all provinces and regions from east to southwest China and mainly cultivated artificially. The leaves are used as medicine to dispel wind and dampness, kill insects, relieve itching, and treat stomach pain.

P. nigrum

Fruits contain 2.5%~3.0% essential oil. The chemical constituents of the essential oil are alpha-pinene, beta-pinene, beta-caryophyllene, alpha-terpinene, phellandrene-3, sesquiterpenes, limonene, phellandrene-2, linalool, delta-cadinene, cubebene, beta-selinene, alpha-selinene. It is originally from Southeast Asia. It is now planted in various provinces and regions in southeastern and southwestern China. The fruit is used in traditional medicine to treat digestive disorders and diarrhea. It can also warm the stomach and intestines and is widely used as a food seasoning.

Piper nigrum

Saururaceae family

Houttuynia cordata

The whole plant contains 0.4%~0.6% essential oil. The chemical constituents of the essential oil are 3-hexenol, beta-caryophyllene, alpha-ocimen e,benzaldehyde, linalool , pinene-4,alpha-pinene, piperitone, terpineol, 2-tridecanone, elemicin, dehydrovomifoliol, linalyl acetate. It is found in south of the Yellow River and in Tibet, Taiwan, Yunnan, and Gansu. It grows in streamsides and wetlands under forests. The whole plant is used in traditional medicine for its heat-clearing, swelling-reducing, and detoxifying effects.

Saururus chinensis

The whole plant contains 0.32% essential oil. The chemical constituents of the essential oil are 3-hexenol, 2-methylcyclopentanol, eucalyptol, linalool, camphor, chrysanthenone, elemicin. It is found in wetlands along rivers and streams in provinces and regions south of the Yangtze River. Folk medicine is also used in Japan and Vietnam. It has a fever- reducing and diuretic effect.

Saururus chinensis

Brassicaceae family

Brassica juncea

Seeds contain 0.21%~0.25% essential oil. The chemical constituents of the essential oil are 3-isothiocyanato-1-propene, 4-isothiocyanato-1-butene, p-sabinene. Originally from Asia, it is now cultivated in various parts of China as a vegetable. It is a common vegetable and the seeds are used as a pungent seasoning.

Chenopodium ambrosioides

Also known as Mexican tea. The whole plant contains 0.4%~1.0% essential oil. The chemical constituents of the essential oil are α-pinene, para-cymene, 1,8-cineole,

Chenopodium ambrosioides

α-terpineol, acetyl terpineol-4-ester, p-cymen-8-ol, ascaridol, and terpinolene. It grows in various regions of China, such as East and South China, and Sichuan, and is used for treating pinworms, roundworms, and tapeworms.

Pelargonium graveolens

Also known as rose-scented geranium, the stems and leaves contain 1.0%~1.4% essential oils. The chemical constituents of the essential oil are geraniol, geranyl formate, citronellol, citronellyl formate, rose oxide, linalool, and benzyl alcohol. Originally from southern Africa, it is now cultivated in various regions of China for use in perfumes and cosmetics.

Passiflora edulis f. *flavicarpa*

Also known as yellow passionfruit, the fruit pulp contains essential oil. The chemical constituents of the essential oil are 3-hexenol, propylcyclohexane, β-myrcene, ethyl acetate, isopropyl 3-methylpropanoate, hexyl butyrate, 4-hydroxybenzyl acetate, hex-3-enyl hexanoate, hexyl hexanoate, 2,6-di-tert-butyl-p-cresol, etc. It is native to South America, particularly Brazil, but is grown in various regions of China, such as Guangdong, Yunnan, Fujian, and Taiwan. The fruit pulp juice can be used to make beverages and ice cream, and the seeds contain fat oil.

Clusiaceae family

Hypericum ascyron

The whole plant contains 0.32%~0.38% essential oil. The chemical constituents of the essential oil are ethyl acetate, 3-hexenal, 3-hexenol, hexanol, 3-hexenoic acid, 1,8-cineole, β-caryophyllene, α-copaene (ZE), β-santalene, α-bisabolol, etc. It is found in Northeast, North, Central and South China, growing in the undergrowth and grasses on mountain slopes. It is used in traditional medicine for treating headaches, vomiting blood, and the seeds are used to treat stomach pain.

Malvaceae family

Abelmoschus moschatus

Seeds contain 0.3% essential oil. The chemical constituents of the essential oil are decyl acetate, α-terpineol, ethyl laurate, t-2-t-6-cedrol, cis-2-t-6-cedrol, ethyl undecylenate, hexadecanal, ethyl cis-2-t-6-cedryl acetate, abelmoschone, ethyl t-6-cedryl acetate, linoleic acid, ethyl linoleate, etc. It is found in Guangdong, Hunan, Jiangxi, and Taiwan, growing in valleys, near gullies or on grassy slopes. The roasted seeds can be used as a substitute for coffee.

Abelmoschus moschatus

Rutaceae family

Boenninghausenia albiflora

The whole plant contains 0.18% essential oil. The chemical constituents of the essential oil are α-pinene, β-pinene, β-myrcene, p-cymene, para-cymenene, 1,8-cineole, β-terpinene, p-cymene, camphor, α-terpineol, caprylic aldehyde, ethyl caprate, β-caryophyllene, γ-terpinene, etc. It is found in the southern parts of the Yangtze River basin, growing on limestone hillsides, in damp forest edges or shrubbery. The whole plant is used in traditional medicine for clearing heat, cooling blood, relaxing muscles, promoting blood circulation, and having anti-inflammatory effects.

Ruta graveolens

The whole plant contains 0.06% essential oil. The chemical constituents of the essential oil are 4-methyl-2-heptanone, 2-nonanone, 2-nonanol, methyl octanoate, 2-undecanone, 2-undecanol, 2-dodecanone, methyl 1-methylheptyl

Ruta graveolens

acetate, 2-tridecanone, and psoralen. It is originally from Europe, but is now grown in both the North and South of China. The whole plant is used in traditional medicine for dispelling wind, reducing fever, promoting diuresis, and is also used to make syrup.

Toddalia asiatica

Leaves contain 0.32% essential oil. The chemical constituents of the essential oil are α-phellandrene, camphene, β-pinene, β-myrcene, para-cymene, limonene, γ-terpinene, β-caryophyllene, β-selinene, β-bisabolene, β-elemene, ethyl p-methane- 3-carboxylate, δ-selinene, and p-mentha-2,8-diene. It is distributed in Guangdong, Guangxi, Hunan, Guizhou, Yunnan, Sichuan, Fujian, Zhejiang, and other provinces and regions, and grows in the jungle. Its root is used in medicine to disperse stasis, reduce swelling, and relieve wind and pain.

Panax ginseng

Araliaceae family

Panax ginseng

Roots contain 1.14%~1.16% essential oil. The chemical constituents of the essential oil are α-cadinol, β-caryophyllene, β-elemene, β-bisabolene, β-selinene, t-farnesol, α-selinene, 3,3-dimethylhexane,heptadecane, coumaric acid, and 15-methyl-2,7-hexadecadiene. It is distributed in eastern Jilin, eastern Liaoning, eastern Heilongjiang, and the Korean Peninsula, growing in deciduous broad-leaved forests at an altitude of 1400~1500 meters. Ginseng is an important tonic, and its flowers also contain essential oils.

Anethum graveolens

Seed contains 1.2%~3.5% essential oil. The chemical constituents of the essential oil are 1.2%~3.5% essential oil, including limonene, trans-anethole, cis-anethole, anethole, carvone, apiol, and 1-methyl-1-vinyl-2,4,5-trimethylbenzene. It is native to Europe and is grown in northeastern China, Gansu, Sichuan, Guangdong, and other regions. Its stems, leaves, and fruits have a dill flavor and can be used as a seasoning. Its fruits are used in medicine to expel wind and strengthen the stomach, and its sprouts can be used as vegetables.

Anethum graveolens

Umbelliferae family

Angelica glauca

Roots contain 0.48% essential oil. The chemical constituents of the essential oil are 6-methyl-2-cyclohexen-1-one, 6-propyl-2-cyclohexen-1-one, pentylbenzene, β-selinene, δ-selinene, furanmethanal, γ-terpineol, and dehydrodendrolasin. It is found in Xinjiang and grows in grassy areas under forests or along river valleys at an altitude of about 1000 meters. Its root is used in traditional medicine as a substitute for ginger, with the effects of relieving wind and dampness, and promoting sweating.

A. sinensis

Fresh leaves contain 0.4% essential oil, while roots contain 0.4%~0.7% essential oil. The chemical constituents of the essential oil are α-pinene, myrcene, limonene, thujone, safranal, para-ethylbenzaldehyde, 3,4-dimethylbenzaldehyde, ligustilide, 1,1,5-trimethyl-2-methylene-2,5-cyclohexadiene-4-one, cubenol, t-beta-bisabolene, 2,4,6-trimethylbenzaldehyde, beta-phellandrene, foshouol, gamma-terpinene, and delta-terpinene, among others. It is mainly cultivated and less commonly found in the wild in Shaanxi, Gansu, Hubei, Sichuan, Yunnan, and Guizhou provinces. The essential oil from the leaves is used in cosmetics to treat melasma. The root, which also contains essential oil, is a well-known medicinal herb. The root essential oil contains haoxiben ester at a concentration of 48%~60%.

Apium graveolens

Seed contains 0.2% essential oil.. The chemical constituents of the essential oil are β-pinene, β-myrcene, limonene, 1-ethenyl-2-hexenyl-cyclopropane, β-caryophyllene, β-phellandrene, α-phellandrene, and α-pinene, among others. It is grown in various regions throughout China and is a vegetable. The seed oil and seed fat are important seasonings, and the whole plant is used in medicine to lower blood pressure and promote diuresis.

Coriandrum sativum

Conioselinum tataricum

Roots and rhizomes contain 1.3% essential oil. The chemical constituents of the essential oil are α-pinene, β-myrcene, α-terpinene, p-cymene, β-ocimene (X), β-ocimene (Y), γ-terpinene, terpinolene, bornyl acetate, eugenol methyl ether, myristicin, elemicin, ligustilide, and β-bisabolene, among others. GaoBen is found in Xinjiang and grows in grasslands. The roots and rhizomes are used medicinally and have the effects of dispersing wind-cold, relieving pain, and drying dampness.

Coriandrum sativum

Fruit and seeds contain 1.0% essential oil. The chemical constituents of the essential oil are α-pinene, β-pinene, para-cymene, γ-terpinene, linalool, terpinen-4-ol, ethyl linaloolate, etc. Originally from the the Mediterranean region of Europe, it is cultivated in most parts of China. The seed essential oil is an important seasoning in Chinese and Western cuisine and is used to flavor vegetables in southern China, known as herbs. The dried fruit is also used in traditional medicine to invigorate the stomach, dispel wind, and relieve phlegm.

Cuminum cyminum

Seeds contain 3.7% essential oil.The chemical constituents of the essential oil are α-pinene, β-pinene, para-cymene, limonene, 1,8-cineole, γ-terpinene, terpineol-4, α-terpineol, propylphenol, carvone, isobutyraldehyde, isobutyl benzoate, etc. An introduced species, it is cultivated in Xinjiang and other regions. The essential oil has the aroma of carvone and is used to prepare food flavors. It is also one of the raw materials for curry.

Cyclorhiza waltonii

Roots contain essential oil. The chemical constituents of the essential oil are α-cedrene, α-pinene, sabinene, eucalyptol, α-terpinene, para-cymene, β-phellandrene, γ-terpinene, thujene, myrcene, etc. A unique species of China, it is distributed in Tibet, Yunnan, and Sichuan and grows on mountain slopes at altitudes of 3500~4600 meters. The root is used in traditional medicine for its antipyretic and detoxifying effects.

Foeniculum vulgare

Fruits contain 3.0%~6.0% essential oil. The chemical constituents of the essential oil are α-pinene, sabinene, β-myrcene, para-cymene, limonene, γ-terpinene, fenchone, estragole, anethole, etc. Originally from the Mediterranean region, it is now cultivated

Foeniculum vulgare

in various provinces and regions of China. The essential oil is mainly used in food, toothpaste, and liquor flavorings, and is also used in traditional medicine for its antibacterial, stomach-warming, and cold- dispersing effects.

Valerian family

Nardostachys jatamansi

Roots and rhizomes contain 5%~6.5% essential oil. The chemical constituents of the essential oil are 9-β-dimethyl ami no-α-bisabolene, 1(10)-dien-9-β-ol, and jatamansone. It is used in traditional Chinese medicine as a digestive aid, for stomach issues, and to relieve pain. It is also used in the perfume industry. It is native to southwestern China and is commonly found growing in mountainous forest areas.

Patrinia rupestris subsp. scabra

Roots contain 0.3% essential oil. The roots and whole plant are used in traditional Chinese medicine for their cooling and drying properties, as well as their ability to stop bleeding. It is found in northeastern and northern China and grows on sunny mountain slopes and embankments.

Valeriana jatamansi

Roots contain 0.5%~0.8% essential oil. The chemical constituents of the essential oil are α-pinene, limonene, and bornyl acetate. It is used in traditional Chinese medicine for its ability to aid digestion, relieve pain, and dispel wind and toxins. It is found in various regions of China, including Shaanxi, Henan, Hubei, Hunan, Sichuan, Guizhou, Yunnan, and Tibet.

V. officinalis

The whole plant contains 0.5%~0.6% essential oil. The chemical constituents of the essential oil are valerenic acid, bornyl acetate, and isovaleric acid. It is used in traditional Chinese medicine for its sedative properties, and the essential oil can be used in various products such as cigarettes, food, and cosmetics. It is found in various regions of China, from the northeast to the southwest, growing in high mountain slopes, forests, and valleys.

V. officinalis var. *latifolia*

Roots contain 2% essential oil. The chemical constituents of the essential oil are isovaleric acid, α-pinene, camphene, sabinene, β-pinene, myrcene, limonene, bornyl acetate, valerenic acid, valeranone, and β-caryophyllene. It is found in Northeast China, Southwest China, and Eastern China, growing on mountain slopes, in forests, or by streams. Its medicinal properties are similar to those of valerian.

Asteraceae family

Ageratum conyzoides

An annual herb. The whole plant contains 0.44% essential oil. The chemical constituents of the essential oil are hexenol, sabinene, bornyl acetate, β-caryophyllene, linalool, β-cubebene, γ-terpinene, and ageratochromene.

Originally from South America, it is widely cultivated in Jiangxi, Fujian, Guangdong, Guangxi, Yunnan, Guizhou, and Sichuan provinces in China, growing on slopes, forest edges, and riverbanks. Its essential oil can be used in perfume production, while the whole plant can be used as green manure and fish poison.

Ajania przewalskii

A perennial herb. The whole plant contains 0.26% essential oil. The chemical constituents of the essential oil are 1,8-cineole, α-selinene, α-selinene isomer, camphor, bornyl acetate, pinene-4, α-pinene, chrysanthenol, cis-ligustilide, cis-2-hinokitiol, and cis-pinosylvin. It is distributed in provinces such as Northwest China and Sichuan. Its smoke is used to repel insects.

Ageratum conyzoides

A. tenuifolia

A perennial herb. The whole plant contains 0.5% essential oil. The chemical constituents of the essential oil are α-pinene, camphene, sabinene, β-pinene, 1,8-cineole, camphor, bornyl acetate, α-terpineol, t-geranyl acetate, and apiole. It is found in various provinces in northwestern China and Tibet. Its essential oil can be used in the production of fennel oil.

Anaphalis margaritacea

The whole plant contains 0.2% essential oil. The chemical constituents of the essential oil are alpha-pinene, beta-cedrene, beta-caryophyllene, alpha-bisabolene, alpha-caryophyllene, gamma-bisabolene, alpha-turmerone, pentadecane, delta-bisabolene, gamma-eudesmol, delta-cadinene, and nerolidol. It is found in various provinces in northwestern China, as well as Yunnan, Hubei, and Hunan, growing on gravelly slopes and in gullies at altitudes between 300 and 3400 meters. The essential oil is used in cosmetics, cigarettes, and traditional medicine for its anti-diarrheal, hemostatic, and vermifuge effects.

Artemisia abaensis

The whole plant contains 0.3% essential oil. The chemical constituents of the essential oil are hexanal, 4-methyl-3-penten-2-one, benzaldehyde, 2,5,5-trimethyl-1,2-cyclohexadiene, 3-carene, 7-octen-4-ol, camphene, 1,8-cineole, benzylalcohol, thujone, butyl-3-hexen-5-ol, alpha-sabinene, 2,6-dimethyl-7-octen-3-one, camphor, bornyl acetate, and alpha- terpineol. It is found in eastern Qinghai, southwestern Gansu, and the Aba region of Sichuan, growing on lakes and roadside slopes at various altitudes. The root has been used in traditional medicine for its anti-wind effect.

A. aksaiensis

The whole plant contains 0.36% essential oil. The chemical constituents of the essential oil are hexanal, chamazulene, eucalyptol, camphor, longifolene, pinus sylvestris alcohol-4, pulegone, coumarin, beta-farnesene, gamma-olivene, etc. It grows on slopes at altitudes of 3000~3800 meters in the autonomous county of

Aksai in western Gansu Province. Its common name is alkali wormwood.

A. anethifolia

The whole plant contains 0.98% essential oil. The chemical constituents of the essential oil are heptanol-4, eucalyptol, hinokitiol, longifolene, pinus sylvestris alcohol-4, rhododenol, para-isopropylphenol, hinokitiol isomers, beta-selinene, etc. It grows on dry mountain slopes in North China, Northwest China, and Heilongjiang Province. Its basal leaves can be used to extract essence, which contains eucalyptol.

A. anethoides

The whole plant contains 0.6% essential oil. The chemical constituents of the essential oil are guaene-3, chamazulene, eucalyptol, gamma--terpinene, guaene-4, alpha-cedrene, alpha-cedrene isomers, camphor, longifolene, pinus sylvestris alcohol-4, kurumol, alpha-terpineol, rhododenol, ethyl acetate-alpha-pinene ester, beta-acacetin, etc. It grows in various provinces and regions in Northeast, North and Northwest China, as well as Shandong, Henan and northern Sichuan, and can tolerate saline-alkaline soil. Essential oil can be extracted from plants grown in Gansu, which contains eucalyptol, and from plants grown in Sichuan, which contains pulegone. Its common name is fennel-leaved wormwood.

A. anomala

Herbs contain 0.21% essential oil. The chemical constituents of the essential oil are aldehydes, 2-hexenal, 1,8-cineole, t-oxides of fenchol (furan type), fenchol, pulegone, camphor, bornyl acetate, pinocarvone, β-linalool, β-caryophyllene, β-elemene, α-curcumene, δ-cadinene,

Artemisia anomala

nerolidol, α-acoradiene, etc. It is found provinces south of the Qinling Mountains, growing at low altitudes along forest edges and in abandoned fields. The whole herb is used in traditional medicine, with effects including blood circulation promotion, menstrual regulation, heat-clearing, fever-alleviating, pain-relieving, and digestion-promoting.

A. argyi

Herbs contain 0.42% essential oil. The chemical constituents of the essential oil are 1,8-cineole, α-thujone, isothujone, sabinene, camphor, bornyl acetate, pinocarveol-4, α-terpineol, eugenol, bisabolene, α-phellandrene, etc. It is found throughout most regions of China except for extremely arid and high-altitude areas, growing on low-altitude hillsides and in abandoned fields. The whole herb is used in traditional medicine for promoting warmth and dispelling dampness, relieving colds, stopping bleeding, and reducing inflammation. The cultivated variety is called qiyi (*Artemisia argyi* var. *argyi*).

Artemisia annua

A. annua

Herbs contain 0.51% essential oil. The chemical constituents of the essential oil are 1,8-cineole, artemisinin, 3,3,6-trimethyl-2-cyclohexenone, camphor, benzyl benzoate, β-caryophyllene, coumarin, β-farnesene, α-copaene, β-phellandrene, p-tert-butylmethoxytoluene, δ-cadinol, etc. It is found throughout China, with plants growing at 1500 meters above sea level in the eastern regions and at 3000~3650 meters above sea level in the northwest and southwest regions. The whole herb is used in traditional medicine for heat-clearing, fever-alleviating, blood-cooling, diuretic, and antiperspirant purposes, and artemisinin extracted from the herb is used as an antimalarial drug.

A. atrovirens

The whole plant contains 0.14% essential oil. The chemical constituents of the essential oil are compounds such as hexanal, 2-hexenal, 7-octen-4-ol, 1,8-cineole, artemisinin, camphor, borneol, linalool, menthol, naphthalene, pinocarvone-4, santolinaalcohol, beta-farnesene, alpha-pinene, beta-caryophyllene, orange flower alcohol, alpha-bisabolol, and yacatholene. It grows in the Qinling Mountains, the Yellow River basin, and in southern China in Sichuan, Yunnan, and other areas. It can also be found in Thailand. It typically grows on low-atitude mountain slopes, along roadsides, and other similar areas.

A. brachyphylla

The whole plant contains 0.48% essential oil. The chemical constituents of the essential oil are 7-octen-4-ol, 3-octene, 1,8-cineole, gamma-terpineol, artemisinin, camphor, borneol, alpha-terpineol, eugenol, beta-caryophyllene, yacatholene, alpha- pinene, and alpha-phellandrene. It is found in eastern Jilin Province in China, and in North Korea. It typically grows in subalpine meadows and on forest edges at altitudes above 1100 meters.

A. calophylla

The whole plant contains 0.13% essential oil. The chemical constituents of the essential oil are 7-octen-4-ol, 1,8-cineole, artemisinin, dec-3-en-1-yl acetate, camphor, borneol, alpha-terpineol, nonanal, hinokitiol, linalool, menthol, artemisia ketone, and nepetalactone. It is found in southern Qinghai, western Guangxi, Sichuan, Yunnan, Guizhou, and eastern Tibet. It typically grows on roadsides and forest edges at altitudes between 1600~3000 meters above sea level.

A. capillaris

New shoots contain essential oil. The whole plant contains 0.3%~0.6% essential oil. The chemical constituents of the essential oil are eugenol, methyl eugenol, β-caryophyllene, α-curcumene, artemisinin, nerolidol, geranyl acetate, 2-methylpropyl propanoate, artemisinic ketone, 2-methylpropyl eugenol, isoeugenol, butyl eugenol, pentyl eugenol, 6,10,14-trimethyl-2-pentadecanone, and other compounds. It grows in provinces across northern and southern China.

The essential oil can be used to make cooling oil and spray, while the whole plant has diaphoretic, diuretic, and antipyretic effects.

A. desertorum

The whole plant contains 0.45% essential oil. The chemical constituents of the essential oil are 3-methylphenol, thujol, geranyl acetate, jasmone, α-curcumene, geranyl propanoate, α-santalene oxide, and α-santalol. It grows in provinces across northern, northwestern, northeastern, and southwestern China, as well as in North Korea, Japan, and India. It grows in grasslands and meadows at altitudes of up to 4000 meters. The essential oil has strong antibacterial and anti-inflammatory effects.

A. dubia

The whole plant contains 0.27% essential oil. The chemical constituents of the essential oil are limonene, 1,8-cineole, linalool, camphor, borneol, cis-chrysanthenol-8, alpha- pinene, alpha-curcumene, yinchenin, and eugenol methyl ether. It grows in grasslands and sparse forests at altitudes below 3500 meters in Inner Mongolia, Gansu, Sichuan, Yunnan, and Tibet, as well as in India and Nepal.

A. dubia var. subdigitata

The whole plant contains 0.5% essential oil. The chemical constituents of the essential oil are hexanal, benzaldehyde, 6-methyl-5-hepten-2-one, cis-chrysanthenol-8, indole, eugenol methyl ether, alpha-curcumene, yinchenin, linalool oxide, 3,4,5-trimethylbenzaldehyde, and linalool oxide isomer. It is found in Inner Mongolia, Hebei, Shanxi, Shaanxi, Ningxia, Gansu, Qinghai, Shandong, and Henan.

A. edgeworthii

The whole plant contains 0.32%~0.36% essential oil. The chemical constituents of the essential oil are germacrene D, 1,8-cineole, 3,3,6-trimethyl-2-geranylgeraniol, alpha-terpineol (furanoid), linalool, sabinene, camphor, 2,3-dihydrobenzofuran, sabinene isomer, pinocarvone-4, cis-chrysanthenol-8, alpha-pinene, cis-p-mentha-2,8-dien-1-ol, heliotropin, beta-caryophyllene, alpha-curcumene, and nerolidol. It grows in Qinghai, Xinjiang, Yunnan, and Tibet at altitudes between 2200 and 4700 meters on mountain slopes. In Qinghai, it is used in traditional medicine as yinchen.

A. fauriei

The whole plant contains 0.38% essential oil. The chemical constituents of the essential oil are chamazulene, 1,8-cineole, linalool, terpinolene, camphor, bornyl acetate, pinocarvone-4, santolinatriene alcohol, and santolinatriene aldehyde. It grows the in coastal areas of Hebei, Shandong, and Jiangsu provinces in China. It also grows in Japan and North Korea. The composition of the essential oil varies in different regions.

A. frigida

The whole plant contains 0.72% essential oil. The chemical constituents of the essential oil are 1,8-cineole, artemisinin, 3,6,6-trimethyl-2-juvenile alcohol, linalool, alpha-thujone, alpha-thujone isomer, camphor, bornyl acetate, and alpha-pinene. It grows in the northern, northeastern, northwestern, and Tibetan regions of China. It also grows in Central Asia and North America. It grows on dry and semi- dry mountain slopes and sand dunes. The whole plant is used as a pain reliever, anti-inflammatory, cough suppressant, and substitute for Artemisia scoparia in traditional medicine. The composition of the essential oil varies in different regions.

A. giraldii

The whole plant contains 0.5%~0.6% essential oil. The chemical constituents of the essential oil are hexanal, 6-methyl-5-hepten-2-one, benzaldehyde, hexenal, alpha-pinene, linalool, 1,8-cineole, isoamyl benzoate, alpha-terpineol, beta- caryophyllene, beta-elemene, alpha-curcumene, artemisia ketone, alpha-eudesmol, and artemisia ketone. It grows in Inner Mongolia, Hebei, Shanxi, Shaanxi, Ningxia, Gansu, and Sichuan provinces in China. It grows on the Loess Plateau at an altitude of 1000~2300 meters. The whole plant is used for its antipyretic, detoxifying, and lung-benefiting effects.

A. gyangzeensis

The whole plant contains 0.36% essential oil. The chemical constituents of the essential oil are n-hexanal, 2-hexenal, champhene, limonene, γ-terpinene, para-cymene, linalool, camphor, para-mentha-8-ol, bornyl acetate, citronellol, β-cedrene, artemisinin, γ-elemene, β-phellandrene, linalyl acetate, etc. It is found in the Jiangzi area of Tibet, Qinghai, and southern Gansu, growing on mountain slopes at an altitude of 3,900 meters.

A. halodendron

The whole plant contains 0.12%~0.42% essential oil. The chemical constituents of the essential oil are β-phellandrene, β-selinene, oxide linalool (furan-type), camphor, pinocamphone-4, α-curcumene, artemisinin, isocarbophyllene, orange flower alcohol, α-sweet myrrh alcohol oxide, α-sweet myrrh alcohol, etc. It is found in the western part of Northeast China, northern China, and various provinces in Northwest China, growing in low-altitude sandy grasslands and rocky deserts.

A. hedinii

The whole plant contains 0.46% essential oil. The chemical constituents of the essential oil are ethyl 3-methylbutyrate, methyl 2-methylbutyrate, cis-linalool oxide (furan-type), linalool, phenylethanol, α-terpineol, phenyl acetate, indole, isomer of methyl salicylate, isomer of methyl benzoate, β-caryophyllene, γ-terpinene, 2,6-di-tert-butyl-p-cresol, orange flower alcohol, 1-nonene-3-ol, etc. It is found in Inner Mongolia and various provinces in Northwest China, as well as India, Nepal, and Kashmir, growing on roadsides, slopes, and forest edges at altitudes of 2000~5000 meters. These plants are used in traditional Chinese medicine for various purposes, such as relieving cough and reducing phlegm, calming asthma, reducing inflammation, and treating external symptoms.

A. indica

The composition of essential oils varies depending on the region. The whole plant contains 0.18%~0.22% essential oil. The chemical constituents of the essential oil are benzaldehyde, 2,2-dimethylhexanal, 1,8-cineole, artemisia ketone, alpha- thujone, alpha-thujone isomer, side-firalcohol, camphor, bornyl acetate, pinene-4, artemisia ketone, patchoulenone, isoborneol, beta-caryophyllene, ylangene, delta-cadinene, and xanthanol, among others. It is distributed throughout the country except for Xinjiang, Qinghai and Ningxia. It grows on slopes at low altitudes on the edge of forests. The whole plant is used medicinally as a substitute for "mugwort" and has the effects of clearing heat, detoxification, hemostasis, and anti-inflammatory.

A. japonica

The whole plant contains 0.23% essential oil. The chemical constituents of the essential oil are 3-hexenal, delta-oxa-bridged [15,10] octane, octane, benzaldehyde, 9-oxa-bridged [6,1,0] nonane, 1,8-cineole, fenchone (furan-type), borneol, camphor, pinene-4, artemisia ketone, patchoulene, alpha-terpineol, 3,4,5-trimethyl-2-cyclopentenone, 2-ethenyl-2,5-dimethyl-4-hexenol, alpha-curcumene, 2,6-di-tert-butyl-4-methylphenol, 2,2-dimethyl-3- (dimethyl-1-propenyl) -cyclopropanecarboxylic acid ethyl ester, and isobornyl benzoate. It is distributed throughout the country in various regions except for Xinjiang, Qinghai and Inner Mongolia. It grows on forest edges, wilderness, mountain slopes, and roadsides at altitudes below 3,300 meters. The whole plant is used medicinally as a substitute for "qinghao"

Artemisia indica

Artemisia japonica

and has the effects of clearing heat, detoxification, relieving summer heat and dampness, hemostasis, anti-inflammatory, and promoting blood circulation.

A. integrifolia

The whole plant contains 0.34% essential oil. The chemical constituents of the essential oil are hexanal, 7-octen-4-ol, 1,8-cineole, benzaldehyde, nonanal, camphor, borneol, naphthalene, yinchenyne, alpha-terpineol, etc. It is mainly distributed in the northern regions of Northeast China and North China. It can also be found in North Korea and eastern Russia. It often grows at the edge of low-altitude forests and roadsides. The essential oil has the potential to be developed as a raw material for making "menthol oil".

A. kawakamii

The whole plant contains 0.90% essential oil. The chemical constituents of the essential oil are 6-methyl-5-hepten-2-one, linalool, isoborneol, cis-ocimene, kaurene, cis-abienol, ethyl isovalerate, eugenol, bornyl acetate, alpha-curcumene, elemol, alpha-pinene, etc. It is a unique species in Taiwan, and is mainly distributed in Taichung, Yilan, Miaoli, Chiayi, etc. The essential oil has the potential to be developed as a raw material for making "menthol oil".

Artemisia lactiflora

A. lactiflora

The whole plant contains 0.41% essential oil. The chemical constituents of the essential oil are beta-caryophyllene, alpha-farnesene, alpha-curcumene, beta-farnesene (Z), zingiberene, beta-farnesene (E), linalool oxide, etc. It is mainly distributed south of the Nanling Mountains, and in Sichuan, Yunnan, Guizhou. It can also be found in Vietnam and Laos. It often grows in mountain valleys at the edge of forests. The whole plant is used in traditional medicine, and is used as a substitute for "Qihao" in Guangdong and Guangxi, with effects such as promoting blood circulation and treating liver and kidney diseases.

A. lancea

The whole plant contains 0.41% essential oil. The chemical constituents of the essential oil are 2-hexenal, 7-octen-4-ol-6-methyl-5-hepten-2-one, 1,2-epoxy-4(14)-pene, linalool, viridiflorol, artemisia ketone, camphor, bornyl acetate, beta-caryophyllene, beta-farnesene, beta-bisabolene, and beta-selinene, among others. It is distributed throughout China except for Xinjiang, Qinghai, Ningxia, and Tibet. It can also be found in Japan, North Korea, and India. It grows in low-altitude forests, wastelands, and other areas. It is used in traditional medicine as a substitute for "ai" and yinchen, with effects that include dispelling cold, warming meridians, stopping bleeding, promoting fetal safety, and reducing inflammation.

A. lavandulaefolia

The whole plant contains 0.39% essential oil. The chemical constituents of the essential oil are 7-octen-4-ol-1, 8-cineole, linalool, alpha-thujone, camphor, bornyl acetate, alpha-pinene, epilavandulyl acetate, l-carvone, geijerene, beta-caryophyllene, beta-farnesene, and beta-bisabolene, among others. It is distributed throughout China except for Xinjiang, Qinghai, Southeast coast, and Taiwan. It can also be found in Japan, North Korea, Mongolia, and Russia. It grows in low-altitude forests, riverbanks, and grasslands. The whole plant is used in traditional medicine and has similar effects to "ai".

A. leucophylla

The whole plant contains 0.32%~0.36% essential oil. The chemical constituents of the essential oil are 7-octen-4-ol, 1,8-cineole, artemisia ketone, 3,6,6-trimethyl-2-menthol, alpha-thujone, camphor, borneol, isoborneol, pinusol-4, alpha-pinene, indole, elemene, alpha-caryophyllene, and alpha-selinene, among others. It is distributed throughout Northeast, North, Northwest, and Southwest China. It can also be found in Mongolia, North Korea, and Russia. It grows along roadsides, forest edges, riverbanks, and lakesides. It is used as a substitute for "ai" and has effects that include dampness elimination, hemostasis, and anti-inflammatory properties.

A. macrocephala

The whole plant contains 0.23% essential oil. The chemical constituents of the essential oil are capric acid, 3-methylhexanoic acid, 1,8-cineole, linalool, nonanal, alpha-terpineol, phenylethanol, camphor, methyl 2,4-decadienoate, borneol, yomogi alcohol, alpha-yomogi alcohol, methyl salicylate, and ethyl salicylate. It is found in northwestern China and Tibet at altitudes ranging from 1500~3400 meters in arid and semi-arid regions. The whole plant is used in traditional medicine and as animal feed in pastoral areas.

A. manshurica

The whole plant contains 0.20% essential oil. The chemical constituents of the essential oil are hexanal, 2-hexenal, oxabicyclo [5.1.0] octane, benzaldehyde, 6-methyl-5-hepten-2-one, oxabicyclo [6.1.0] nonane, citronellene, 1,8-cineole, camphor, borneol, yomogi alcohol-4, naphthalene, anethole, bornyl acetate, eugenol, chrysanthenone, and 2,6-di-tert-butyl-p-cresol. It is found in Heilongjiang, Jilin, Liaoning, and northern Hebei provinces in China, growing in low-altitude humid areas and on forest edges.

A. mongolica

The whole plant contains 0.18% essential oil. The chemical constituents of the essential oil are hexanal, 2-hexenal, benzaldehyde, 7-octanol, 1,8-cineole, 2,5,5-trimethyl-2,6-heptadien-4-one, alpha-terpineol, isomer of alpha-terpineol, camphor, borneol, yomogi alcohol-4, and pulegone. It is widely cultivated throughout China, except in Hainan, Yunnan, and Tibet, mainly growing at middle to low altitudes.

A. moorcoftiana

Essential oil components vary depending on the production area. The whole plant contains 0.85% essential oil. The chemical constituents of the essential oil are artemisia ketone, 2,6,6-trimethyl-2-ene-2-ol, cis-pinocarveol, camphor, bornyl acetate, menthol-4, pulegone, myristicin, nerolidol, and pinene. It is grown in Gansu, Ningxia, Qinghai, Sichuan, and Yunnan in China, as well as in Pakistan and Kashmir. It grows in sub-alpine mountains, meadows, and grasslands at altitudes of 3000~4800 meters.

A. myriantha

Essential oil components vary depending on the production area. The whole plant contains 0.16% essential oil. The chemical constituents of the essential oil are hexanal, 2-hexenal, α-thujene, α-pinene, β-pinene, myrcene, lauryl alcohol, 6-methyl-5-hepten-2-one, germacrene, limonene, 1,8-cineole, γ-terpinene, linalool, borneol, camphor, bornyl acetate, naphthalene, α-terpineol, β-caryophyllene, β-elemene, α-curcumene, and spathulenol. It is grown in Shanxi, Gansu, Qinghai, Sichuan, Guizhou, Yunnan, and Guangxi in China, as well as in India, Bhutan, and Nepal. It grows along roadsides and in thickets at altitudes of 1000~2000 meters. The whole plant is used in traditional medicine for its anti-inflammatory properties.

A. myriantha var. pleiocephala

Essential oil components vary depending on the production area. The whole plant contains 0.47% essential oil. The chemical constituents of the essential oil are 2-heptenal, α-pinene, sabinene, 1,8-cineole, α-terpineol, linalool-4, thujone, α-terpinolene, 1,3-cyclohexadiene-7, α-curcumene, chamazulene, artemisinin, nerolidol, yucampol, and artemisinin lactone. It is distributed in Qinghai and southwestern provinces of China, as well as northern India, Bhutan, and Nepal, growing on mountain slopes at middle to high altitudes.

A. occidentalisichuanensis

The whole plant contains 0.56% essential oil. The chemical constituents of the essential oil are 2-heptenal, 2,5,5-trimethyl-1,3,6-heptatriene, pinene, 7-octen-4-ol, 6-methyl-5-hepten-2-one, sabinene, 1,8-cineole, linalool, camphor, bornyl acetate, α-terpineol, carvone, β-caryophyllene, β-farnesene, chamazulene, β-pinene, juniperene, and yucampol. It is found in western Sichuan province of China and its essential oil can be used as a raw material for making perfumes.

A. ordosica

The whole plant contains 0.64% essential oil. The chemical constituents of the essential oil are 2-heptenal, α-pinene, β-pinene, limonene, β-myrcene (Z), β-myrcene (E), thujone, α-terpinolene, nerolidol, ar-curcumene, chamazulene, linalool, β-caryophyllene, and α-santalene. It grows in the desert and dry grasslands at an altitude of below 1500 meters in northern China and northwestern provinces. It is used by locals in Inner Mongolia for anti-inflammatory, hemostatic, anti-rheumatic, and antipyretic purposes, and also as feed for livestock in pastoral areas.

A. palustris

The whole plant contains 1.20% essential oil. The chemical constituents of the essential oil are alpha-pinene, 6-methyl-5-hepten-2-one, 1,8-cineole, alpha-terpinene, artemisia ketone, alpha-thujone, camphor, bornyl acetate, and pinene-4-ol. It is found in the western part of the three northeastern provinces and northern Hebei and Inner Mongolia in China. It grows in low-altitude grasslands and has the potential for developing natural essential oils.

A. parviflora

The whole plant contains 0.30%~0.42% essential oil. The chemical constituents of the essential oil are hexanal, 2-hexenal, 8-oxabicyclo[5.1.0]octane, 6-methyl-5-hepten-2-one, 9-oxabicyclo [6.1.0] nonene, 1,8-cineole, artemisia ketone, linalool, camphor, bornyl acetate, pinene-4-ol, indole, 2,6-di-tert-butyl-p-cresol, nerol, and others. It is found in Gansu, Qinghai, Shaanxi, Hubei, Sichuan, Yunnan, and Tibet in China, as well as in Afghanistan, India, Nepal, and Myanmar. It grows in grassy areas and on slopes at altitudes of 2000 to 3100 meters. The whole plant is used in traditional medicine for its antipyretic, detoxifying, hemostatic, and dampness-removing effects and can be used as a substitute for "qinghao" (*Artemisia annua*).

A. phyllobotrys

The whole plant contains 0.21%~0.38% essential oil. The chemical constituents of the essential oil are 1,8-cineole, 3-hexenyl butyrate, linalool, alpha-thujone, alpha-thujone isomer, camphor, bornyl acetate, pinene-4-ol, cuparene, eudesma-4(14), 11-diene, alpha-pinene, beta-linalool, beta-ionone, elemol, beta-eudesmol, alpha-curcumene, and others. It is found in southern Qinghai and western Sichuan in China, growing in high-altitude grasslands at altitudes of 3000 to 3900 meters.

A. princeps

The whole plant contains 0.13%~0.20% essential oil. The chemical constituents of the essential oil are aldehydes, ketones, alcohols, and other compounds. It grows throughout most of China, except for arid regions, and can also be found in Japan and North Korea. Mugwort is used in traditional Chinese medicine to treat various conditions, such as colds, menstrual irregularities, and inflammation.

A. rubripes

The whole plant contains 0.40%~0.50% essential oil. The chemical constituents of the essential oil are terpenoids, ketones, alcohols, and other compounds. It grows in various regions of China, as well as in Russia, Japan, and North Korea. Red-stem wormwood is used in traditional Chinese medicine as a substitute for mugwort and has warming, pain-relieving, and hemostatic properties.

A. sacrorum

The essential oil components vary depending on the place of origin. The whole plant contains 0.88% essential oil. The chemical constituents of the essential oil are hexanal, benzaldehyde, 1,8-cineole, camphor, eucalyptol, methyl eugenol, beta-caryophyllene, beta-farnesene, beta-bisabolene, chamazulene, yomogi ketone, orange blossom alcohol, yomogi enone, and butyl cinnamate. It is distributed throughout China and the Eurasian continent, growing in grasslands, deserts, and along the edges of forests and roads at altitudes below of 4000 meters. The basal leaves and young flower heads are used in traditional medicine, and the usage is the same as that of *A. vulgaris*. It is also known as Tu Yin Chen or North Yin Chen.

A. scoparia

The essential oil components vary depending on the place of origin. The whole plant contains 0.88% essential oil. The chemical constituents of the essential oil are hexanal, benzaldehyde, 1,8-cineole, camphor, eucalyptol, methyl eugenol, beta-caryophyllene, beta-farnesene, beta-bisabolene, chamazulene, yomogi ketone, orange blossom alcohol, yomogi enone, and butyl cinnamate. It is distributed throughout China and the Eurasian continent, growing in grasslands, deserts, and along the edges of forests and roads at altitudes below 4000 meters. The basal leaves and young flower heads are used in traditional medicine, and the usage is the same as that of *A. vulgaris.* It is also known as Tu Yin Chen or North Yin Chen.

A. selengensis

The whole plant contains 0.52% essential oil. The chemical constituents of the essential oil are octanol-4, 1,8-cineole, camphor, alpha-terpineol, alpha-terpineol isomer, bornyl acetate, pinene, sabinene, pinocarveol-4, alpha-pinene, and cedrol. It is distributed throughout the country,except for the coastal areas of Southeast China, Tibet, and Northwest China, and it also grows in Mongolia and Siberia. It grows near lakes and marshes. The whole plant is used in traditional medicine to treat hepatitis, and has anti-inflammatory, hemostatic, antitussive, and expectorant effects.

A. sieversiana

The essential oil components vary depending on the place of origin. The whole plant contains 0.30%~0.40% essential oil. The chemical constituents of the essential oil are octanol-4,1,8-cineole, hexanoic acid-3-hexenyl ester, linalool, alpha-terpineol, alpha-terpineol isomer, camphor, bornyl acetate, pinene, limonene, menthol, nerol, jasmine ketone, methyl salicylate, alpha-pinene, and dimethyl phthalate. It is distributed in the high mountain regions of Northeast China, North China, Southwest China, as well as in India and Pakistan. It grows in dry and semi-arid regions. The whole plant is used in traditional medicine as an anti-inflammatory and hemostatic agent. The essential oil has sunscreen effects, and the seeds are used to make fruit jelly and noodle dishes in pastoral areas.

A. pubescens

The whole plant contains essential oil. The chemical constituents of the essential oil are 1,8-cineole, linalool, camphor, bornyl acetate, coumarin, alpha-curcumene, chamazulene, artemisinin, and t-cadinol. It grows in grasslands and pastures at low to medium altitudes in Northeast, North, Sichuan, and Northwest

China. It is also found in Mongolia and Japan, and its spring shoots can be used as a substitute for Artemisia annua. In addition, it is used as fodder for livestock in pastoral areas.

A. qinlingensis

The whole plant contains essential oil. The chemical constituents of the essential oil are 7-octen-4-ol, 1,8-cineole, camphor, borneol, pinene-4-ol, eugenol methyl ether, coumarin, beta-eudesmol, and xanthanolide. It grows on mountain slopes and roadsides at medium to low altitudes in southwestern Hunan, southern Shaanxi, and eastern Gansu provinces in China. The whole plant is used in traditional medicine as a substitute for *Artemisia argyi*, and natural camphor can be extracted from it.

A. robusta

The whole plant contains essential oil. The chemical constituents of the essential oil are hexanal, 2-hexenal, 3,3-dimethyl-6-methylene-1,4,6-heptatriene, pinene, octenol, octanol-3, beta-selinene, 1,8-cineole, camphor, borneol, dihydrochamazulene, bornyl acetate, beta-caryophyllene, gingerene, p-cymene-2,6-ditert-butyl, delta-cadinene, yucacamphor, and beta-eudesmol. It grows on mountain slopes and roadsides at altitudes of 2200~3500 meters in Sichuan and Yunnan provinces in China, as well as in northern India. Artemisia roxburghiana is another plant in the Asteraceae family that contains essential oils throughout the plant. It contains compounds such as 5,5-dimethyl-2-furanone, 1,8-cineole, linalool, alpha-cedrene, camphor, bornyl acetate, piperitone, eucalyptol, coumarin, and beta-caryophyllene. It grows in dry river valleys and roadsides at middle to high altitudes in Shaanxi, Qinghai, Hubei, and southwestern provinces of China, as well as in Afghanistan, India, and Nepal. The essential oil can be developed as a raw material for the production of fragrances.

A. roxburghiana var. *purpurascens*

The whole plant contains essential oil. The chemical constituents of the essential oil are aldehydes, benzaldehyde, 1,8-cineole, α-terpineol, α-terpineol isomer, bornyl acetate,camphor, californone, longifolene, pinanol-4, kurumene, eugenol methyl ether, artesunate, sclareol, nerolidol, etc. It is distributed in western

Sichuan and Tibet. It is also found in Nepal, Pakistan, Kashmir, northern India, and Thailand.The essential oil is rich in α-terpineol and can be developed into a resource for α-terpineol.

Atractylodes lancea

Beicangzhu: Rhizomes contain 1.03%~2.24% essential oil. The chemical constituents of the essential oil are atractylone, costunolide, β-eudesmol, atractylol, β-caryophyllene, β-eudesmol isomer, elemol, β-pinene, beta-sitosterol A, etc. It is found in northeast and north China, as well as Shandong, Henan, Shaanxi, and other regions on mountain slopes and grasslands. The essential oil can be used to make Atractylodes hard fat, which can be used to formulate perfumes with the scent of Atractylodes or used medicinally for spleen-strengthening, digestion-promoting, and dampness-eliminating effects. Nancangzhu: Rhizomes contain 3.30%~6.90% essential oil. The chemical constituents of the essential oil are β-pinene, artemisone, cis-and trans-chrysanthenol, β-eudesmol, and achillin, among others. The essential oil can be used to make perfumes and the rhizomes are used in traditional Chinese medicine to treat various ailments and are similar to those of *A. argyi*, which is also known as Chinese mugwort. It grows in mountainous areas, thickets, and grassy areas in Zhejiang, Jiangsu, Jiangxi, Shandong, Hubei, and Sichuan provinces in China.

A. macrocephala

Rhizomes contain 0.62%~1.42% essential oil. The chemical constituents of the essential oil are β-pinene, artemisone, cis- and trans- chrysanthenol, β-eudesmol, and achillin, among others. The rhizomes are used in traditional Chinese medicine to replenish Qi, invigorate the spleen, and dry dampness. It is native to China and is cultivated in Eastern and Southwestern China.

A. japonica

Rhizomes contain 1.6% essential oil. The chemical constituents of the essential oil are Costunolide, β-eudesmol, atractylol, β-caryophyllene, β-eudesmol isomer, elemol, β-pinene, β-caryophyllene oxide, atractylone, atractylon, etc. It is found in northeast China, growing under oak forests on mountain slopes or in shrubs. The essential oil has the same medicinal use as Bei Chang Shu.

A. simulans

The whole plant contains 0.31% essential oil. The chemical constituents of the essential oil are 1,8-cineole, phenylethanol, β-thujone, camphor, pinocarvone-4, cis-chrysanthenol-8, pulegone, bornyl acetate, isobornyl acetate, α-curcumene, 2-methylbutyric acid bornyl ester, and yucampol. It grows in the low mountain slopes and wastelands of central and southern China and the southwestern provinces.

A. songarica

The whole plant contains 0.56% essential oil. The chemical constituents of the essential oil are α-pinene, sabinene, β-myrcene, limonene, cis-β-ocimene (Z), β-ocimene (E), γ-terpinene, 3,4-dimethyl-2,4,6-octatriene, pinocarvone-4, 3-hexenyl butyrate, yinchenin, linalool oxide, and linalool. It grows in the gravelly hills of the northern Junggar Basin in Xinjiang, and is also distributed in Central Asia's regions.

A. speciosa

The whole plant contains 0.40%~0.50% essential oil. The chemical constituents of the essential oil are 1,8-cineole, artemisia ketone, 3-hexenyl butyrate, α-thujone, α-thujone isomer, cis-sabinene hydrate, camphor, borneol, pinocarvone-4, α-terpineol, artemisone, and α-santalol. It grows in Sichuan, Qinghai, Yunnan, and Tibet at an altitude of 3500~3800 meters on gravelly slopes and roadsides. The essential oil can be extracted to obtain α-thujone.

A. sphaerocephala

The whole plant contains 0.67% essential oil. The chemical constituents of the essential oil are β-caryophyllene, limonene, β-ocimene, γ-pinene, pinocarvone-4, 3-nonanal, α-thujene, thujopsene, nerol, α-bisabolol oxide B, α-bisabolol, and others. It is found in various provinces in North and Northwest China, growing in deserts and sand dunes at altitudes between 1000~2500 meters. The plant is used in medicine as an anti-inflammatory and anthelmintic agent.

A. stolonifera

The whole plant contains 0.27% essential oil. The chemical constituents of the essential oil are 7-octen-4-ol, 6-methyl-5-hepten-2-one, camphene, 1,8-cineole, linalool, α-selinene, camphor, bornyl acetate, pinocarvone-4, pulegone, myristicin, β-thujene, β-caryophyllene, α-bisabolol, and nerol. It is found in Northeast, North and East China, Shandong, Jiangsu, Anhui, Zhejiang, and Hubei provinces, as well as Japan, Korea, and Central Asia. It usually grows at low altitudes near the edge of forests, and in moist areas. The essential oil can be used to make liniments.

A. subulata

The whole plant contains 0.56% essential oil. The chemical constituents of the essential oil are 2-hexenal, 7-octen-4-ol, 1,8-cineole, α-pinene, γ-pinene, camphor, bornyl acetate, pinocarvone-4, α-pinene, β-thujene, β-phellandrene, and α-phellandrene. It is found in Northeast and North China, as well as in Japan and Korea in East Asia. It usually grows on low-altitude slopes, forest edges, riverbanks, and swamp edges. The plant has the potential to be used for the development of medicinal essential oils.

A. sylvatica

The whole plant contains 0.22% essential oil. The chemical constituents of the essential oil are hexanal, 2-hexenal, 7-octen-4-ol, 1,8-cineole, linalool, camphor, borneol, pinene-4, alpha-pinene, artemisia ketone, indole, beta-bourbonene, beta-caryophyllene, valencene, beta-eudesmol, alpha-santalol, etc. It is distributed throughout China except for Xinjiang, Fujian, Guangdong, and Hainan. It is also found in North Korea, Mongolia, and eastern Russia. It grows in shaded areas on the edge of forests and shrubs.

A. verlotorum

The essential oil components vary depending on the place of picking. The whole plant contains 0.57% essential oil. The chemical constituents of the essential oil are 7-octenol-4, 3-octanol, 1,8-cineole, camphor, artemisia ketone, kesserol, beta-caryophyllene, jinhehuanene, alpha-caryophyllene, valencene, etc. It is found everywhere except for extremely dry and cold regions. It is distributed in tropical and subtropical regions worldwide. It grows at low to medium altitudes.

A. vestita

The whole plant contains 0.30%~0.40% essential oil. The chemical constituents of the essential oil are sesquiterpenes, 1,8-cineole, 3,6,6-trimethyl-2-borneol, alpha-thujone, alpha-thujone isomer, borneol, para-isopropylphenol, pulegone, nerolidol, etc. It is found in northwestern and southwestern China, western Hubei, and northwestern Guangxi. It grows on mountain slopes, grasslands, and shrubs at medium to low altitudes. The whole plant is used medicinally for its heat-clearing, anti-inflammatory, and diuretic properties.

A. vulgaris

The whole plant contains 1.20% essential oil. The chemical constituents of the essential oil are 7-Octen-4-ol, 1,8-cineole, butyl-3-hexenyl ester-alpha-thujone, alpha-thujone isomer, hinokitiol, camphor, bornyl acetate, pine oil alcohol-4, methoxyeugenol, beta-caryophyllene, alpha-caryophyllene, beta-farnesene, nerolidol, etc. It is distributed in Shaanxi, Gansu, Qinghai, Xinjiang, Sichuan, Mongolia, and Eurasia. It grows on roadsides, grasslands, and slopes at an altitude of 1500~3500 meters.

A. waltonii

The whole plant contains 0.75% essential oil. The chemical constituents of the essential oil are 1,8-cineole, linalool, camphor, withered alcohol, meta-anisaldehyde, methoxyeugenol, inula acetylene, linalyl acetate, inula acetylene ketone, jin hehuanol, etc. It is distributed in southern Qinghai, western Sichuan and Tibet. It grows in shrubs, mountain slopes, and grasslands at altitudes of 3000 to 4300 meters. The essential oil has strong antibacterial ability and has the potential to be developed into products. Locals in Tibet use it for incense.

A. xerophytica

The whole plant contains 0.54% essential oil. The chemical constituents of the essential oil are 6-methyl-5-hepten-2-one, myrcene, 1,8-cineole, camphor, 3,6,6-trimethyl-2-menthenol, alpha-thujone, alpha-thujone isomer, cis-hinokitiol, camphor, pine oil alcohol-4, nerolidol, trans-hinokitiol, nerolidol, etc. It is distributed in Inner Mongolia and provinces in the northwest. It grows in the Gobi, semi-arid grasslands, and semi-fixed sand dunes. It is a medicinal essential oil resource with development potential.

A. yunnanensis

The whole plant contains 0.26% essential oil. The chemical constituents of the essential oil are hexanal, 2-hexenal, benzaldehyde, 7-octen-4-ol, 1,8-cineole, linalool, alpha-thujone, alpha-thujone isomer, camphor, bornyl acetate, pine oil alcohol-4, ta-ojinnanol, benzofuran, coumarin, etc. It is distributed in Qinghai, Sichuan, and Yunnan. It grows in dry river valleys or limestone areas below 3700 meters. Its essential oil has the potential to be developed into medicinal essential oil.

Aster ageratoides

The whole plant contains 0.22%~0.30% essential oil. The chemical constituents of the essential oil are α-ocimene, linalool, δ-terpinene, β-terpinene, α-germacrene, β-caryophyllene, α-caryophyllene, β-farnesene, α-farnesene, δ-selinene, β-elemene, γ-terpinene, δ-cadinene, γ-terpineol, bornyl alcohol, labdane, and lavalcohol. It is found in Guangdong and Guizhou provinces in China. The essential oil is rich in sesquiterpenes and has a pleasant fragrance, making it suitable for use in perfume production. It is also known as Fei Ji Cao in Chinese.

Blumea balsamifera

The whole plant contains essential oil. The chemical constituents of the essential

Blumea balsamifera

oil are camphor, l-menthol, β-caryophyllene, elemene, and β-eudesmol, among others. Its leaves and roots are used in traditional medicine to promote blood circulation, dispel wind, and relieve diarrhea, abdominal pain, and other ailments. It is also the source of "artemisinin", an important antimalarial drug. It grows in Yunnan, Guizhou, Guangxi, Guangdong, Fujian, and Taiwan.

Conyza canadensis
The whole plant contains 0.1%~0.3% essential oil. The chemical constituents of the essential oil are α-pinene, β-pinene, linalool, limonene, geraniol, caryophyllene, cis-thujone and trans-thujone, β-thujone, nerolidol, and olivene. Its decoction is used in traditional medicine to inhibit dysentery and paratyphoid fever. It is native to North America but is now distributed in both northern and southern China, growing in the wild, in wastelands, and in fields.

Cyathocline purpurea
The whole plant contains essential oil. The chemical constituents of the essential oil are 1-octen-3-ol, ethyl octanoate, vanillin, butyl vanillate, isobutyl vanillate, and pentyl vanillate, among others. It is used in traditional medicine to kill insects and has anti-inflammatory, detoxifying, heat-clearing, and hemostatic effects. It grows in the undergrowth of mountain slopes, grasslands, and fields near water in Yunnan, Sichuan, Guizhou, and Guangxi provinces in China.

Eupatorium odoratum
The whole plant contains 0.12%essential oil. The chemical constituents of the essential oil are α-pinene, sabinene, β-pinene, α-ocimene, cubebene, β-caryophyllene, β-farnesene, γ-terpinene, δ-selinene, and guaiol. Originally from America, it is now distributed in Yunnan and Hainan provinces in China. The plant is used in traditional medicine to treat external injuries and bleeding, as well as unknown toxicities.

Parthenium argentatum
Leaves contain 1.0% essential oil. The chemical constituents of the essential oil are isobutanol, α-pinene, camphene, sabinene, β-pinene, 1,8-cineole, menthone, methyl salicylate, and ethyl salicylate. Originally from America, it is now

Eupatorium odoratum

introduced to the southern part of China. This plant is a raw material that is rich in rubber.

Stemmacantiha uniflora

The whole plant contains 0.02/100 mg essential oil. The chemical constituents of the essential oil are tridecane, cubebene, cis-caryophyllene, selinene, β-caryophyllene, pentadecane, γ-terpinene, α-eudesmol, δ-juniperene, oxide-caryophyllene, hexadecane, β-eucalyptol, pentadecanal, and 6-hydroxy-3-[(3-methyl-2-butenyl)oxy]-2-cyclohexen-1-one. It grows in mountainous areas, slopes, foot of the mountain, and field edges. Its roots are used in traditional medicine to clear heat and detoxify.

Saussurea purpurascens

The whole plant contains essential oil. The chemical constituents of the essential oil are linalool, β-pinene, chamazulene, γ-juniperene, β-cedrene, δ-juniperene, 4-methyl-2,6-tert-butylphenol, hexadecane, β-acacia alcohol, β-acacia aldehyde, 2,6-di-tert-butyl-p-benzoquinone, cypressene alcohol, β-eucalyptol, δ-juniperol, heptadecane, 1-tridecene, and 1,2-dihydro-2,2,4- trimethyl-quinoline. It grows in Northeast China, Northwest China, Eastern China, and Southern China at altitudes of 300~1800 meters on hillsides, grasslands, and roadside ditches. It is also distributed in Japan and Korea.

Seriphidium nitrosum

The whole plant contains 0.43% essential oil. The chemical constituents of the essential oil are ethyl 3-methylbutanoate, eucalyptol, α-thujone, α-thujone isomer, camphor, borneol, and pine oil alcohol-4. It grows in the western part of Inner Mongolia, the northwestern part of Gansu, and Xinjiang, altitudes below 1500 meters in desert grasslands. Its essential oil has the ability to kill and control pests in warehouses, and can be developed as a natural resource for α-thujone.

S. transiliense

The whole plant contains 0.36%~0.48% essential oils, which exhibit activity against pests in storage. The chemical constituents of the essential oil are 1,8-cineole, α-thujone, α-thujone isomers, cis-pinaneol, camphor, borneol, pinocarveol-4, kuraridin, and cis-pinaneol. It grows in the northern part of Xinjiang, China.

Tagetes erecta

The whole plant contains 0.10% essential oil. The chemical constituents of the essential oil are limonene, β-ocimene, β-ocimene-X, isopinocamphone, 4-methyl-6-hepten- 3-one, caryophyllene oxide, decanal, tagetone, linalool, t-beta-caryophyllene, ethyl-3-hexenoate, perilla aldehyde, citral, borneol, α-terpineol, eudesma-4(15),7-dien-11-ol, and others. It is native to South America, specifically

• *Tagetes erecta*

Adenosma glutinosum

Limnophila rugosa

treat rheumatism, bone pain, bruises, eczema, and insect bites.

A. indianum

The whole plant contains 0.40% essential oil. The chemical constituents of the essential oil are α-pinene, β-pinene, limonene, p-cymene, 1,8-cineole, phellandrene-3, linalool, carvone, bornyl alcohol, α-terpineol, musk tenuiflorum, t-cadinol, d-cadinenone, α-cubebene, β-caryophyllene, cedrol, isoborneol, methyl isoeugenol, α-gurjunene, acetoxydihydrocadalene, α-amyrin, and patchouli alcohol, among others. It is found in Guangdong, Guangxi, Yunnan and other provinces and regions, growing on dry slopes at altitudes of 200~600 meters.

Scrophulariaceae family

Limnophila rugosa

The whole plant contains 0.20%~0.43% essential oil. The chemical constituents of the essential oil are aicaonene, t-anethole, cis-anethole, anethole, α-copaene, linalool, among others. It is found in Guangdong, Yunnan and Taiwan, growing near water. The whole plant is used medicinally and has a slightly pungent taste. It is used to clear heat and relieve surface symptoms, and can be used to treat colds. It is also used by some ethnic minorities in Yunnan as a pickling seasoning.

Acanthaceae family

Justicia gendarussa

The whole plant contains 0.42% essential oil. The chemical constituents of the essential oil are 2-ethyl furan, 3-methyl butanol, 1,3,5-cycloheptatriene, acetaldehyde, methyl pyrrole, 3-hexenal, 3-hexenol, 3,5-dimethylpyrrole, benzaldehyde, 1-hexen-3-ol, 3-octanone, linalool, 3-(2-methylpropyl)-2-cyclohexenone, methyl 2-hydroxybenzoate, α-ionone, and β-ionone, among others. It is found in southern and southwestern China, as well as in tropical regions of Asia. It grows in roadside shrubs and forests.

Verbenaceae family

Caryopteris forrestii

The whole plant contains essential oils ranging from 0.24%~2.00%. The chemical constituents of the essential oil are α-pinene, β-pinene, β-caryophyllene,

Justicia gendarussa

α-phellandrene, p-cymene, limonene, β-myrcene, sabinene, bornyl acetate, and other terpenes. It grows in Sichuan, Yunnan, Guizhou, and Tibet at an altitude of 1700~3000 meters on hillsides, roadsides, and wastelands. The plant has traditionally been used as a medicine to dispel wind, relieve surface-level symptoms, alleviate cough, and alleviate pain. It is also used as a source of limonene.

C. incana

The whole plant contains 0.24%~2.00% essential oil. The chemical constituents of the essential oil are α-pinene, sabinene, β-pinene, α-phellandrene, β-caryophyllene, p-cymene, limonene, t-β-ocimene, β-elemene, α-phellandrene, β-myrcene,

Caryopteris incana

α-terpinene, β-caryophyllene, α-copaene, and other terpenes. It grows in Jiangsu, Anhui, Zhejiang, Jiangxi, Hunan, Hubei, Fujian, Guangdong, and Guangxi at an altitude of hillsides and roadsides. The plant has traditionally been used as a medicine to dispel wind, relieve surface-level symptoms, alleviate cough, and alleviate pain. It can also be used for treating snake bites.

C. trichosphaera

The whole plant contains 0.24% essential oil.The chemical constituents of the essential oil are α-pinene, β-pinene, p-cymene, limonene, α-cubebene, β-elemene, β-caryophyllene, sabinene, and other terpenes. It grows in western Sichuan, Deqin and Zhongdian in Yunnan, and Changdu in Tibet at an altitude of 2700~3300 meters on hillsides, shrubs, and river valleys. The plant has traditionally been used as a medicine to dispel wind, relieve surface-level symptoms, alleviate cough, and alleviate pain. It can also be used as a source of limonene.

Lantana camara

Branches and leaves contain 0.02% essential oil. The chemical constituents of

the essential oil are 1,8-cineole, γ-terpinene, linalool, camphor, borneol, cubenol, α-pinene, β-caryophyllene, γ-terpinene, δ-juniperene, nerolidol, α-ylangene, and β-elemene. It is native to the tropical regions of America and is grown in Taiwan, Guangdong, and Guangxi in China. The roots, leaves, and flowers are used in traditional medicine for their effects in clearing heat and toxins, relieving pain and swelling, and relieving itching.

Lamiaceae family

Agastache rugosa

The whole plant contains 0.28% essential oil. The chemical constituents of the essential oil are β-caryophyllene, β-pinene, α-terpineol, α-humulene, β-terpineol, ylangene, δ-terpineol, and linalool. It is widely distributed throughout China and commonly cultivated. The essential oil is a precious natural fragrance with a long-lasting aroma, often used in perfumes. The stems and leaves are aromatic and can be used as a digestive aid, treating colds, headaches, and fever. The water decoction of the plant has inhibitory effects on spirochetes.

Elsholtzia cyprianii

The whole plant contains 0.81% essential oil (dry weight). The chemical constituents of the essential oil are octanol-3, 1-octenol-5, benzaldehyde, linalool, phenylethyl ketone, naphthyl methyl ketone, t-pinene, β-caryophyllene, β-oxo-santalene, musk thyme phenol, and oxidized p-cymene. It is found in Shaanxi, Henan, Anhui, Hubei, Hunan, Guizhou, Sichuan, Guangxi, and Yunnan provinces in China, growing along roadsides, riverbanks, and forest edges at altitudes of 400~2900 meters. The whole plant is used in traditional medicine for treating colds and boils, among other conditions.

Agastache rugosa

Leonurus japonicus

Leonurus japonicus

The whole plant contains 0.083% essential oil. The chemical constituents of the essential oil are 1-octen-3-ol, octanol-3, linalool, cubenene, cis-piperitol, trans-piperitol, elemol, gamma-terpineol, delta-selinene, piperitenone oxide, and plant sterols. It is distributed throughout China, growing in mountains, wilderness, riverbanks, grasslands, and creeks. It can be used in traditional medicine to regulate menstruation, promote blood circulation, dispel stasis, diurese, and reduce swelling.

L. macranthus

The whole plant contains 0.05% essential oil. The chemical constituents of the essential oil are 1-octen-3-ol, cubenene, beta-bourbonene, trans-piperitol, beta-bisabolene, elemol, gamma-terpineol, acacetin propionate, and plant sterols. It is distributed in Liaoning, Jilin, and Hebei provinces, growing on grassy slopes and bushes at an altitude of 400 meters. The essential oil has certain activity and can be used to treat bronchitis by reducing wheezing.

L. sibiricus

The whole plant contains 0.125% essential oil. The chemical constituents of the essential oil are 1-octen-3-ol, linalool, cubenene, cis-piperitol, trans-piperitol, elemol, gamma-terpineol, delta-selinene, piperitenone oxide, and plant sterols. It is distributed in Inner Mongolia, Shanxi, Shaanxi, Hebei provinces, and Central Asia. It grows on sandy grasslands and in pine forests. It has similar effects as large- flowered motherwort.

Mentha piperita

The chemical constituents of the essential oil are menthol, menthone, alpha-pinene, l imonene, 1,8-cineole, gamma-terpinene, and pulegone. Originally from Europe, it has been introduced to Beijing and Nanjing. It can be used for extracting raw materials of menthol.

M. spicata

The whole plant contains 0.60%~0.90% essential oil. The chemical constituents of the essential oil are limonene, menthone, linalool, menthol, dihydrocarvone, carvone, carvone oxide, beta-bourbonene, beta-pinene, terpinol-4, trans- dihydrocarvone, and acetyl carvacrol. Originally from the Canary Islands and Madeira Islands, it has been introduced to Hebei, Jiangsu, Zhejiang, Guangdong, Guangxi, Sichuan, Yunnan, and Guizhou. It can be used in

Mentha spicata

traditional medicine to treat colds, heatstroke, headaches, and externally to treat heat rash, eczema, sores, haemorrhoids, and bleeding.

Micromeria biflora

The whole plant contains 0.60% essential oil. The chemical constituents of the essential oil are β-pinene, limonene, 1,8-cineole, menthone, menthol, isomenthone, isopulegone, eugenol methyl ether, myristicin, and linalool. It is mainly found in Yunnan and Guizhou provinces in China, growing on slopes, grasslands, and riverbanks at altitudes between 3000~4000 meters. The plant is used in traditional medicine for its anti-inflammatory and antipyretic effects and is commonly used to treat colds, coughs, diarrhea, and abdominal pain.

Elsholtzia ciliata

The whole plant contains 0.08% essential oil. The chemical constituents of the essential oil are α-pinene, β-pinene, α-phellandrene, para-cymene, γ-terpinene, terpineol-4, thymol, carvacrol, t-α-terpinene, sabinene, β-bisabolene, and β-caryophyllene. It is found throughout China except in Xinjiang and Qinghai provinces, growing on roadsides, hillsides, wastelands, forests, and riverbanks. The plant is used in traditional medicine to treat paralysis, tuberculosis, hemorrhage, colds, and sores.

E. splendens

The whole plant contains 0.70% essential oil. The chemical constituents of the essential oil are α-pinene, 7-octen-4-ol, para-cymene, t-ocimene, β-terpineol, terpineol-4, thymol, β-caryophyllene, β-cyclocitral, and β-farnesene. It is found in several provinces in China, including Liaoning, Hebei, Henan, Shandong, Jiangsu, Zhejiang, Jiangxi, and Guangdong, growing in grasslands, ditches, hillsides, and roadsides. The plant is used in traditional medicine to treat colds, fever, coughs, and inflammation.

E. stauntoni

Leaves contain 1.30% essential oil. The chemical constituents of the essential oil are α-pinene, 1,8-cineole, benzaldehyde, linalool, bornyl alcohol, t-p-menth-8-en-1-ol, cis-p-menth-2-en-1-ol, δ-3-carene, and α-terpineol. It grows in valleys, streamsides, and along rivers at altitudes of 700~1600 meters in Hebei, Henan, Shanxi, Shaanxi, and Gansu. The essential oil can be used to treat dysentery, gastroenteritis, and colds.

Lavandula angustifolia

The whole plant contains 2.0%~2.30% essential oil. The chemical constituents of the essential oil are α-pinene, camphene, β-pinene, β-ocimene, limonene, 1,8-cineole, linalool, lavandulol, camphor, terpinen-4-ol, and acetate linalyl. It is native to the Mediterranean region and is now cultivated in many parts of China. The essential oil is an important ingredient in cosmetics, soaps, and floral water. It has anti- inflammatory, antiseptic, analgesic, and diuretic effects.

L. latifolia

The whole plant contains 1.20%~1.70% essential oil. The chemical constituents of the essential oil are α-pinene, camphene, β-pinene, 2-octanone, limonene, 1,8-cineole, linalool, camphor isobornyl acetate, bornyl acetate, and benzoic

Lavandula angustifolia

acid. It is native to France, Italy, Spain, Bulgaria, and the Balkan Peninsula, and has been introduced to Jiangsu and Shandong in China. Its uses are similar to those of lavender and it is commonly used in the fragrance industry.

Mosla chinensis

The whole plant contains 0.40%~0.60% essential oil. The chemical constituents of the essential oil are α-pinene, camphene, sabinene, β-pinene, β-ocimene, myrcene, geraniol, β-caryophyllene, α-terpineol, and β-eudesmol. It grows on grassy slopes and under forests below 1400 meters in East China, Central and South China, and Southwest China. The stems and leaves are used in traditional Chinese medicine to treat colds, heat stroke, vomiting, abdominal pain, diarrhoea, bruises, and eczema when applied externally.

M. dianthera

The whole plant contains 0.64% essential oil (Lamiaceae). The chemical constituents of the essential oil are linalool, limonene, camphor, menthone, resveratrol, t-patchoulene, 2,6-dimethyl-6-(4-isopentenyl)-2-cyclohexene-[3,1,1] hept-2-ene, β-patchoulene, squalene, t-β-caryophyllene, cis-β-caryophyllene, and more. It is found in East, Central, and Southwest China, as well as Shaanxi, growing on slopes, roadsides, and near water. The whole plant is used in folk medicine to treat colds, fever, heatstroke, headaches, nausea, and other ailments.

Mosla dianthera

M. formosana

The whole plant contains 0.18%~2.28% essential oil. The chemical constituents of the essential oil are carvone, elemol, resveratrol, umbellulone, patchouketone, α-cedrene, limonene, and pinene. This plant is found in Taiwan.

M. scabra

The whole plant contains 2.57%~3.50% essential oil. The chemical constituents of the essential oil are linalool, 2-methyl-5-(1-methylethyl)-2,5- cyclohexadiene-1,4-dione, eugenol, methyl vanillate, ethyl apiol, ethyl isoferulate, cubenene, β-patchoulene, α-bergamotene, α-patchoulene, β-caryophyllene isomers, α-humulene (Z,E), carvacrol, and carvacrol isomers. It is found in East, Central, Sichuan, Gansu, Shaanxi, and Liaoning, growing on slopes and roadsides. The whole plant is used in medicine to relieve heat, dispel wind and dampness, reduce swelling, and detoxify.

Nepeta cataria

The whole plant contains essential oil. The chemical constituents of the essential oil are 7-octenol-4, limonene, β-ocimene, α-ocimene, linalool, phenethyl ketone, patchouli ketone, α-citral, β-caryophyllene, α-humulene, β-selinene, α-bisabolene, nerolidol, and β-eudesmol. It grows in Xinjiang, Gansu, Shaanxi, Henan, Shanxi, Shandong, Hubei, Guizhou, Sichuan, and Yunnan, mostly in shrubs beside homes at an altitude of 2500 meters or lower. The whole plant is used medicinally to treat colds, headaches, sore throats, and conjunctivitis.

Ocimum basilicum

The whole plant contains essential oil. The chemical constituents of the essential oil are α-ocimene, linalool, carvone, p-cymene, eugenol methyl ether, eugenol, α-humulene, and α-bisabolene. It grows in Xinjiang, Jilin, Hebei, Zhejiang, Anhui, Jiangxi, Hubei, Hunan, Guangdong, Guangxi, Taiwan, and Yunnan, and is mostly cultivated. As an important cultivated aromatic plant, its essential oil can be used to make soap, toothpaste, and cosmetics.

O. gratissimum

The whole plant contains essential oil. The chemical constituents of the essential oil are eugenol, cubenene, β-selinene, β-caryophyllene, β-bergamotene, β-ocimene, γ-elemene, γ-elemene isomer, β-selinene isomer, γ-terpinene, and δ-terpinene.It is cultivated in Guangdong,

Ocimum gratissimum

Guangxi, Fujian, Zhejiang, Jiangsu, Shanghai, and other regions. The essential oil of this plant is an important resource for isolating eugenol and can also be used to make daily fragrance raw materials.

O. gratissimum var. *suave*

The whole plant contains 0.64%~1.27% essential oil. The chemical constituents of the essential oil are β-elemene, α-elemene, pinene-4, eugenol, β-caryophyllene, and β-bisabolene. It is cultivated in Jiangsu, Zhejiang, Fujian, Taiwan, Guangdong, Guangxi, and Yunnan provinces of China. The essential oil extracted from this plant can be used as a raw material in the food, pharmaceutical, and chemical industries. It is also a resource plant for isolating eugenol.

Origanum marjorana

The whole plant contains 0.20%~0.50% essential oil. The chemical constituents of the essential oil are α-pinene, camphene, sabinene, β-pinene, myrcene, 1,8-cineole, cis-thujanol (furan type), t-thujanol (furan type), linalool, camphor, borneol, pinene-4, α-terpineol, and methyl chavicol. It is native to the Mediterranean coast and Western Asia. Currently, it is cultivated in Guangdong, Guangxi, and Shanghai provinces of China. In France, Germany, and other countries, it is used to make herb-based alcoholic beverages and as a spicy seasoning for meat processing.

Perilla frutescens

Perilla frutescens

The whole plant contains 0.3% essential oil. The chemical constituents of the essential oil are 6-methyl-5-hepten-2-one, cis-thujanol (furan type), t-thujanol (furan type), linalool, geraniol, β-citral, p-cymene, α-citral, ethyl isoeugenol, β-caryophyllene, α-bisabolene (*Z*, *E*), and nerolidol. It is cultivated throughout China. The essential oil extracted from this plant is used as a raw material for isolating citral, and it can also be used directly as a fragrance in soap and food products. The whole plant is used in traditional Chinese medicine to treat wind-cold, cough, phlegm, asthma, constipation, etc. The seed oil can regulate blood concentration.

P. frutescens var. *crispa*

The whole plant contains 0.20%~0.26% essential oil. The chemical constituents of the essential oil are limonene, linalool, perilla aldehyde, and perilla alcohol, among others. It is cultivated throughout China and is used in traditional medicine to treat symptoms such as fever, chills, chest and abdominal distension, and indigestion.

Pogostemon cablin

The whole plant contains 1.50% essential oil. The chemical constituents of the essential oil are patchouli alcohol, benzaldehyde, eugenol, cinnamaldehyde, patchouli camphor, patchouli pyridine, beta-bourbonene, alpha-guaiene, and alpha-patchoulene, among others. It is grown in Fujian, Taiwan, Guangdong, and Guangxi. The essential oil has a strong and intense aroma, making it an excellent fixative. The plant is also used in traditional medicine to treat stomach disorders and fever.

Prunella vulgaris

The whole plant contains 0.31% essential

Perilla frutescens var. *crispa*

Pogostemon cablin

oil. The chemical constituents of the essential oil are alpha-pinene, beta-pinene, linalool, alpha-terpineol, 1,8-cineole, fenchol, 1-nonanol-4, pulegone, 1-menthene-8-ol, piperitone, ethyl-1-menthene-8-carboxylate, ethyl cinnamate, delta-cadinene, and ethyl salicylate, among others. It grows almost everywhere

Prunella vulgaris

in China, including forests, grasslands, wetlands, field ridges, and roadsides. The plant is used in traditional medicine to treat eye inflammation, goiter, lymph node swelling, breast hyperplasia, and high blood pressure.

Rosmarinus officinalis

The whole plant contains 0.48%~0.52% essential oil. The chemical constituents of the essential oil are alpha-pinene, camphor, beta-pinene, verbenone, 1,8-cineole, bornyl acetate, camphene, and beta-caryophyllene, among others. It is native to Europe and North Africa, but has been introduced to China. The plant is used in traditional medicine to improve digestion and induce sweating, and its essential oil has moderate inhibitory effects against Staphylococcus aureus and Escherichia coli.

Salvia sclarea

The whole plant contains 0.10%~0.12% essential oil. The chemical constituents of the essential oil are perilla ketone, β-caryophyllene, α-terpinolene, linalool, α-pinene, methyl chavicol acetate, estragole acetate, isomer of estragole acetate, bornyl acetate, β-pinene, and β-phellandrene. Originally from Europe, it is now cultivated in Shaanxi, Henan, Hebei, and other areas in China. The dried flowers are widely used in daily life, food, and also in the perfume, liquor and tobacco fragrance industries.

Teucrium viscidum

The whole plant contains essential oil. The chemical constituents of the essential oil are sabinene, β-phellandrene, 1,8-cineole, cubebene, β-phellandrene, β-ocimene, β-pinene, α-terpineol, α-pinene, fenchene, β-phellandrene, ylangene, β-caryophyllene, and nerol. It is found in various provinces south of the Yangtze River in China, as well as in Japan, Korea, and Southeast Asia. The whole plant is used in traditional medicine for cooling blood, detoxifying, promoting blood circulation and treating injuries, and snake bites.

Thymus mongolicus

The whole plant contains 0.20%~0.50% essential oil. The chemical constituents of the essential oil are α-pinene, perilla ketone, 7-octen-4-ol, β-caryophyllene,

2-carene, p-cymene, 1,8-cineole, γ-terpinene, linalool, camphor, bornyl acetate, para- tert-butylbenzyl alcohol, thymol, methyl thymol acetate, isomer of thujone, β-selinene, 2-hydroxy-5-methoxyacetophenone, and ylangene. It is found in Gansu, Qinghai, Shaanxi, Shanxi, Shandong, Hebei, and Inner Mongolia in China, growing in mountainous areas, valleys, and grasslands at altitudes of 1100~3600 meters. The essential oil is used in making toothpaste, body powder, soap, and detergents, and has strong antibacterial properties.

T. quinquecostatus

The whole plant contains 0.21%~0.58% essential oil. The chemical constituents of the essential oil are linalool, alpha-pinene, camphene, para-cymene, 1-octen-3-ol, muscone methyl ether, bornyl acetate, pinene, sabinene, thymol, and p-cymen-8-ol. It grows on slopes at altitudes of 600~900 meters in Shandong, Liaoning, Hebei, Henan, and Shanxi. The essential oil is used in cosmetics, while the whole plant is used in traditional medicine for cough suppression, anti-inflammatory effects, and treatment of colds and joint pain. It is also used to extract thymol and linalool.

Zingiberaceae family

Alpinia blepharocalyx

Seeds contain 0.11% enssential oil. The chemical constituents of the essential oil are alpha-terpineol, beta-pinene, 2-hydroxy-3-methylpentanoic acid methyl ester, para-cymene, cis-ocimene (furan type), trans-ocimene (furan type), linalool, citral, alpha-terpineol, cubenene, geraniol, ethyl acetate, and benzyl acetate. It grows in sparse forests at altitudes of 100~1000 meters in the southern and western parts of Yunnan. The seeds are used in traditional medicine to warm the stomach and invigorate the spleen, and to treat cold and pain in the abdomen, vomiting of sour fluids, and diarrhea caused by cold and dampness.

A. oblongifolia

Seeds contain 0.08%~0.11% essential oil. The chemical constituents of the essential oil are alpha-pinene, beta-pinene, limonene, ocimene, linalool, beta-citral, geraniol, alpha-citral, ethyl acetate, 3-phenyl-2-propen-1-ol, camphene, sabinene, and ethyl cinnamate. It grows in the southeastern and southwestern provinces of China, at altitudes of 100~2500 meters in the lower mountain forests. The essential oil of the seeds has strong antibacterial activity, and the seeds are used in traditional medicine to treat stomach pain and distension, belching, and diarrhea.

A. galanga

Rhizomes contain essential oil. The chemical constituents of the essential oil are kaeempferene, 1,8-cineole, camphor, p-cymene, terpinene-4-ol, methyl cinnamate, pentadecane, cis-p-methoxycinnamate, and trans-p-methoxycinnamate. It is found in various provinces in China, including Guangdong, Guangxi, and Yunnan. The plant is widely cultivated and the rhizome is often used as a food seasoning and in medicine to treat pain and colds.

A. guinanensis

Seeds contain 0.20% essential oil. The chemical constituents of the essential oil are cineol, 1,8-cineole, camphor, t-geraniol, nerolidol, cycloheptanol, 2-heptanol, bornyl acetate, salicyl alcohol, etc. It is found in Long'an County, Guangxi, growing in scrub on limestone slopes. The

Alpinia oblongifolia

Alpinia galanga

Alpinia guinanensis

Alpinia japonica

seed essential oil has strong antibacterial activity.

A. henryi

Seeds contain 0.09% essential oil. The chemical constituents of the essential oil are linalool, cineol, 1,8-cineole, borneol, camphor, 4-phenyl-2-butanone, α-cubebene, 4-phenyl-3-buten-2-one, δ-juniperene, salicyl alcohol, salicyl alcohol, etc. It is found in Guangdong, Yunnan, and Guangxi. It grows in dense forests in the mountains. The seeds and fruits are used in folk medicine in Guangdong, Guangxi, and Yunnan to substitute for white cardamom.

A. japonica

Seeds contain 0.10% essential oil. The chemical constituents of the essential oil are alpha-pinene, camphene, beta-pinene, 1,8-cineole, borneol, camphor, terpineol-4, α-terpineol, bornyl acetate, eucalyptol, ortho-propyltoluene, sabinene, yu-chun-woodene, α-cubebene, orange blossom alcohol, etc. It is found in southeastern, southwestern, and southern provinces of China. It grows in humid shady areas under forests. The seed essential oil has antibacterial activity and is used in folk medicine for treating cold and abdominal pain, rheumatism, and hemorrhage caused by physical strain.

A. hainanensis

Seeds contain 0.12% essential oil. The chemical constituents of the essential oil are camphene, 1,8-cineole, borneol, camphor, geraniol, terpineol-4, α-terpineol, 3-phenyl-3-buten-2-one, 3-phenyl-2- butanone, β-elemene, α-cubebene, β-pinene, orange blossom alcohol, carotene alcohol, juniper alcohol, and acacia alcohol. It is found in Guangdong and Guangxi, growing in sparse forests or under dense forests in the mountains. The seeds are used in folk medicine for treating coldness and pain in the heart and abdomen, indigestion with cold and dampness, and vomiting and diarrhoea.

Alpinia hainanensis

A. maclurei

Seeds contain 0.09% essential oil. The chemical constituents of the essential oil are β-pinene, camphene, terpinolene, 5-(1-methylethyl)-2-cyclohexenone, p-isopropylbenzaldehyde, bornyl acetate, 2-propenyl-3-methyl-4-cyclohexenol, α-cubebene, ethyl phenylacetate, etc. It is distributed in Guangdong, Guangxi, Yunnan, and also in Vietnam. It grows in sparse forests or dense forests on mountains. It is used traditionally as a substitute for Yi Zhi herb.

A. malaccensis

Seeds contain 0.03% essential oil. The chemical constituents of the essential oil

Alpinia maclurei

Alpinia oxyphylla

Alpinia tonkinensis

are 1,8-cineole, geraniol, 4-phenyl-3-buten-2-one, capric acid, methyl salicylate, nerol, dodecanoic acid, β-sitosterol, α-sitosterol, myristic acid, palmitic acid, etc. It is cultivated in Tibet, Yunnan, Xishuangbanna and Guangdong, and distributed in Southeast Asia. It grows under under evergreen broad-leaved forests. The seeds are traditionally used in medicine with pain-relieving, digestive, and vomiting effects, commonly used for the treatment of gastrointestinal diseases. It is also known as *Alpinia hainanensis*.

A. oxyphylla

Seeds contain 2.00% essential oil. The chemical constituents of the essential oil are α-cedrene, β-pinene, α-pinene, β-myrcene, α-terpinene, 3-carene, camphene, pinocarvone-4, linalool oxide, chamazulene, etc. Cultivated in Guangdong, Guangxi, and in recent years in Yunnan and Fujian. It grows in damp places under forests. The seeds are traditionally used in medicine with warming spleen, stopping diarrhea, promoting salivation, warming the stomach, consolidating essence, and shrinking urine effects.

A. tonkinensis

Seeds contain 0.14% essential oil. The chemical constituents of the essential oil are α-pinene, camphene, β-pinene, limonene, elemene, 2,6-dimethyl-2,7-

octadien-6-ol, trans-nerolidol, bornyl acetate, 3-phenylpropanoic acid, muscone, methyl salicylate, etc. It is distributed in Guangxi. The seeds are traditionally used as a substitute for Alpinia katsumadai in medicine.

A. zerumbet

Seeds contain 0.30%~0.50% essential oil. The chemical constituents of the essential oil are α-pinene, β-pinene, para-cymene, 1,8-cineole, α-terpineol, α-cubebene, t-germacrene, cis-t-nerolidol, etc. It is distributed in Southeast to Southwest China. The seeds are used in traditional medicine to dry dampness, dispel cold, invigorate the spleen, and warm the stomach.

Alpinia zerumbet

Amomum aurantiacum

Seed contains 0.08%~1.20% essential oil. The chemical constituents of the essential oil are α-pinene, β-pinene, linalool, terpineol, β-elemene, nerolidol, patchouli alcohol, etc. It is found in Yunnan, growing on slopes and under forest at an altitude of 600 meters. The fruit is used in traditional medicine and has an aromatic and stomach-invigorating effect.

A. austrosinense

Seeds contain 0.60%~0.80% essential oil. The chemical constituents of the essential oil are α-pinene, β-pinene, 1,8-cineole, limonene, linalool, α-terpineol, nerolidol, etc. It is found in Guangxi and Hunan, growing on mountains at an altitude of 450 to 1000 meters. The whole plant is used in traditional medicine to treat rheumatism, bone pain, bruises, and stomach cold.

A. longiligulare

Seeds contain 1.47% essential oil. The chemical constituents of the essential oil are limonene, camphor, isoborneol, borneol, α-terpineol, bornyl acetate, α-santalene, nerolidol, etc. It is found in Chengbian, Yaxian, and Danxian of Hainan Province, and cultivated in Xuwen and Suixi of Guangdong Province. It grows in dense forests in valleys. The fruit can be used as an aromatic and stomach-invigorating agent, but its effectiveness is not as good as that of cardamom.

A. maximum

Seeds contain 0.50% essential oil. The chemical constituents of the essential oil are 1,8-cineole, linalool, bornyl cinnamate, camphor, isobornyl acetate, bornyl acetate, t-ocimenone (furan type), pinene-4, campholenic alcohol, α-pinene, cineol, and others. It is found in Yunnan, Tibet, Guangdong, and Guangxi. It grows in shaded, moist areas The flesh of the fruit is edible, and it can also be used as a substitute for sand ginger and cardamom.

A. muricarpum

Leaves contain 0.08% essential oil. The chemical constituents of the essential oil are α-pinene, β-pinene, linalool, α-terpineol, bornyl cinnamate, camphor, and others. It is found in Guangdong, Guangxi, and also in the Philippines. It grows in dense forests at an altitude of 300~1000 meters. The essential oil of the leaves has moderate antibacterial activity.

A. odontocarpum

Seeds contain 1.20%~1.40% essential oil. The chemical constituents of the essential oil are α-pinene, β-pinene, β-ocimene, 1,8-cineole, γ-terpinene, linalool, terpinen-4-ol, α-terpineol, γ-terpinene, and others. It is found in Guangxi. It grows in sparse forests at an altitude of 1500 meters. The fruit can be used medicinally as a substitute for cardamom.

A. subulatum

Seeds contain 0.20%~0.30% essential oil. The chemical constituents of the essential oil are α-pinene, β-pinene, β-ocimene, 1,8-cineole, linalool, pinene-4, α-terpineol, zingiberene, and others. It is found in Tibet, Yunnan, and Guangxi. It grows in shaded, moist forests at an altitude of 300~1300 meters. The fruit is a traditional Chinese medicine with stomach-strengthening and pain-relieving effects and can also be used as a seasoning.

A. tsaoko

Seeds contain 0.70%~0.90% essential oil. The chemical constituents of the essential oil are α-phellandrene, α-selinene, linalool-4, α-terpineol, citral, geraniol, 2-phenyl-2-butenal, bornyl acetate, α-citral, 2-decanal, 1-ethyl-3-methylbenzene, 2-methylcinnamaldehyde, methyl eugenol, dodecanal, and nerolidol. It belongs to the ginger family and is commonly used as a seasoning ingredient and food preservative. It grows wild in sparse forests or is mostly cultivated in Yunnan, Guangxi, Guizhou, and other provinces of China. Tsaoko is also used in traditional medicine for dispelling cold and dampness, relieving abdominal pain caused by internal obstruction.

A. villosum

Seeds contain essential oil. The chemical constituents of the essential oil are 2.50%~3.00%. The chemical constituents of the essential oil are α-phellandrene, karenene, β-ocimene, limonene, camphene, terpinen-4-ol, cubebene, and nerolidol. It belongs to the ginger family and grows wild in mountainous forests, but is mostly cultivated in Fujian, Guangdong, Guangxi, and Yunnan provinces of China. Cardamom is a traditional aromatic and digestive medicine in China and is used for promoting digestion, relieving flatulence, and resolving food

Amomum tsaoko

Amomum villosum

Curcuma aromatica

stagnation. The essential oil extracted from the leaves has similar medicinal effects to those of the seeds.

A. villosum var. *xanthioides*

Green cardamom is a variety of cardamom. Rhizomes contain 3.00% essential oil. The chemical constituents of the essential oil are limonene, camphene, terpinen-4-ol, cubebene, cubebol, α-cubebene, α-phellandrene, β-phellandrene, karenene, linalool, and α-terpineol. It grows in southern Yunnan and is also cultivated in Guangdong Province of China. Green cardamom has similar medicinal properties to those of cardamom.

Curcuma aromatica

Rhizomes contain 6.10% essential oil. The chemical constituents of the essential oil are 1-α-turmerone, 1-β-turmerone, camphor, curcumenol, ar-turmerone, germacrone, ar-curcumene, and others. It is commonly used in traditional medicine to promote digestion, relieve depression, cool the blood, and disperse blood stasis. It is native to the southeastern and southwestern regions of China.

C. wenyujin

Rhizomes contain essential oil. The chemical constituents of the essential oil are 1,8-cineole, camphor, borneol, eucalyptol, curzerene, ar-turmerone, germacrone, curcumene, curcumenol, curcumin, and others. The essential oil of this variety has been found to have a certain therapeutic effect on cervical cancer. It is grown in well-drained sandy soil. It is cultivated in Zhejiang, China.

C. kwangsiensis

Rhizomes contain essential oil. The chemical constituents of the essential oil are alpha-pinene, beta-pinene, 1,8-cineole, camphor, linalool, curzerene, ar- turmerone, germacrone, curcumene, curcumenol, curcumin, and others. Like the other varieties, the essential oil of this plant has been found to have a certain therapeutic effect on cervical cancer. It is native to Guangxi and Yunnan, China.

Curcuma kwangsiensis

C. longa

Rhizomes contain essential oil. The chemical constituents of the essential oil are alpha-pinene, beta-pinene, 1,8-cineole, camphor, curzerene, ar-turmerone, germacrone, curcumene, curcumenol, curcumin, and others. The essential oil of this plant has also been found to have a certain therapeutic effect on cervical cancer, and the rhizomes are used in traditional medicine to promote blood circulation, relieve hypertension and gynecological diseases. It is native to Taiwan, Fujian, Guangdong, Guangxi, Yunnan, and Tibet.

C. phaeocaulis

Rhizomes contain essential oil. The chemical constituents of the essential oil are α-pinene, camphene, β-pinene, 1,8-cineole, borneol, camphor, bornyl acetate, caryophyllene, eugenol, ar-turmerone, zingiberene, curcumenol, ar-curcumene, and dehydrocurdione. It is found in Taiwan, Fujian, Jiangxi, Guangdong, Guangxi, Sichuan, Yunnan and other provinces and regions. It grows wild in shaded areas of forests and is cultivated. The essential oil has antibacterial activity, and the rhizomes are used in traditional medicine to relieve gas,

Curcuma phaeocaulis

promote blood circulation, relieve indigestion, stop bleeding, and treat stubborn indigestion, bruises, and injuries.

C. zedoaria

Rhizomes contain essential oil. The chemical constituents of the essential oil are 1,8-cineole, 3,6-dimethyl-1,6-octadien-3-ol, camphor, isoborneol, borneol, pinosylvin-4, α-pinen-4-ol, (+)-cis-kessane-4-ol, α-cedrene, β-cedrene, ar-turmerone, zedoarondiol, gigantol, and curcumenol. It is found in Fujian, Guangdong, Guangxi, Taiwan, Sichuan, Yunnan and other provinces and regions. It grows wild in valley forests and is cultivated. The rhizomes are used in traditional medicine for their gas-relieving, blood circulation-promoting, digestion-relieving, and pain-relieving effects.

Etlingera littoralis

Leaves contain essential oil. The chemical constituents of the essential oil are α-pinene, camphene, β-pinene, 1,8-cineole, linalool, α-terpineol, α-terpinyl acetate, citronellal, bornyl acetate, and nerolidol. It is found in Guangdong and Malaysia. It grows in shaded areas at altitudes of 200~300 meters. The leaf

Etlingera littoralis

essential oil has a Chinese medicine-type aroma and is not yet fully developed.

Hedychium coronarium

Flowers contain essential oil. The chemical constituents of the essential oil are α-pinene, sabinene, β-pinene, β-caryophyllene, 1,8-cineole, d-limonene, β-phellandrene, methyl benzoate, linalool, cis-ocimene, and trans-ocimene. It is found in various provinces in China, including Taiwan, Guangdong, Hunan, Guangxi, Yunnan, and Sichuan. It grows in forested areas and is also cultivated. The flower has a pleasant fragrance and is an important aromatic flower in southern China.

Hedychium coronarium

Zingiber officinale

Rhizomes contain essential oil. The chemical constituents of the essential oil are sesquiterpenes, α-pinene, camphene, sabinene, β-pinene, myrcene, β-phellandrene, 1,8-cineole, and terpinen-4-ol. It is found in various provinces in China, including the southeastern and southwestern regions. The essential oil of ginger can be used in food and cosmetics, while the rhizome is a commonly used seasoning and has medicinal properties such as antibacterial, anti-inflammatory, and diaphoretic effects.

Liliaceae family

Allium sativum

Bulbs contain 0.50% essential oil. The chemical constituents of the essential oil are allyl disulfide, methyl allyl trisulfide, 3-vinyl-1,4-dimethyl-1,2-dithiin, diallyl, 3-vinyl-1,2-dithiin-5-cyclohexene, diallyl trisulfide, methyl allyl tetrasulfide,3-vinyl-1,2-dithiin-4-cyclohexene, etc. It is originally from western Asia and Europe. It is now planted in various regions of China. It can be used as a vegetable and seasoning. Its essential oil is mostly used in

Zingiber officinale

canned food, which can prevent arteriosclerosis. The essential oil has bactericidal effects and is used in medicine for diuretic, expectorant, sedative, and anti-inflammatory effects.

Araceae family

Acorus calamus

The whole plant contains 1.50%~3.20% essential oil. The chemical constituents of the essential oil are beta-asarone, cis-isoelemicin, trans-asarone,calamenene, beta-pinene, asaraldehyde, etc. It is distributed in all provinces of China.It grows near water, marshes, lakes, and floating islands below an altitude of 260 meters. Its rhizomes have the effects of resolving phlegm, opening the orifices, invigorating the spleen, and diuresis. Its essential oil can be used to blend liquor spices.

A. gramineus

The whole plant contains 0.30% essential oil. The chemical constituents of the essential oil are beta-caryophyllene, beta-eudesmol, isoeugenol methyl ether, alpha-caryophyllene, methyl eugenol, spathulenol, nerolidol, 2,10,11-trimethyl-2,4,11-dodecatrien-6-one, etc. It is distributed in Zhejiang, Jiangxi, Hubei, Henan, Guangdong, Guangxi, Shaanxi, Gansu, Sichuan, and other provinces and regions. It grows in wetlands near water below 1800 meters and is often cultivated.

A. macrospadiceus

It is an herbaceous plant. The whole plant contains 1.50%~2.0% essential oil. The chemical constituents of the essential oil are alpha-pinene, beta-pinene, beta-selinene, pseudo-limonene, p-cymene, elemene, asarone, methyl eugenol, myristicin, and alpha-terpineol. It is found in Jiangdong, Rongshui County, Guangxi and has antibacterial and antifungal properties. Its oil is also used to extract methyl eugenol.

Sparganiaceae family

Sparganium stoloniferum

It is a herbaceous plant. Rhizomes contain 0.04%~0.06% essential oil. The chemical constituents of the essential oil are furfural, furfuryl alcohol, 5-methyl furfural, caproic acid, 2-acetyl pyrrole, phenethyl alcohol, 3-ethylphenol, 2-hydroxy-5-methyl phenyl ketone, 2,3-dihydrobenzofuran, 5-hexyl-2(2H)-furanone, elemicin, hydroquinone, 8-hydroxy-3-methyl-3,4-dihydro-1H-2-

benzopyran-1-one, beta-caryophyllene, dihydrodehydrodiconiferyl alcohol, and dehydrodiconiferyl alcohol. It is found in the Northeast, the Yellow River Basin, the middle and lower reaches of the Yangtze River, and Tibet. Its rhizomes are used in traditional Chinese medicine to treat stagnation of Qi and blood, abdominal pain, rib-side pain, amenorrhea, postpartum abdominal pain, bruises, and injuries.

Iridaceae family

Iris pallida

It is a rhizomatous perennial plant. Rhizomes contain 0.50%~0.80% essential oil. The chemical constituents of the essential oil are irone, benzoic acid, decanal, phenyl ethyl ketone, acetaldehyde, butyldienolide, eugenol, myristic acid, methyl benzoate, benzaldehyde, and phenethyl alcohol. It is native to Europe and is commonly cultivated in gardens in China. Its oil is used in the preparation of cosmetics and fragrances.

Poaceae family

Cymbopogon citratus

It is a herbaceous plant. The chemical constituents of the essential oil are myrcene, limonene, linalool, citral, geraniol, citronellal, nerol, neral, and geranyl acetate. It is cultivated in Fujian, Taiwan, and Yunnan. Its oil is used in the production of soap fragrances and is a common ingredient in many household products, such as cleaning solutions and insect repellents.

C. distans

The whole plant contains 0.20% essential oil. The chemical constituents of the essential oil are pulegone, para-cymene, linalool, acetic acid, and citronellene. It is commonly found in the mountainous grasslands of southwestern China, Gansu, and Shaanxi. The essential oil extracted from the plant has antibacterial and disinfectant properties and can be used to make soap. It can also be used to extract pulegone, which is used in the synthesis of menthol.

C. citratus

The whole plant contains 0.37%~0.40% essential oil. The chemical constituents of the essential oil are alpha-pinene, camphene, beta-pinene, citronellol, eucalyptol, citral, geraniol, and acetic acid. It is native to Southeast Asia and is cultivated in southern China. The essential oil extracted from the plant is used to produce citronellol and geraniol for use in various cosmetic fragrances.

C. winterianus

The whole plant contains 1.20%~1.40% essential oil. The chemical constituents of the essential oil are citronellal, geraniol, citral, geranyl acetate, limonene, eugenol methyl ether, isobutanol, butanedione,

Cymbopogon citratus

piperitone, and beta-caryophyllene. It is native to Sri Lanka and is cultivated in southern China. The essential oil extracted from the plant is widely used in the fragrance industry. It can be used to produce citronellal and geraniol, and is also used directly in soap making.

1.5 Fragrant Flower Essential Oil Resources

In addition to commonly used fragrant flower essential oil resources in China, such as *Osmanthus fragrans* and *Jasminum grandiflorum* in the Oleaceae family, many are introduced and naturalized species, such as *Magnolia grandiflora* in the Magnoliaceae family, *Michelia alba* and *M. champaca* in the Magnoliaceae family, and *Rosa rugosa* in the Rosaceae family. However, there is great potential for other fragrant flower essential oil resources that have yet to be developed. The table contains information about the chemical components and distribution of essential oils in several plant species, including *Syringa reticulate* subsp. *amurensis*, *Plumeria rubra* 'Acutifolia', *Gardenia jasminoides*, *Gardenia jasminoide* var. *fortuniana*, and *Luculia pinceana*. The table includes details about the yield of essential oil, the chemical components present in the oil, the areas where the plants grow, and the medicinal and cosmetic uses of the oils.

Magnoliaceae family

Magnolia denudata

Buds contain essential oil 0.29%~0.67%. The chemical constituents of the essential oil are pinene, beta-linalool, alpha-terpineol, 1,8-cineole, para-cymene, beta-caryophyllene, alpha-phellandrene, linalool, geranyl acetate, methyl salicylate, ethyl benzoate, t-nerolidol, beta-eudesmol, etc. Native to Zhejiang, Anhui, Jiangxi, southern Hunan, and northern Guangdong. Currently, it is widely planted south of the Yellow River. Buds (bracts) are used in traditional Chinese medicine to treat headaches, bone pain, toothaches, etc.

M. grandiflora

Fresh flowers contain essential oil. The chemical constituents of the essential oil are methyl acetate, α-phellandrene, camphene, pinene, β-phellandrene, 6-methyl-5-hepten-2-one, 1,8-cineole, methyl benzoate, linalool, phenethyl alcohol, methyl octanoate, bornyl acetate, nerolidol, α,β-citral, methyl undecanoate, jasmone, methyl laurate, etc. It is native to southeastern United States. It is planted in all provinces south of the Yangtze River in China. It is a common aromatic

Magnolia grandiflora

tree species in gardens. The leaves are used in traditional medicine to treat hypertension, and the tree is also used for lumber.

M. purpurella

Fresh flowers contain essential oil. The chemical constituents of the essential oil are pinene, β-phellandrene, limonene, α-terpineol,cis-furanoid oxide of linalool, trans-furanoid oxide of linalool, linalool, α-elemene, β-caryophyllene,methyl benzoate, phenethyl alcohol, 2-methyl-6-methylen-1,7-octadien-3-one, etc. It is found in Wangcheng County, Hunan Province, China. It grows in mountainous areas at an altitude of 150~300 meters. It is a unique tree species in China that can be used as an aromatic tree for garden landscaping.

M. sargentiana

Flower buds contain 0.30% essential oil. The chemical constituents of the essential oil are linalool, trans-caryophyllene, caryophyllene oxide, methoxy-eugenol, safrole, fargesol, γ, β, α-eudesmol, etc. It is found in Sichuan and Yunnan, China. It grows in high-altitude forests. Flower buds contain essential oil.

M. sprengeri

Flower buds contain 0.20%~0.30% essential oil. The chemical constituents of the essential oil are linalool, trans-caryophyllene, geraniol, caryophyllene oxide, fargesol, α, β, γ-eudesmol, safrole, etc. It is found in Hubei Province, China. It grows in mixed mountain forests. Flower buds and branches contain essential oil.

Manglietia chevalieri

Fresh flowers contain essential oil. The chemical constituents of the essential oil are phenol-3, hinokitiol, beta-pinene, lemonene, 1,8-cineole, linalool, naphthalene, elemicin, beta-caryophyllene, etc. It is originally from Vietnam and Laos. It is introduced and cultivated in Guangdong and Guangxi as excellent garden and landscaping trees, and used for construction materials.

M. fordiana

Fresh flowers contain essential oil. The chemical constituents of the essential oil are lemonene, beta-pinene, rhododendrol, and orange blossom alcohol, etc. It is a tree species used for timber in various provinces in south and southwest China.

Manglietia hainanensis, also known as Hainan Magnolia, has fresh flowers that contain essential oils, including beta-pinene, limonene, alpha-pinene, phytol, phenethyl alcohol, neryl aldehyde, elemicin, and methyl benzoate. It is a unique species in Hainan and is used as an excellent tree for garden landscaping and shipbuilding timber.

M. hainanensis

Fresh flowers contain essential oil. The chemical constituents of the essential oil are beta-pinene, lemonene, alpha-pinene, phytol, orange blossom alcohol,

rhododendrol, elemicin, and methyl salicylate. It is an endemic species in Hainan, used as excellent garden and landscaping trees, and for shipbuilding timber.

M. megaphylla

Fresh flowers contain essential oil. The

Manglietia hainanensis

Manglietia megaphylla

chemical constituents of the essential oil are pinene, beta-pinene, eucalyptol, and alpha-pinene. It is used as a timber tree in southern Guangxi and southwestern Yunnan provinces.

M. moto

Fresh flowers contain essential oil. The chemical constituents of the essential oil are 1,8-cineole, bornyl alcohol-4, alpha-terpineol, cubenene, beta-pinene, nerolidol, etc. It is native to Hunan, Guangdong, and Fujian. The wood has a fine texture and is used in fine woodworking.

Michelia alba

Fresh flowers contain 0.30% essential oil. The chemical constituents of the essential oil are methyl acetate, cis-linalool oxide, trans-linalool oxide, linalool, eugenol methyl ether, ethyl cinnamate, phenylethanol, etc. It is native to Indonesia. It is now widely planted in Guangdong, Guangxi, Yunnan, Sichuan, and Fujian. The leaves contain mainly camphor, and the flowers are an important source of fragrant flowers. The branches and leaves also contain 0.20%~0.28% essential oil.

Manglietia moto

Michelia alba

M. balansae

Fresh flowers contain essential oil. The chemical constituents of the essential oil are butyl acetate, ethyl butyrate, methyl 3-methylbutyrate, ethyl hexanoate, 2-methylpropyl-2-hexanoate, ethyl 1-ethoxyethyl acetate, etc. It is native to Guangdong, Guangxi, Hainan, Fujian, and southern Yunnan. It is a garden and landscaping tree species that can be used to develop floral fragrance resources.

Michelia champaca

M. champaca

Fresh flowers contain essential oil. The chemical constituents of the essential oil are heptanol-3, hexanol-3, phenyl ethanol (furan type), phenylpropanol (pyran type), methyl benzoate, phenylethyl alcohol, phenylpropyl acetate, jinhuanol, t-nerolidol, alpha-ionone, etc. It is native to Southeast Asia. It is now planted in Guangdong, Yunnan, and south of the Yangtze River. It can be used to extract essential oil and oleoresin.

M. figo

Flowers contain essential oil. The chemical constituents of the essential oil are ethyl 2-methylpropionate, butyl acetate, 1,3-butanediol, methyl hexanoate, 2-methypropyl hexanoate. It is widely

Michelia figo

planted in gardens and by local people in southern China. It is a tree species commonly used for garden and home landscaping.

M. foveolata

Flowers contain essential oil. The chemical constituents of the essential oil are 2-pentanol, 2-methyl-2-butene-1-ol, methyl 2-methylpropanoate, ethyl 2-ethoxyacetate, cis-3-hexenol, alpha-pinene, beta-pinene, methyl benzoate, phenethyl alcohol, methyl 2-octenoate, 1-methoxy-3,7-dimethyl-2,6-octadiene, etc. It is mainly found in eastern Guizhou, southern Yunnan, Jiangxi, Guangdong, southern Hunan and Vietnam, and is used as a tree species for wood materials.

Michelia macclurei

M. macclurei

Flowers contain essential oil. The chemical constituents of the essential oil are methyl 2-methylpropanoate, methyl 2-methylbutyrate, alpha-pinene, humulene, beta-pinene, methyl benzoate, 2-methyl-6-methylene-1,7-octadien-3-one, 1,2-dimethoxybenzene, limonene, etc. It is mainly found in Guangdong and Guangxi, and grows in forests below 1000 meters above sea level. It is a beautiful and fragrant flower used as a tree species for garden and landscaping.

M. macclurei var. sublanea

Flowers contain essential oil. The chemical constituents of the essential oil are methyl

Michelia macclurei var. *sublanea*

2-methylbutyrate, heptanal, humulene, beta-pinene, limonene, methyl benzoate, linalool, 1,2-dimethoxybenzene, ethyl salicylate, gamma-olena, beta-bisabolene, beta-caryophyllene, etc. It is mainly found in southwestern Guangdong and southern Guangxi, and is used as a tree species for garden and landscaping. The flower is beautiful and fragrant.

M. maudiae

Flowers contain 0.18% essential oil. The chemical constituents of the essential oil are α-pinene, camphene, β-pinene, 1,8-cineole, linalool, terpineol-4, α-terpineol, β-phellandrene, γ-terpinene, nerolidol, etc. It is native to Guangdong and Hainan, growing in dense forests at an altitude of 600~1000 meters. It can be developed as an ornamental tree in gardens.

M. mediocris

Flowers contain essential oil. The chemical constituents of the essential oil are 5,5-dimethyl-2-furanone, para-cymene, 1,8-cineole, t-oxidized linalool, cis-

Michelia maudiae

oxidized linalool, 1,2-dimethoxybenzene, fenchone, methyl benzoate, etc. It is native to Guangdong and Guangxi, also found in Vietnam and Cambodia. It grows in mixed forests at an altitude of 400~1000 meters. It can be used as an ornamental tree in gardens.

M. guangxiensis

Flowers contain essential oil. The chemical constituents of the essential oil are butyl acetate, ethyl butyrate, methyl 2-methylpropionate, methyl 2-methyl-2-propionate, ethyl caproate, ethyl 2-methylpropionate, etc. It is found in Miao'er Mountain in Guangxi and can be used as an ornamental tree in gardens.

M. yunnanensis

Flowers contain essential oil. The chemical constituents of the essential oil are limonene, camphor, bornyl acetate, methyl jasmonate, cypressene, borneol, carvone, p-cymene, etc. It is found in pine and fir forests or shrubs with red soil in Yunnan. The flowers can be used to extract essential oils and the leaves can be crushed and used as a seasoning.

M. odora

Flowers contain essential oil. The chemical constituents of the essential oil are methyl 2-methyl butyrate, ethyl acetate1-ethoxy, 2-heptanone-5-methyl-2-hexanol-4-methyl-2-heptanone, ethyl-4-hexenoate, cis-linalool oxide (furan type), t-linalool oxide (furan type), decanone-2, decanol-2, pentadecane, and others. It is native to mountain forests and sparse forests at altitudes of 100~1000 meters in Guangdong, Guangxi, Jiangxi, and Hainan, and is suitable for use as a garden ornamental and roadside tree due to its straight trunk and beautiful, fragrant flowers.

Michelia guangxiensis

Woonyoungia septentrionalis

Fresh flowers contain essential oil. The chemical constituents of the essential oil are ethyl 1-ethoxyethyl acetate, cis-furanoid oxide of linalool, trans-furanoid oxide of linalool, benzyl hex-3-enoate, methyl anthranilate, etc. It is found in northeast Guangxi and southeast Guizhou, China. It grows in limestone forests. It is a tall tree species.

Annonaceae family

Artabotrys hexapetalus

Flowers contain 0.75% essential oil. The chemical constituents of the essential oil are ethyl 2-methyl propanoate, butyl butyrate, ethyl butyrate, methyl 2-methyl propanoate, ethyl 2-methyl propionate, ethyl-2-methyl butyrate, 2-methyl-2-

Michelia odora

propenoic acid, 2-methyl propanoic acid, isobutyl butyrate, and 3,3-dimethyl-2-propenoic acid tert-butyl ester, among others. It is commonly found in cultivated areas in Zhejiang, Taiwan, Fujian, Jiangxi, Guangdong, Guangxi, and Yunnan, and is a fragrant and ornamental tree commonly seen in gardens in southern China, but rarely seen in the wild.

Cananga odorata

Flowers contain 0.5%~1.0% essential oil. The chemical constituents of the essential oil are para-methoxybenzyl alcohol, methyl benzoate, linalool, benzyl acetate, methyl salicylate, ethyl benzoate, isoeugenol, gamma-terpinene, and benzyl benzoate, among others. It is native to Indonesia, the Philippines, and Malaysia, and is cultivated in Taiwan, Fujian, Guangdong, Guangxi, and Yunnan, where many varieties exist. Its essential oil is called cananga oil and is a high-value fragrance ingredient.

Artabotrys hexapetalus

Cananga odorata

Desmos chinensis

Desmos chinensis

Fresh flowers contain 0.24% essential oil. The chemical constituents of the essential oil are benzaldehyde, umbellulone, 1,8-cineole, benzyl alcohol, gamma-terpinene, acetophenone, linalool, cinnamaldehyde, cubenol, beta-selinene, delta-selinene, benzyl benzoate, etc. It grows in hills, mountain slopes, edge of forests, or low-altitude wilderness. It is a fragrant and ornamental tree commonly used in gardens. Its extract can be used in perfumes and fragrances, and can also be used to produce alcohol mash.

Fissistigma shangtzeense

Fresh flowers contain 0.04%~0.05% essential oil. The chemical constituents of the essential oil are 1,8-cineole, oxidized linalool (furan type), 2,4-di-tert-butylphenol, linalool, oxidized linalool (pyran type), alpha-terpineol, geraniol, citronellol, 2,5-octadienoic acid methyl ester, selinene, nerol, and myristic acid methyl ester. It grows in mountainous forests of Guangxi and Yunnan. The flowers are fragrant and have the potential to be a natural fragrance resource.

Nymphaeaceae family

Nelumbo nucifera

Fresh flowers contain essential oil. The chemical constituents of the essential oil are 1,4-dimethylbenzene, beta-selinene, alpha-selinene, heptadecane, heptadecyne, benzothiazole, alpha-terpineol, linalool, benzyl benzoate, etc. It is widely grown in various provinces and regions in China for ornamental purposes. There are many cultivated varieties.

Nelumbo nucifera

Nymphaea 'Fragrant Hybrid'

Fresh flowers contain essential oil. The chemical constituents of the essential oil are benzyl alcohol, 6,9-heptadecadiene, 2-heptadecanone, 8-heptadecene,

Nymphaea 'Fragrant Hybrid'

Chloranthus spicatus

n-pentadecane, n-hexadecanoic acid, nonadecane, chlorophenol, cyclohexane, 9,12,15-octadecadienoic acid, 1,9-tetradecanedione, 9,12-octadecadienoic acid, 2-pentadecanone, 4- (2,6,6-trimethyl-1-cyclohexenyl) -3-buten-2-one, etc. Water lilies are distributed in various provinces of China, and there are a huge number of varieties.

Chloranthaceae family

Chloranthus spicatus

Fresh flowers contain essential oil. The chemical constituents of the essential oil are alpha-pinene, kessane, guaiene, beta-pinene, laurene, cis-beta-ocimene, trans-beta-ocimene, 2-methyl-6-methylene-1,7-octadien-3-one, isokessane, nerol, methyl jasmonate, etc. It grows in Yunnan, Sichuan, Guizhou, Guangdong, and Fujian, mainly in mountainous forests and cultivated for ornamental purposes. It is a common fragrant flower in the south that can be used to make scented tea.

Caryophyllaceae family

Dianthus caryophyllus

Flowers contain essential oil. The chemical constituents of the essential oil are eugenol, linalool, alpha-pinene, 3-hexenol, benzyl alcohol, benzaldehyde, alpha-terpineol, methyl benzoate, phenethyl alcohol, benzyl benzoate, benzyl salicylate, and methyl jasmonate. Carnations are commonly grown as ornamental plants in gardens.

Lythraceae family

Lawsonia inermis

Fresh flowers contain essential oil. The chemical constituents of the essential oil are hexanal, 2-hexenal, 3-hexenol, 2-hexenol, ethyl 3-hexenoate, cis-furanoid linalool oxide, trans-furanoid linalool oxide, linalool, alpha-pinene, suadene, linalyl acetate, and methyl benzoate. It grows in tropical and subtropical regions of Guangdong, Guangxi, Yunnan, Fujian, Jiangsu, and Zhejiang provinces in China as an ornamental plant, and its leaves are used as a natural red dye.

Lawsonia inermis

Onagraceae family

Oenothera stricta

Flowers contain essential oil. The chemical constituents of the essential oil are cis-furanoid linalool oxide, trans-furanoid linalool oxide, linalool, 6-undecanone, benzothiazole,3,4-dihydro-2,5-dimethyl-2H-pyran-2-aldehyde, R-(-)-actinidiolide, and nonanoic acid. Native to South America, it is also grown in Northeast China. The fragrant flowers can be extracted to make perfume.

Thymelaeaceae family

Daphne tangutica

Flowers contain essential oil. The chemical constituents of the essential oil are furfuraldehyde, furfuryl alcohol, benzaldehyde, benzyl alcohol, cis-oxides of linalool (furan type), t-oxides of linalool (furan type), linalool, nonanal, phenethyl alcohol, cis-oxides of linalool (pyran type), t-oxides of linalool (pyran type), phenethyl alcohol, nonanal, benzothiazole, 3-phenyl-2-propenoic acid

Pittosporum tobira

ethyl ester, benzyl benzoate, and methoxy eugenol. It can be found in Qinghai, Gansu, Tibet, Hubei, Sichuan, and Shaanxi provinces of China, growing in the forest undergrowth and rock crevices at altitudes between 1400 and 3900 meters above sea level. The flower extract can be used to make fragrances.

Pittosporaceae family

Pittosporum tobira
Flowers contain essential oil. The chemical constituents of the essential oil are 2-methyl-2-butene-1-ol, benzaldehyde, benzyl alcohol, cis-oxides of linalool (furan type), t-oxides of linalool (furan type), linalool, ethyl benzoate, indole, 2,6-di-tert-butyl-p-cresol, and nerolidol. It is commonly cultivated in provinces south of the Yangtze River in China as a fragrant plant in gardens.

Myrtaceae family

Cleistocalyx operculatus
Flower buds contain 0.18% essential oil. The chemical constituents of the essential oil are α-pinene, β-pinene, eucalyptol, β-ocimene (Z), β-ocimene (E), sabinene, γ-terpinene, α-copaene, δ-cadinene, beta-elemene, δ-selinene, jin-hua-ol, cis-chrysanthemol, and 3,6,8,8-tetramethyl-1,2,3,4,5,6,7,8-octahydro-7-methyleneazulene. It is commonly found near water in Guangdong, Guangxi, and Yunnan provinces of China. The fragrant flower buds can be used to make tea and are believed to aid digestion. The branches and leaves also contain essential oils.

Cleistocalyx operculatus

Syzygium aromaticum

Flower buds contain 15.30% essential oil. The chemical constituents of the essential oil are eugenol, benzaldehyde, caryophyllene, alpha-pinene, methyl salicylate, and eugenol acetate. It is an important aromatic medicinal plant that can be used to relieve coughs and pain. The leaves also contain essential oils. It is originally from Tanzania and now cultivated in China.

Rosaceae family

Rosa mairei

Flowers contain essential oil. The chemical constituents of the essential oil are 1,1-diethoxyethane, cis-verbenol oxide, terpineol, phenylethanol, methyl salicylate, alpha-terpineol, ethyl phenol, citral, nerol, geraniol, eugenol methyl ether, dihydro-beta-ionone, eugenol, and tetradecanal. The fresh flowers have potential as a natural fragrance material. It grows in Yunnan, Sichuan, Tibet,

Rosa rugosa

and Guizhou in China, at altitudes of 2300~4180 meters on sunny slopes.

R. rugosa

Fresh flowers contain essential oil. The chemical constituents of the essential oil are cinnamyl alcohol, rose oxide, citral, geraniol, ethyl acetate, eugenol, methyl eugenol, and ionones. The flower paste and essential oils are used in food and high-end perfumes, while the petals are used for food and tea. It is native to northern China, Japan, and the Korean Peninsula, and is now widely cultivated throughout China.

R. sertata

Flowers contain 0.03% essential oil. The chemical constituents of the essential oil are cis-rose oxide, t-rose oxide, geraniol, linalool, geranyl acetate, eugenol, methyl eugenol, elemicin, γ-elemene, 13-ketone-2, 13-ol-2, cis-acaciaspirin, t-acaciaspirin, rhodinol, and p-cymene-8-ol. It is extensively cultivated in northwest China and Xinjiang for the extraction of its essential oil or absolute for perfume blending.

Robinia pseudoacacia

Fresh flowers contain essential oil. The chemical constituents of the essential oil are7-octen-4-ol, cis-linalool oxide (furanoid), α-bisabolol, linalool, phenethyl alcohol, cis-linalool oxide (pyranoid), geraniol, santalol, nerol, ethyl eugenol, cedrol, and γ-decalactone. It is native to North America, but is cultivated worldwide, including in China, and is used to make a variety of flower-scented fragrances. Additionally, the flowers are known for their hemostatic properties and can be used to treat internal bleeding.

Humulus lupulus

Dried flowers of hops contain 0.90%~1.0% essential oil. The chemical constituents of the essential oil are beta-myrcene, methyl 2-methyl propanoate, methyl 6-methyl heptanoate, methyl caprate, methyl nonanoate, 2-decylketone, methyl decanoate, beta-caryophyllene, methyl dodecanoate, alpha-caryophyllene, methyl undecanoate, and methyl cinnamate. Hops are wild in northern Xinjiang and cultivated in northeastern, northern, and Shandong regions of China. The dried flowers are used in beer production, and the female inflorescences are used in medicine to aid digestion.

Elaeagnaceae family

Elaeagnus oxycarpa

Fruits contain 0.2%~0.4% essential oil. It is found in northern and northwestern China, including Inner Mongolia, growing in sandy areas at altitudes of 400 to 600 meters. The unique and long-lasting fragrance of the plant can be used in cosmetics and soap perfumes, and the leaves, roots, and fruit are used in traditional medicine.

Rutaceae family

Citrus aurantium

Fresh flowers contain 0.28% essential oil. It is found in the Yangtze River Basin and southern China. The flowers, leaves, and fruit all contain essential oils that can be used in food and cosmetics, and the fruit is used in traditional medicine.

C. maxima

Flowers contain essential oil. The chemical constituents of the essential oil are phellandrene, β-pinene, limonene, α-ocimene, linalool. It is found in southern

Citrus aurantium

Citrus maxima

China. The fruit, wood, and leaves all contain essential oils that can be used in perfumes, and the fruit can also be used to make pectin.

C. limon

Also known as lemon. Fruit peels contain 1.56% essential oil. It is found in Yunnan, Guangxi, Guangdong, Fujian, Sichuan, and Guizhou in both wild and cultivated forms. The essential oils found in the fruit, leaves, and flowers can be used in soap perfumes, and the fruit is used to make refreshing desserts and drinks.

Meliaceae family

Aglaia duperreana

Flowers contain 0.2%~0.4% essential oil. The chemical constituents of the essential oil are beta-caryophyllene, trans-linalool oxide, cis-linalool oxide, linalool, indole, alpha-bergamotene, alpha-pinene, beta-selinene, beta-ocimene, gamma-elemene, beta-elemene, and cis-dihydrocoumarin. The flower is a natural fragrance material unique to China and is often used in perfumery. It is commonly found in southern China, particularly in Guangdong, Guangxi, and Fujian.

A. odorata

Flowers contain essential oil. The chemical constituents of the essential oil are alpha-pinene, beta-caryophyllene, trans-linalool

Aglaia duperreana

oxide, cis-linalool oxide, methyl salicylate, indole, alpha-cubebene, beta-ocimene, trans-beta-farnesene, sabinene, gamma-terpinene, and linalool. The flower is often used in cosmetics, soap, and as a fixative in perfumes. It is another fragrant flower commonly found in China, particularly in Fujian, Guangdong, Guangxi, Yunnan, and Sichuan.

Aglaia Odorata

Oleaceae family

Jasminum grandiflorum

Flowers contain essential oil. The chemical constituents of the essential oil are benzyl alcohol, para-cresol, linalool, benzyl acetate, 1H-indole, eugenol, methyl jasmonate, isoeugenol, alpha-farnesene, nerol, benzyl benzoate, and hexenol-3. The flower is a highly valued natural fragrance material used in high-end perfumes and cosmetics. It is native to the Mediterranean coast and Morocco, but is widely cultivated around the world.

J. officinale

Flowers contain essential oil. The chemical constituents of the essential oil are benzyl acetate, benzyl benzoate, linalool, nerol, geraniol, benzyl alcohol, benzaldehyde, eugenol, and methyl jasmonate. It is cultivated in Yunnan, Sichuan, and Tibet. It is used in the production of high-end cosmetics and perfumes due to its pleasant aroma.

Jasminum grandiflorum

· 429 ·

J. pentaneurum

Flowers contain essential oil. The chemical constituents of the essential oil are benzyl alcohol, cis-oxides of benzyl alcohol, trans-oxides of benzyl alcohol, linalool, methyl benzoate, methyl jasmonate, nerol, and 3-(4,8-dimethyl-3,7-nonadien-1-yl) furan, making it a potential natural fragrance material. It is cultivated in Guangdong and Guangxi.

Jasminum sambac

J. sambac

Flowers contain essential oil. The chemical constituents of the essential oil are cis-3-hexenol acetate, cis-3-hexenol, benzyl alcohol, methyl benzoate, benzyl acetate, farnesene, benzyl benzoate, juniperene, and para-methoxybenzyl methyl ether. It is used in the production of perfumes, soaps, and other cosmetic products and is a common garden plant. It is in the Oleaceae family that is cultivated in Guangdong, Guangxi, Yunnan, Fujian, Sichuan, and Guizhou.

Ligustrum sinense

Flowers contain essential oil. The chemical constituents of the essential oil are benzaldehyde, limonene, 1,8-cineole, phenylethyl alcohol, methyl eugenol, linalool, benzyl acetate, nerol, camphor, and N-phenylacetamide. It is native to southern provinces in China, such as the Yangtze River basin, and is often cultivated in mountainous areas and along roadsides. Its leaves and extracts are used for medicinal purposes to reduce inflammation and relieve pain. Its extracts are also used in the production of fragrances for soaps and cosmetics.

Ligustrum sinense

Osmanthus fragrans

As an ornamental tree for its fragrant flowers. Fresh flowers contain essential oil. The chemical constituents of the essential oil are linalool oxide (furanoid), cis-linalool oxide (furanoid), linalool, nonanol, α-terpineol, 6-ethyl-2,2,6-trimethyltetrahydropyran-3-ol, nerol, dihydro-β-ionone, methyl salicylate, and more. It is widely planted in the lower reaches of the Yangtze River, Guangxi, Guangdong, Yunnan, and other areas. It has two varieties:

O. fragrans var. *thunbergii* (Golden Osmanthus) and *O. fragrans* var. *aurantiacus* (Orange Osmanthus). Golden Osmanthus has golden-yellow flowers and contains essential oils such as methyl heptenone, 6-diethoxyacetyloctane, 5-benzyloxyvalerolactone, cyclohexene-3-methanol, menthone, ethoxymethanal, linalool oxide (furanoid), cis-linalool oxide (furanoid), hexyl hexanoate, kouloin, α-ionone, and dihydro-β-ionone. It is mainly grown as an ornamental tree. Orange Osmanthus has orange- red flowers and contains essential oils such as β-ocimene (*E*), cis-linalool oxide (furanoid), trans-linalool oxide (furanoid), linalool, cis-linalool oxide (pyranoid), trans-linalool oxide (pyranoid), α-ionone, dihydro-β-ionone, 5-hexyldihydro-2(3H)-furanone, and β-ionone. It is also grown as an ornamental tree and has a lighter fragrance compared to Silver Osmanthus and Golden Osmanthus. The flowers of Osmanthus fragrans can also be consumed directly or used to make cosmetics and food flavors.

Osmanthus fragrans

Syringa reticulata subsp. *amurensis*

Flowers contain 4.0% essential oil. The chemical constituents of the essential oil are cis-oxides of linalool, trans-oxides of linalool, linalool, phenylethanol, 2,6,6-trimethyl-2-vinyl-5-hydroxypyrane, ethyl caprate, δ-selinene, nerolidol, and ethyl eugenol. It is mainly distributed in Northeast China, Hebei, Ningxia and other provinces, and grows on riverbanks and in mixed forests at the edge of the forest. The bark and branches of the plant are used to treat coughs and asthma, while the flower oil can be used in cosmetics and fragrances.

Plumeria rubra 'Acutifolia'

Apocynaceae family

Plumeria rubra 'Acutifolia'

Flowers contain essential oil. The chemical constituents of the essential oil are benzaldehyde, tricyclic [3,2,1,0,1,5] octane, octane, methyl benzoate, linalool, eucalyptol, citral, nerol, methyl salicylate, and other compounds. It is native to tropical regions in the Americas and is cultivated as a tropical and subtropical aromatic ornamental plant in China. It is also used as a raw material in Cantonese herbal tea. The red variety of Plumeria rubra is commonly found in China, but there are also all-white varieties.

Rubiaceae family

Gardenia jasminoides

Fresh flowers contain essential oil. The chemical constituents of the essential oil are cis-3-hexenol, methyl benzoate, ethyl-3-hexenoate, linalool, isobutyric acid-cis-3-hexenyl ester, butyric acid-cis-3-hexenyl ester, koumarin, isobutyric acid-cis-3-hexenyl ester, tiglic acid-benzyl ester, and cis-3-hexenyl benzoate, among other compounds. It is native to southern, southwestern, and central China and is commonly found as a fragrant ornamental plant in gardens. Its essential oil and extracts are used to make various fragrances.

Gardenia jasminoides var. *fortuniana*

Flowers contain essential oil. The chemical constituents of the essential oil are ethyl propyl ether, methyl butyl ether, ethyl butyl ether, methyl-2-butenoate, ethyl-2-methyl cyclopentane carboxylate, 6-methyl-5-hepten-2-one, gamma-terpinen-7-al (furan type), linalool, butyric acid-3-hexenyl ester, gamma-terpinen-7-al (pyran type), 2-methyl-2-penten-1-ol, 1-methyl-2-(2-propyl) cyclohexane, and hexyl cyclohexane, among other compounds. It is a common fragrant ornamental flower found in gardens in China.

Luculia pinceana

Fresh flowers contain essential oil. The chemical constituents of the essential oil are 2,4-dimethyl-3-pentanone, alpha-pinene, camphene, beta-myrcene, beta-pinene, limonene, t-terpinen-4-ol (furan type),

Gardenia jasminoides

Gardenia jasminoides var. *fortuniana*

Lonicera confusa

Lonicera japonica

cis-terpinen-4-ol (furan type), linalool, camphor, borneol, methyl salicylate, ylangene, beta-caryophyllene, 2,6-di-tert-butyl-4-methylphenol, and beta-cubebene, among other compounds. It is native to mountain slopes and shrubs above 1300 meters in Yunnan and Guangxi provinces in China and is a fragrant flower with potential for further development.

Caprifoliaceae family

Lonicera confusa

Flowers contain essential oil. The chemical constituents of the essential oil are 3-hexenol, 2-methylbutyrate, 6-methyl-5-hepten-2-one, benzyl alcohol, α-thujene, cis-oxide linalool type, t-oxide linalool type, methyl benzoate, linalool, phenethyl alcohol, cis-oxide fenchol type, methyl anthranilate, α-cubebene, etc. It is found in Sichuan, Guangdong, Guangxi, Hunan, Guizhou, Yunnan, and other provinces in China. It grows in sparse forests and shrublands. The flowers are used in traditional medicine to treat colds, dysentery, and diarrhea, while the leaves are used for rheumatism.

L. japonica

Flowers contain 0.4% essential oil. The chemical constituents of the essential oil are carene, 1-hexene, cis-hexenol-3, cis-oxide fenchol, t-oxide fenchol, fenchol, α-pinene, nerolidol, nerol, terpineol, benzyl alcohol, β-phenylethanol, thymol, eugenol, etc. It is native to Northeast Asia, widespread in China, including Liaoning, Shaanxi, Hunan, Yunnan, and Guizhou. It grows in shrublands and sparse forests, extensively cultivated. It can be used in traditional Chinese medicine as a cooling and detoxifying agent, also a popular garden ornamental plant.

Asteraceae family

Wedelia trilobata

Fresh flowers contain essential oil. The chemical constituents of the essential oil are α-pinene, carene, sabinene, β-pinene, β-myrcene, β-caryophyllene, p-cymene, methyl benzoate, camphor, bornyl acetate, isoamyl phenol, musk xylene, t-4-hydroxy-3-methyl-6-isopropyl-2-cyclohexenone, t-4-hydroxy-3-methyl-6-isopropyl-2-cyclohexenone isomer, α-patchoulene, β-bisabolene, etc. It is an introduced species in China, commonly found in southern regions. It is planted for ground cover and ornamental purposes.

Solanaceae family

Cestrum nocturnum

Flowers contain 0.3%~0.6% essential oil. The chemical constituents of the essential oil are ethyl acetate, 1-ethoxy-2-methylpropane, ethoxybutane, pentyl

Wedelia trilobata

Cestrum nocturnum

acetate, benzaldehyde, benzyl alcohol, methyl benzoate, phenethyl alcohol, 2-methoxy-4-(2-propenyl) phenol, phenylmethyl acetate, methyl salicylate, methyl 2-aminobenzoate, 4-methyl-6-hepten-3-one, α-cubebene, etc. It is native to South America, now widely introduced around the world. It is planted in Guangdong, Guangxi, Fujian, and Yunnan in China. It is used as a fragrant ornamental plant in gardens. The flowers bloom at night and have disinfectant and antiseptic properties.

Lamiaceae family

Clerodendrum philippinum var. *simplex*

Flowers contain essential oil. The chemical constituents of the essential oil are ethoxy pentane, acetic acid, ethoxy ethyl ester, benzaldehyde, 7-octen-4-ol-6-methyl-5-hepten-2-one, 1,8-cineole, phenethyl aldehyde, cis-phenol oxide (furan type), t-phenol oxide (furan type), phenol, benzyl alcohol, methyl 2-hydroxybenzoate, nerolidol, linalool, benzyl benzoate, and others. The leaves and roots are used in traditional medicine to dispel wind, promote blood

Clerodendrum philippinum var. *simplex*

circulation, and strengthen bones and muscles.

Zingiberaceae family

Hedychium coronarium

Flowers contain essential oil. The chemical constituents of the essential oil are linalool, pinene, 1,8-cineole, α-terpineol, β-terpineol, cis-nerolidol (furanoid), t-nerolidol (furanoid), methyl benzoate, benzyl alcohol, α-phellandrene, indole, methyl jasmonate, α-bisabolene, pentyl benzoate, 3-(4,8-dimethyl-3,7-nonadienyl) furan, etc. It is found in Taiwan, Guangdong, Hunan, Guangxi, Yunnan, Sichuan and other provinces and regions. It grows in forests or is cultivated, prefers wet areas near water, now mostly seen in cultivation. It is a commonly grown fragrant flower in Southern China.

Amaryllidaceae family

Narcissus tazetta var. *chinensis*

Flowers contain essential oil. The chemical constituents of the essential oil are 1,3-dimethoxy-2-propanol, ethyl-3-methyl-2-butenoate, α-phellandrene, β-phellandrene, limonene, 1,8-cineole, β-terpineol, cis-nerolidol (furanoid), t-nerolidol (furanoid), undecane, benzyl alcohol, methyl benzoate, α-terpineol, ethyl phenylacetate, ethyl benzoate, ethyl phenylpropionate, ethylgeranate, etc. Originally from the warm coastal areas of

Narcissus tazetta var. *chinensis*

Asia, it is now mostly cultivated by humans, with Fujian Province having the largest plantations. The flowers have an elegant fragrance and are an important fragrant flower for the Spring Festival, but they should not be kept indoors for too long.

Asparagaceae family

Polianthes tuberosa

Fresh flowers contain essential oil. The chemical constituents of the essential oil are α-phellandrene, benzaldehyde, pinene, β-phellandrene, limonene, 1,8-cineole, methyl benzoate, 2-hydroxy methyl benzoate, indole, methyl anthranilate, β-myrcene, 6-methyl-5-hepten-2-one, etc. Originally from Mexico, it is now cultivated in various regions of China. The extract can be used to make high-quality fragrance materials, and it is also a commonly used fragrant flower for decoration.

Cymbidium sinense

Orchidaceae family

Cymbidium sinense

Flowers contain essential oil. The chemical constituents of the essential oil are heptane, 3-methyl-3-pentanol, 3,4-dimethyl-3-ketone, 2,4-dimethyl-3-pentanol, 3,4-dimethylhexanol, decane, 1,8-cineole, α-terpineol, methyl benzoate, 3-phenylpropanal, cubenol, dihydro-β-ionone, β-ionone, 2,6-di-tert-butyl-4-methylphenol, cis-t-jasmone, etc. It is cultivated in various regions of China, and there are many varieties and subspecies. It is a common fragrant flower for gardens and halls, with a pleasant and lasting aroma.

Dendrobium chrysotoxum

Flowers contain essential oil. The chemical constituents of the essential oil are kairomone-3, benzaldehyde, 1,7,7-trimethyl bicyclo [2.2.1] hept-2-ene, 3,7-dimethyl-1,3,7-octatriene, phenylacetaldehyde, cyclopentane pentanol, linalool-2, camphor, borneol, ethyl octanoate, and citronellol. It is found in the southwestern regions of China, and it grows epiphytically on trees. It is cultivated in various botanical gardens and is known for its beautiful flowers and pleasant fragrance, making it a popular ornamental plant for gardens.

Dendrobium chrysotoxum

Chapter II

Chemical Composition and Chemical Analysis of Plant Essential Oils

Section 1

Chemical composition of plant essential oils

Essential oils are aromatic, volatile liquids obtained from plants via steam distillation. These aromatic oils contain the "essence of" the plant's fragrance, which consists of organic volatile compounds, generally of low molecular weight below 300.

The composition of essential oils is highly species-specific. The variation in the chemical composition of essential oil can change from plant to plant, even in the same species. Essential oils are stored in specialized structures such as secretory glands, cavities, channels, and glandular trichomes. Their composition usually varies in the different organs of an individual species. These differences are closely related to the organ function. In addition, the chemical composition is also influenced by the stage of plant development, geographical origin, drying method, and distillation method.

Essential oils are complex mixtures, in individual oils, up to 400 chemicals or even more can be identified when proper analytical equipment and methods are used. According to their chemical structures, these compounds can be divided into four major groups: terpenoids, aromatic, aliphatic derivatives, and miscellaneous. The current chapter provides a general overview of the chemical composition and structural elucidation of essential oils.

Terpenoids

The terpenoids are, by far, the most important group of natural products concerning essential oils. They are defined as substances composed of isoprene (2-methylbutadiene) units. In nature, terpenoids occur predominantly as hydrocarbons, alcohols and their ethers, aldehydes, ketones, carboxylic acids, esters and glycosides. Data on volatile oils containing terpenoid constituents isolated from plant materials are given in Table 2-1. Volatile oils in which the main components are aromatic and derived from the shikimate pathway are listed in Table 2-2.

1.1 Biosynthetic pathway of terpenoids

Terpenoids form a large and structurally diverse family of natural products derived from C_5 isoprene units. Typical structures contain carbon skeletons represented by $(C_5)_n$, and are divided into monoterpenes (C_{10}), sesquiterpenes (C_{15}), diterpenes (C_{20}), sesterterpenes (C_{25}), triterpenes (C_{30}), and tetraterpenes (C_{40}) depending on their carbon units. Although terpenoids are extraordinarily diverse, all originate through the condensation of the universal five-carbon precursors, isopentenyl diphosphate (IPP) and dimethylallyl diphosphate (DMAPP). In higher plants, two independent pathways located in separate intracellular compartments are involved in the biosynthesis of IPP and DMAPP. In the cytosol, IPP is derived from the classic mevalonic acid (MVA) pathway that starts with the condensation of acetyl-CoA (Figure 2-1), whereas in plastids, IPP is formed from pyruvate and glyceraldehyde 3-phosphate via the methylerythritol phosphate (MEP) pathway (Figure 2-2). Research indicates that the cytosolic pool of IPP serves as a precursor of farnesyl diphosphate (FPP, C_{15}) and, ultimately, sesquiterpenes and triterpenes, whereas the plastidial pool of IPP provides precursors of geranyl diphosphate (GPP, C_{10}) and geranylgeranyl diphosphate (GGPP, C_{20}) and, ultimately, monoterpenes, diterpenes, and tetraterpenes.

Table 2-1 Volatile oils containing principally terpenoids derived from MEP pathway

Oils	Plant source	Plant part used	Oil content (%)	Major constituents with typical (%) composition	Uses, notes
Bergamot	*Citrus aurantium* ssp. *bergamia* (Rutaceae)	fresh fruit peel (expression)	0.5	limonene (42) linalyl acetate (27) γ-terpinene (8) linalool (7)	flavoring, aromatherapy, perfumery also contains the furocoumarin bergapten (up to 5%) and may cause severe photosensitization
Camphor oil	*Cinnamomum camphora* (Lauraceae)	wood	1 ~ 3	camphor (27 ~ 45) cineole (4 ~ 21) safrole (1 ~ 18)	soaps
Caraway	*Carum carvi* (Umbelliferae/Apiaceae)	ripe fruit	3 ~ 7	(+)-carvone (50 ~ 70) limonene (47)	flavor, carminative, aromatherapy
Cardamom	*Elettaria cardamomum* (Zingiberaceae)	ripe fruit	3 ~ 7	α-terpinyl acetate (25 ~ 35) cineole (25 ~ 45) linalool (5)	flavor, carminative, ingredient of curries, pickles
Chamomile (Roman chamomile)	*Chamaemelum nobile* (*Anthemis nobilis*) (Compositae /Asteraceae)	dried flowers	0.4 ~ 1.5	aliphatic esters of angelic, tiglic, isovaleric, and isobutyric acids (75 ~ 85) small amounts of monoterpenes	flavoring, aromatherapy blue colour of oil is due to chamazulene

(continued)

Oils	Plant source	Plant part used	Oil content (%)	Major constituents with typical (%) composition	Uses, notes
Citronella	*Cymbopogon winterianus* *C. nardus* (Graminae/Poaceae)	fresh leaves	0.5 ~ 1.2	(+)-citronellal (25 ~ 55) geraniol (+)-citronellol (10 ~ 15) geranyl acetate (8) (20 ~ 40)	perfumery, aromatherapy, insect repellent
Coriander	*Coriandrum sativum* (Umbelliferae/Apiaceae)	ripe fruit	0.3 ~ 1.8	(+)-linalool (60 ~ 75) γ-terpinene (5) α-pinene (5) camphor (5)	flavor, carminative
Dill	*Anethum graveolens* (Umbelliferae/Apiaceae)	ripe fruit	3 ~ 4	(+)-carvone (40 ~ 65)	flavor, carminative
Eucalyptus	*Eucalyptus globulus* *E. smithii* *E. polybractea* (Myrtaceae)	fresh leaves	1 ~ 3	cineole (= eucalyptol) (70 ~ 85) α-pinene (14)	flavor, antiseptic, aromatherapy
Eucalyptus (lemon-scented)	*Eucalyptus citriodora* (Myrtaceae)	fresh leaves	0.8	citronellal (65 ~ 85)	perfumery
Ginger	*Zingiber officinale* (Zingiberaceae)	dried rhizome	1.5 ~ 3	zingiberene (34) β-sesquiphellandrene (12) β-phellandrene (8) β-bisabolene (6)	flavoring the main pungent principles in ginger (gingerols) are not volatile

(continued)

Oils	Plant source	Plant part used	Oil content (%)	Major constituents with typical (%) composition	Uses, notes
Juniper	Juniperus communis (Cupressaceae)	dried ripe berries	0.5 ~ 2	α-pinene (45 ~ 80) myrcene (10 ~ 25) limonene (1 ~ 10) sabinene (0 ~ 15)	flavoring, antiseptic, diuretic, aromatherapy juniper berries provide the flavoring for gin
Lavender	Lavandula angustifolia L. officinalis (Labiatae/Lamiaceae)	fresh flowering tops	0.3 ~ 1	linalyl acetate (25 ~ 45) linalool (25 ~ 38)	perfumery, aromatherapy inhalation produces mild sedation and facilitates sleep
Lemon	Citrus limon (Rutaceae)	dried peel from fruit (expression)	0.1 ~ 3	(+)-limonene (60 ~ 80) β-pinene (8 ~ 12) γ-terpinene (8 ~ 10) citral (= geranial + neral) (2 ~ 3)	flavoring, perfumery, aromatherapy terpeneless lemon oil is obtained by removing much of the terpenes under reduced pressure; this oil is more stable and contains 40% ~ 50% citral
Lemon-grass	Cymbopogon citratus (Graminae/Poaceae)	fresh leaves	0.1 ~ 0.3	citral (= geranial + neral) (50 ~ 85)	perfumery, aromatherapy
Matricaria (German chamomile)	Matricaria chamomilla (Chamomilla recutica) (Compositae/Asteraceae)	dried flowers	0.3 ~ 1.5	(−)-α-bisabolol (10 ~ 25) bisabolol oxides A and B (10 ~ 25) chamazulene (1 ~ 15)	flavoring, dark blue colour of oil is due to chamazulene

(continued)

Oils	Plant source	Plant part used	Oil content (%)	Major constituents with typical (%) composition	Uses, notes
Orange (bitter)	*Citrus aurantium* ssp. *amara* (Rutaceae)	dried peel from fruit (expression)	0.5 ~ 2.5	(+)-limonene (92 ~ 94) myrcene (2)	flavoring, aromatherapy, the main flavor and odour comes from the minor oxygenated components; terpeneless orange oil is obtained by removing much of the terpenes under reduced pressure; this oil contains about 20% aldehydes, mainly decanal
Orange (sweet)	*Citrus sinensis* (Rutaceae)	dried peel from fruit (expression)	0.3	(+)-limonene (90 ~ 95) myrcene (2)	flavoring, aromatherapy, the main flavor and odour comes from the minor oxygenated components; terpeneless orange oil is obtained by removing much of the terpenes under reduced pressure; this oil contains about 20% aldehydes, mainly octanal and decanal
Orange flower (Neroli)	*Citrus aurantium* ssp. *amara* (Rutaceae)	fresh flowers	0.1	linalool (36) β-pinene (16) limonene (12) linalyl acetate (6)	flavoring, perfumery, aromatherapy
Peppermint	*Mentha x piperita* (Labiatae/Lamiaceae)	fresh leaves	1 ~ 3	menthol (30 ~ 50) menthone (15 ~ 32) menthyl acetate (2 ~ 10), menthofuran (1 ~ 9)	flavoring, carminative, aromatherapy

(continued)

Oils	Plant source	Plant part used	Oil content (%)	Major constituents with typical (%) composition	Uses, notes
Pine	*Pinus palustris* or other *Pinus* species (Pinaceae)	needles, twigs		α-terpineol (65)	antiseptic, disinfectant, aroma-therapy
Pumilio pine	*Pinus mugo* ssp. *pum-ilio* (Pinaceae)	needles	0.3 ~ 0.4	α- and β-phellandrene (60) α- and β-pinene (10 ~ 20) bornyl acetate (3 ~ 10)	inhalant, the minor components bornyl acetate and borneol are mainly responsible for the aroma
Rose (attar of rose, otto of rose)	*Rosa damascena, gallica, R. alba,* and *R. R. centifolia* (Rosaceae)	fresh flowers	0.02 ~ 0.03	citronellol (36) geraniol (17) 2-phenylethanol (3) C14 –C23 straight chain hydrocarbons (25)	perfumery, aromatherapy
Rosemary	*Rosmarinus officinalis* (Labiatae/Lamiaceae)	fresh flower-ing tops	1 ~ 2	cineole (15 ~ 45) α-pinene (10 ~ 25) camphor (10 ~ 25) β-pinene (8)	perfumery, aromatherapy
Sage	*Salvia officinalis* (Labiatae/Lamiaceae)	fresh flower-ing tops	0.7 ~ 2.5	thujone (40 ~ 60) camphor (5 ~ 22) cineole (5 ~ 14) β-caryophyllene (10) limonene (6)	aromatherapy, food flavoring

(continued)

Oils	Plant source	Plant part used	Oil content (%)	Major constituents with typical (%) composition	Uses, notes
Sandalwood	*Santalum album* (Santalaceae)	heartwood	4.5 ~ 6.3	sesquiterpenes: α-santalol (50) β-santalol (21)	perfumery, aromatherapy
Spearmint	*Mentha spicata* (Labiatae/Lamiaceae)	fresh leaves	1 ~ 2	(−)-carvone (50 ~ 70) (−)-limonene (2 ~ 25)	flavoring, carminative, aromatherapy
Tea tree	*Melaleuca alternifolia* (Myrtaceae)	fresh leaves	1.8	terpinen-4-ol (30 ~ 45) γ-terpinene (10 ~ 28) α-terpinene (5 ~ 13) p-cymene (0.5 ~ 12) cineole (0.5 ~ 10) α-terpineol (1.5 ~ 8)	antiseptic, aromatherapy, an effective broad spectrum antiseptic widely used in creams, cosmetics, toiletries
Thyme	*Thymus vulgaris* (Labiatae/Lamiaceae)	fresh flowering tops	0.5 ~ 2.5	thymol (40) p-cymene (30) linalool (7) carvacrol (1)	antiseptic, aromatherapy, food flavoring
Turpentine oil	*Pinus palustris* and other *Pinus* species (Pinaceae)	distillation of the resin (turpentine) secreted from bark		(+)- and (−)-α-pinene (35:65) (60 ~ 70) β-pinene (20 ~ 25)	counter-irritant, important source of industrial chemicals residue from distillation is colophony (rosin), composed chiefly of diterpene acids (abietic acids)

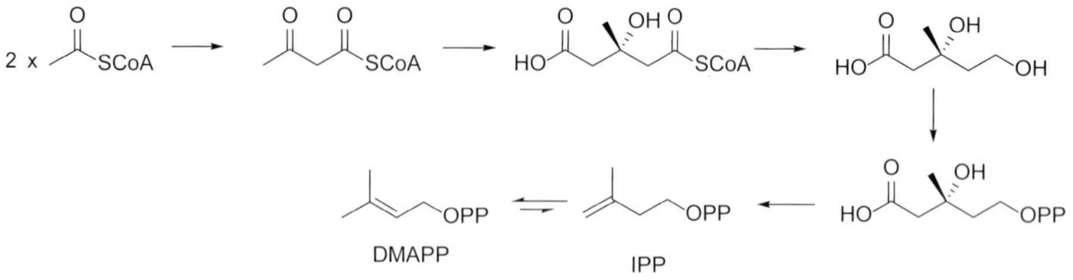

Figure 2-1 Mevalonic acid (MVA) pathway

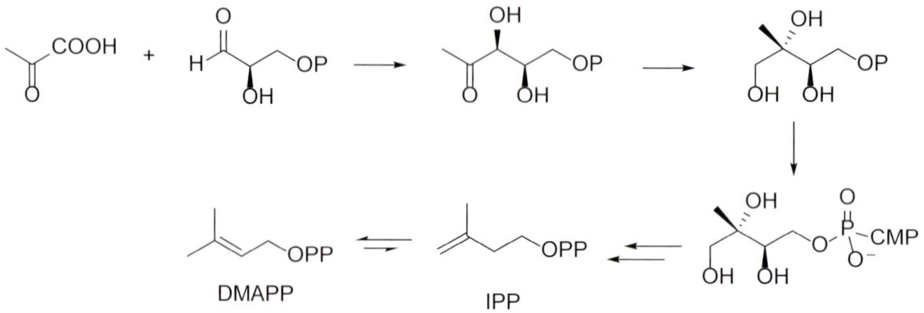

Figure 2-2 Methylerythritol phosphate (MEP) pathway

Linear monoterpenoid GPP is formed by the combination of DMAPP and IPP via the enzyme prenyl transferase. Linalyl diphosphate (LPP) and neryl diphosphate (NPP) are isomers of GPP, and are formed from GPP by ionization to the allylic cation, which allows a change in the attachment of the diphosphate group (to the tertiary carbon in LPP) or a change in the stereochemistry at the double bond (to Z in NPP). These three compounds (GPP, LPP, and NPP), with relatively modest changes, can give rise to a range of linear monoterpenoids found as components of essential oils (Figure 2-3). The resulting compounds may be hydrocarbons, alcohols, aldehydes, or perhaps esters, especially acetates.

Cyclic monoterpenoids, including monocyclic and bicyclic systems, can be created by cyclization reaction, NPP and LPP are more propitious precursors than GPP in terms of stereochemistry for generating the monocyclic menthane system, which produces a carbocation (known as menthyl or α-terpinyl) that has the menthane framework. This carbocation can be further rearranged to yield a range of bicyclic scaffolds, including borneol, camphor, camphene, pinene, and terpinene, etc. (Figure 2-4).

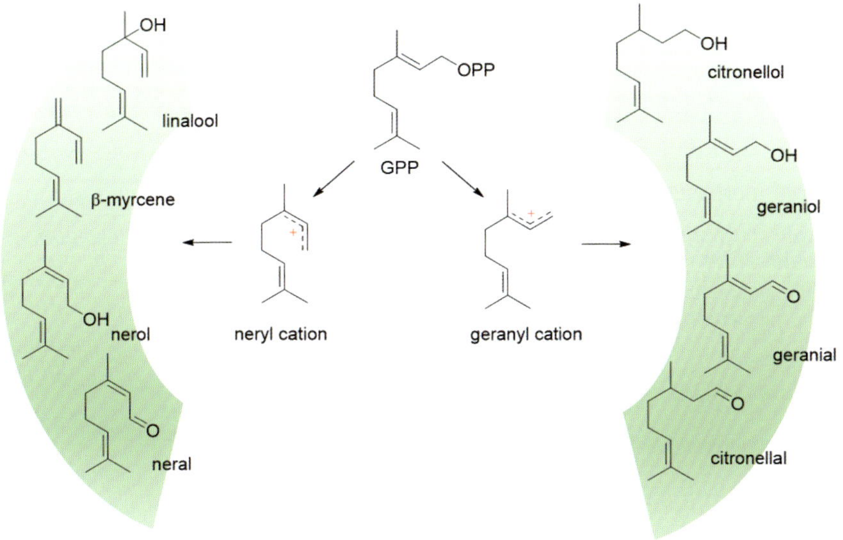

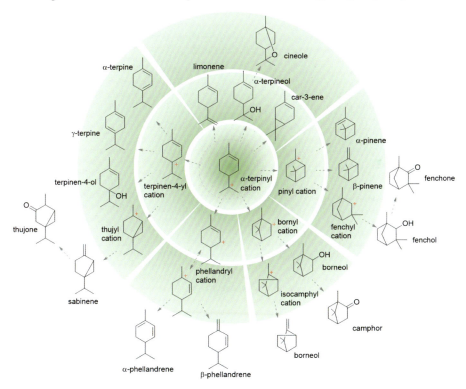

Figure 2-3 Linear monoterpenoids: conversion of geranyl diphosphate

Figure 2-4 Cyclic monoterpenoids from cation rearrangement of α-terpinyl cation

Sesquiterpenoids are formed from three C5 units. Addition of a further C5 IPP unit to GPP in an extension of the GPP synthase reaction leads to the fundamental sesquiterpenoids precursor farnesyl diphosphate (FPP). FPP can then give rise to linear and cyclic sesquiterpenoids. Because of the presence of five more carbons in the starting substrate, the number of possible cyclization modes is increased, and a huge range of mono-, bi-, and tri-cyclic structures are available from enzymatic dissociation of the sesquiterpene FPP (Figure 2-5).

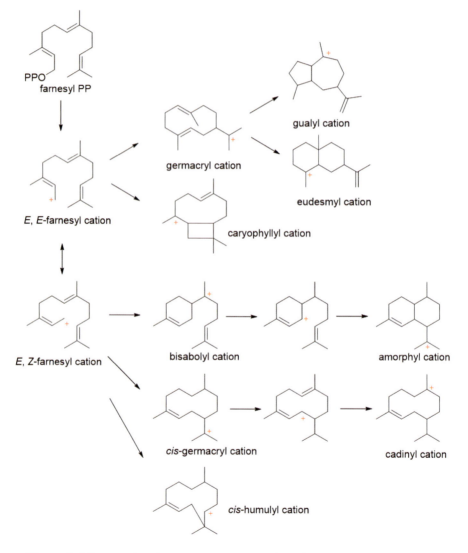

Figure 2-5 Intramolecular rearrangements of C_{15} farnesyl C_1 allylic cation

1.2 Monoterpenoids

Monoterpenoids are highly fragrant molecules that are naturally occurring and widely distributed in nature. There are over 400 different structures of these molecules, which can be extracted from the leaves, flowers, and fruits of various plants. Some examples of linear monoterpenoids include β-myrcene and configurational isomers of β-ocimene. These molecules can be found in essential oils derived from plants such as basil, bay, hops and petitgrain.

β-myrcene   (Z)-β-ocimene   (E)-β-ocimene

Unsaturated monoterpene alcohols and aldehydes are highly valued in the world of perfumery. (R)-(−)-Linalool, acquired from rose, neroli (orange flowers), and spike (lavender) oils, has a woody scent, whereas the (S)-(+)-enantiomer has a sweeter lavender fragrance. Citral, widely used in the perfume industry, is a combination of the (E,Z)-isomers geranial and neral, which releases a strong and delightful aroma similar to that of lemon peels. These compounds serve as crucial components in various fragrances, contributing to their complex and unique scents.

rac. linalool   geranial (E)   neral (Z)

The majority of monocyclic terpenes are derived from the *cis-trans*-isomers of p-menthane. *Trans*-p-menthane, which is present in the oil of turpentine, serves as an important precursor for this process. Another important monocyclic terpene is limonene, which is an unsaturated hydrocarbon found abundantly in various essential oils. Its versatility has led to its widespread use in different industries.

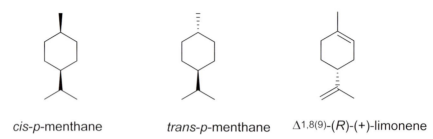

cis-p-menthane   trans-p-menthane   Δ1,8(9)-(R)-(+)-limonene

p-Menthan-3-ol has the ability to form four pairs of enantiomers. Among these, (−)-menthol is the primary component of peppermint oil, and is commonly utilized as a fragrance and flavoring agent in the perfumery industry. This compound exhibits mild anesthetic, antipruritic, antiseptic, carminative, cooling, and stomach-calming properties, and can be applied as an antipruritic and in nasal inhalers. The versatility of (−)-menthol makes it a useful ingredient for various applications.

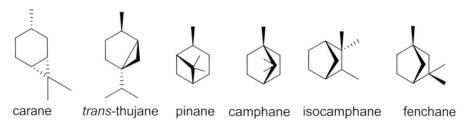

(-)-menthol   (-)-isomenthol   (+)-neomenthol   (-)-neomenthol

The bicyclic cyclobutane pinane, carane, and thujane, as well as bicyclo[2.2.1] heptanes like camphane, isocamphane, and fenchane, are the prominent frameworks of naturally occurring bicyclic monoterpenes.

carane   trans-thujane   pinane   camphane   isocamphane   fenchane

(+)-3-Carene, also known as 3,7,7-trimethylbicyclo[4.1.0]hept-3-ene, is present as a constituent of the oil of turpentine from the tropical pine species Pinus longifolia, and can also be found in certain species of fir, juniper, and citrus. In contrast, thujane derivatives are more widespread in plants. Thujol [(-)-thujan-3α-ol] and its 4-epimer (+)-isothujol are present in species of Artemisia, Juniperus, and Thuja. Meanwhile, (+)-4(10)-Thujene, which is widely recognized as (+)-sabinene, is present in the oil of savin obtained from fresh tops of Juniperus sabina (Cupressaceae). This implies that these chemicals are highly prevalent in many plant species and contribute to their distinctive fragrances.

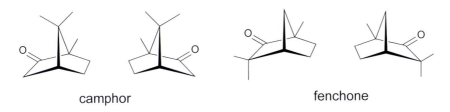

(+)-3-carene   (-)-thujol   (+)-3-thujanone   (+)-4(10)-thujene

The oil of turpentine, extracted on a large scale from the wood of diverse pine trees (*Pinus caribaea*, *P. palustris*, *P. pinaster*), is composed of over 70% α-pinene and up to 20% β-pinene. This indicates that the oil of turpentine has a high concentration of these two chemicals with distinct fragrant properties, which are abundantly found in pine trees.

Naturally occurring camphanes are composed of various compounds, including borneols that feature an endo hydroxy group, isoborneols that exhibit an exo OH, and 2-camphanone (2-bonanone), commonly referred to as camphor. Borneo camphor, also known as (+)-borneol, is extracted from the camphor tree Cinnamomum camphora (Lauraceae) and the roots of ginger-like Curcuma aromatica (Zingiberaceae), both of which are grown in Eastern Asia. On the other hand, (−)-isoborneol is isolated from Achillea filipendulina (Asteraceae). This demonstrates that camphanes can be obtained through various natural sources and have distinctive chemical properties.

camphor   fenchone

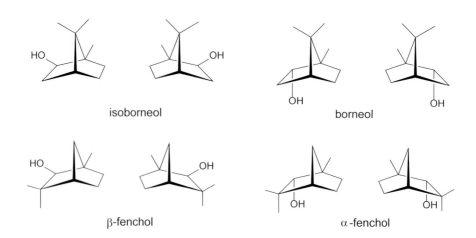

isoborneol   borneol

β-fenchol   α-fenchol

(+)-Camphor, commonly known as Japan camphor, is the primary component of the camphor tree. This compound is also present in other plant families, including the leaves of rosemary Rosmarinus officinalis and sage Salvia officinalis (Labiatae). It produces the typical odor associated with spherical molecules, and serves as an analeptic, topical analgesic, topical antipruritic, antirheumatic, antiseptic, carminative, and counterirritant. Hence, (+)-camphor has diverse applications. To obtain (+)-camphor on a large scale, the crushed wood of fully grown camphor trees is steam-distilled, after which the (+)-camphor may crystallize partially from the distillate.

Fenchane derivatives are present in several essential oils as fenchones and fenchols. For instance, the oil of fennel, extracted from the dried fruit of Foeniculum vulgare (Umbelliferae), consists of up to 20% (+)-fenchone and is typically associated with limonene, phellandrene and α -pinene. This demonstrates that fenchane derivatives contribute to the fragrant profile and composition of fennel oil and other essential oils.

Sesquiterpenoids

The (E,E)-isomer of α-farnesene plays a significant role in the flavors and natural coatings of fruits such as apples, pears, and others. When combined with (E)-β-farnesene, it can be found in various essential oils, including those obtained from chamomile, citrus, and hops. In addition to this, aldehydes, such as α-sinensal and β-sinensal, which are derived from α- and β-farnesene, contribute to the flavor of the oil extracted from the fresh peel of ripe fruits of *Citrus sinensis* (Rutaceae). This implies that these compounds play an essential role in determining the fragrance and taste of different fruits and oils.

α-farnesene (*E,E*)   β-farnesene (E)

The formal ring closure of C-1 and C-6 of farnesane is responsible for creating a cyclohexane ring in bisabolanes, which represents a widely recognized class of monocyclic sesquiterpenes. The oil of ginger, extracted from the rhizome of *Zingiber officinalis*, is composed mainly of (−)-zingiberene (20%~40%), β-sesquiphellandrene, and (+)-β-bisabolene. Fragrant sesquiterpenes such as (+)-α- and (+)-β-bisabolol are detected in the essential oils of various plants, including chamomile and bergamot oil from unripe fruits of *Citrus aurantium* var. bergamia grown in southern Italy. These compounds also contribute to the fragrance of these plants.

(-)-zingiberene   β-bisabolene   (+)-α-bisabolol   (+)-β-bisabolol

Germacranes are created through the ring closure of C-1 and C-10 of farnesane. One example of a 1(10),4-germacradiene is 1(10),4-germacradien-6-ol, which is found as a glycoside in *Pittosporum tobira*. These compounds can undergo COPE rearrangements, resulting in elemadienes, such as in the case of shyobunol that is present in the oils of galbanum and kalmus. Therefore, some of the isolated elemane derivatives may be considered as artifacts that have arisen from germacranes.

There are more than 300 naturally occurring germacranes that have been identified. Examples of these include germacrene B [1(10)-*E*,4-*E*,7(11)-germacratriene] which is found in the peel of *Citrus junos*, germacrene D which is present in bergamot oil (*Citrus bergamia*, Rutaceae), and germacrone

[1(10)-*E*,4-*E*,7(11)-germacratrien-8-one], derived from germacrene B. Germacrone delivers a pleasant flowery to herby fragrance, and is a component detected in the essential oil of myrrh (*Commiphora abyssinica*, Burseraceae) as well as in the essential oils of *Geranium macrorhizum* (Geraniaceae) and *Rhododendron adamsii* (Ericaceae).

germacrane

1(10),4-germa-
cradien-6-ol

shyobunol

Currently, there are approximately 50 known elements, including β-elemenone from the oil of myrrh, which is the COPE rearrangement product of germacrone, (-)-bicycloelemene from peppermint oils of various origins (e.g., *Mentha piperita* or *Mentha arvensis*), and β-elemol. β-elemol is not only a minor component of Javanese oil of citronella but is also present in the elemi oil, extracted from the Manila elemi resin of the *Canarum luzonicum* tree (Burseraceae), and delivers a fragrance similar to that of pepper and lemon.

germacrene B   (-)-germacrene D   (-)-bicycloelemene   (-)-β-elemol

The formation of more than 30 naturally occurring humulanes involves the ring closure of C-1 and C-11 of farnesane, which occurs not only formally but also in biogenesis via farnesyl diphosphate, resulting in the sesquiterpene skeleton. Furthermore, the leaves of *Lindera strychnifolia* (Lauraceae)contain both regioisomeric α- and β-humulene.

| farnesane | humulane | α-humlene | β-humulene |

Epoxyhumuladienes which are derived from α-humulene, as well as (-)-humulol and (+)-humuladienone, are prominent constituents of the essential oils of hops (*Humulus lupulus*, Cannabaceae), cloves (*Caryophylli flos*, Caryophyllaceae), and ginger (*Zingiber zerumbeticum*, Zingiberaceae). These compounds are closely related to humulene.

| (-)-2,3-epoxy-6,9-humuladiene | (-)-6,7-epoxy-6,9-humuladiene | (-)-humulol | (+)-humula-2,9-dien-6-one |

There are around 30 naturally occurring caryophyllanes derived from humulanes, in which the C-2 and C-10 atoms form a cyclobutane ring. (-)-β-Caryophyllene, which is typically found as a mixture with its cis isomer isocaryophyllene, is present in clove oil extracted from the dried flower buds of cloves (*Caryophylli flos*, Caryophyllaceae), as well as in the oils of cinnamon, citrus, eucalyptus, sage, and thyme. Clove oil is known for its pleasant sweet, spicy, and fruity aroma, and is used not only in perfumery and chewing gum flavoring, but also as a dental analgesic, carminative, and counterirritant.

In sesquiterpenes, the eudesmane bicyclic skeleton is formed by the closure of the carbon atoms C-1, C-2, C-7, and C-10, with a *trans*-decalin core structure having corresponding numbering of the ring positions. Over 500 eudesmanes, previously known as selinanes, have been documented in the literature up to now.

Some well-known eudesmane derivatives found in flavors and fragrances include α- and β-selinene, which can be extracted from the oils of *Cannabis sativa* var. indica (Moraceae), celery (*Apium graveolens*, Umbelliferae), and hops (*Humulus lupulus*, Moraceae); (+)-α- and (+)-β-eudesmol, which can be found in certain oils of eucalyptus (*Eucalyptus macarthuri*); (-)-epi-γ-eudesmol with a woody aroma, which is present in the oil of geranium (*Pelargonium odoratissimum* and allied species) from North Africa; and (+)-γ-eudesmol, which is almost odorless and can be found in various essential oils. Additionally, (+)-β-costus acid and (+)-β-costol are among the constituents of the essential oil obtained from the roots of *Aucklandia costus* (Asteraceae), which is used in traditional Chinese and Japanese medicine to treat stomach ailments.

(-)-α-leudesmene   (-)-β-leudesmene   (+)-α-eudesmol   (+)-α-eudesmol

(-)-γ-leudesmene   (-)-*epi*-γ-eudesmol   (+)-β-costus acid   (+)-β-costus acid

4,5-*seco*-cadinane-derived antimalarials are found in the traditional Chinese medicinal herb *Artemisia annua* (Asteraceae), commonly known as qinghao. One such antimalarial is artemisinin, also known as qinghaosu, which is a 3,6-peroxide of the acylal formed by 4,5-seco-cadinane-5-aldehyde-12-oic acid. Dihydroqinghaosu and the 11(13)-dehydro derivative artemisitene are the active ingredients, which are currently used as semisynthetic esters and ethers (e.g. artemether) to treat malaria. These peroxides are believed to eliminate singlet oxygen, which damages the membrane of the pathogens and disrupts their nucleic acid metabolism.

4,5-*seco*-cadinane   (+)-qinghaosu   R=H: (+)-dihydroqinghaosu
R=CH₃: (+)-artemether

The formation of bonds from C-1 to C-10 and C-2 to C-6 of farnesane leads to the formal production of the bicyclic skeleton found in over 500 guaianes that have been isolated from higher plants. The numbering system for guaianes is based on that of decalin. Guaianes are sometimes referred to as proazulenes because their naturally occurring derivatives often undergo dehydration to form terpenoid azulenes (guaia-1,3,5,7,9-pentaenes) upon heating or steam distillation. An example of this is the deep blue-violet oily guaiazulene (guaia-1,3,5,7,9-pentaene) obtained during the extraction of oils from chamomile and guaiac wood from Guajacum species (Zygophyllaceae). When the genuine yellow 15-stearoyloxyguaia-1,3,5,7,9,11-hexaene in the milky juice of the delicious fungus Lactarius deliciosus is enzymatically decomposed, it turns from orange to greenish, and produces violet lactaroviolin (guaia-1,3,5,7,9,11-hexaen-4-aldehyde).

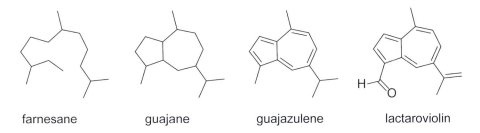

farnesane   guajane   guajazulene   lactaroviolin

The formation of bonds between C-1 and C-6, as well as C-1 and C-11, results in the conversion of farnesane to the bicyclic skeleton of himachalane. The numbering system used for himachalanes is taken from that of farnesane. Several himachalanes, such as α-himachalene and himachalol, are present in the essential oil of cedar wood extracted from *Cedrus deodara* (Pinaceae). 2,7-Cyclohimachalanes are also known as longipinanes, and can be found in various oils of pine wood and some Asteraceae. Examples of longipinanes include 3-longipinene from Pinus species (Pinaceae), and 3-longipinen-5-one from *Chrysanthemum vulgare* (Asteraceae).

Longifolanes, which biogenetically and formally derive from farnesane, are distinguished from longipinanes by the cleavage of the C-3–C-4 bond and the formation of bonds between C-1–C-6, C-2–C-4, C-3–C-7, and C-1–C-11 to produce the tricycle structure. Examples of longifolanes include the isomers longicyclene and longifolene, which are widely found in essential oils. Longifolene can be found in Indian turpentine oil to a considerable extent (up to 20%), which is commercially produced from the Himalayan pine *Pinus longifolia* (Pinaceae) for the synthesis of a widely used chiral hydroboration agent.

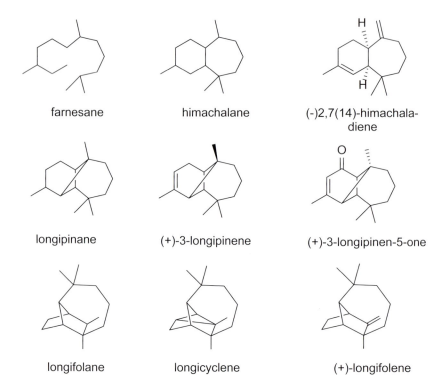

farnesane himachalane (-)2,7(14)-himachala-diene

longipinane (+)-3-longipinene (+)-3-longipinen-5-one

longifolane longicyclene (+)-longifolene

Formally connecting the bonds C-1–C-6 and C-6–C-10 in farnesane leads to the spiro[4,5]decane basic skeleton of acorane. This class of sesquiterpenes is named after the Acorus species. One example is (-)-4-Acoren-3-one, which has been isolated from *Acorus calamus* (Calamus, Araceae) and from the carrot *Daucus carota* (Umbelliferae). The oil of calamus (oil of sweet flag) extracted from the rhizome of *Acorus calamus*, which has a warm and spicy odor and a pleasant bitter taste, is mainly used in perfumery and as a minor ingredient (possibly carcinogenic) in vermouth, some flavored wines, and liqueurs. (+)-3,7(11)-Acoradiene is a constituent of juniper *Juniperus rigida*; its enantiomer is found in *Chamaecyparis nootkatensis* (Cupressaceae).

farnesane   acorane   (+)-3,7(11)-acoradiene   (-)-4-acroren-3-one

Aromatic compounds

A number of important essential oil components are comprised of aromatic compounds. The aromatic ring may contain one or more substituted functional groups or side chains. Among the aromatic alcohols are benzyl, phenylethyl, phenylpropyl, cinnamyl, and cuminyl. Cinnamaldehyde is the primary constituent in cinnamon bark oil (*Cinnamomum zeylanicum*; Lauraceae), widely used as a spice and flavoring. Fresh bark has been found to have abundant levels of cinnamyl acetate, and cinnamaldehyde is released from it through fermentation processes, which form part of the commercial bark preparation, likely involving enzymatic hydrolysis and the participation of reversible aldehyde-alcohol oxidoreductase.

In contrast, cinnamon leaves contain high levels of eugenol and a lower concentration of cinnamaldehyde. Eugenol is also the main constituent in clove oil (*Syzygium aromaticum*; Myrtaceae), which has been used for many years as a dental anesthetic and flavoring agent. The side chain of eugenol is derived from that of the cinnamyl alcohols through reduction, but it differs in the placement of the double bond. This variation is explained by resonance forms of the allylic cation, and the addition of hydride (from NADPH) can produce either allylphenols (such as eugenol) or propenylphenols (such as anethole).

Protonation or possibly phosphorylation may facilitate the loss of hydroxyl from a cinnamyl alcohol, although there is no evidence to support the latter. Myristicin, found in nutmeg (Myristica fragrans; Myristicaceae), is another example of an allylphenol found in flavorings. Myristicin has a history of being used as a mild hallucinogen through the ingestion of ground nutmeg. It is believed that myristicin is metabolized in the body via an amination reaction, creating an amphetamine-like derivative. Anethole is the primary constituent in oils from aniseed (*Pimpinella anisum*; Umbelliferae/Apiaceae), star anise (*Illicium verum*; Illiciaceae), and fennel (*Foeniculum vulgare*; Umbelliferae/Apiaceae). The use of propenyl components from flavorings such as cinnamon, star anise, nutmeg, and sassafras (*Sassafras albidum*; Lauraceae) has been reduced in commercial applications since these constituents have been identified as weak carcinogens in animal laboratory tests. In the case of safrole, the primary component of sassafras oil, it has been shown to arise from side-chain hydroxylation followed by sulfation, resulting in an agent that binds to cellular macromolecules. Further data on volatile oils containing aromatic constituents isolated from these and other plant materials are given in Table 2-2.

cinnamaldehyde   cinnamyl acetate   anethole   estragole

eugenol   myristicin   elemicin

Cymenes refer to benzenoid menthanes, which are not commonly found in their o-isomeric form in nature. *m*-Cymene is present in the essential oil of blackcurrant (*Ribes nigrum*, Saxifragaceae), while *p*-cymene is found in essential oils of cinnamon, cypress, eucalyptus, thyme, turpentine, and others. Both isomers are used as fragrances in the perfumery industry. Carvacrol is extracted from the oils of marjoram, origanum, summer savoy, and thyme, and serves as a disinfectant. Thymol is predominantly found in the oil of thyme (*Thymus vulgaris*, Labiatae) and the essential oil obtained from the seeds of *Orthodon angustifolium* (Labiatae); it is used as a topical antiseptic and antihelmintic. *p*-Cymen-8-ol was discovered in the frass of the woodworm *Hylotrupes bajulus* (Cerambycodae). Cuminaldehyde, with its strong and persistent odor, is extracted from various essential oils such as eucalyptus and myrrh, and is used in perfumery. *p*-cymene, along with the phenol derivatives thymol and carvacrol found in thyme (*Thymus vulgaris*; Labiatae/Lamiaceae), belong to a small group of aromatic compounds produced in nature from isoprene units, rather than through the much more typical routes involving acetate or shikimate. These compounds all possess the carbon skeleton characteristic of monocyclic monoterpenes, and their structural similarities to limonene and other oxygenated monoterpenes like menthone or carvone indicate pathways involving additional dehydrogenation reactions.

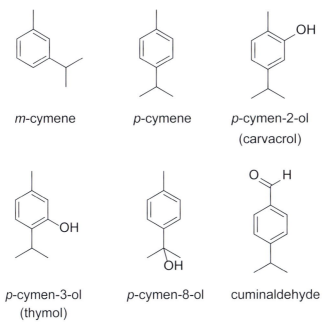

m-cymene   *p*-cymene   *p*-cymen-2-ol
                                     (carvacrol)

p-cymen-3-ol   *p*-cymen-8-ol   cuminaldehyde
(thymol)

Table 2-2 Volatile oils containing principally aromatic compounds

Oils	Plant source	Plant part used	Oil content (%)	Major constituents with typical (%) composition	Uses, notes
Aniseed (Anise)	Pimpinella anisum (Umbelliferae/Apiaceae)	ripe fruit	2 ~ 3	anethole (80 ~ 90) estragole (1 ~ 6)	flavoring, carminative, aromatherapy
Star anise	Illicium verum (Illiciaceae)	ripe fruit	5 ~ 8	anethole (80 ~ 90) estragole (1 ~ 6)	flavoring, carminative, fruits contain substantial amounts of shikimic and quinic acids
Cassia	Cinnamomum cassia (Lauraceae)	dried bark, or leaves and twigs	1 ~ 2	cinnamaldehyde (70 ~ 90) 2-methoxycinnamal-dehyde (12)	flavoring, carminative, known as cinnamon oil in USA
Cinnamon bark	Cinnamomum Zeylanicum (Lauraceae)	dried bark	1 ~ 2	cinnamaldehyde (70 ~ 80) eugenol (1 ~ 13) cinnamyl acetate (3 ~ 4) eugenol (70 ~ 95)	flavoring, carminative, aromatherapy
Cinnamon leaf	Cinnamomum zeylanicum (Lauraceae)	leaves	0.5 ~ 0.7	eugenol (70 ~ 95)	flavoring

(continued)

Oils	Plant source	Plant part used	Oil content (%)	Major constituents with typical (%) composition	Uses, notes
Clove	Syzygium aromaticum (Eugenia caryophyllus) (Myrtaceae)	dried flower buds	15 ~ 20	eugenol (75 ~ 90) eugenyl acetate (10 ~ 15) β-caryophyllene (3)	flavoring, aromatherapy, antiseptic
Fennel	Foeniculum vulgare (Umbelliferae/ Apiaceae)	ripe fruit	2 ~ 5	anethole (50 ~ 70) fenchone (10 ~ 20) estragole (3 ~ 20)	flavoring, carminative, aromatherapy
Nutmeg	Myristica fragrans (Myristicaceae)	seed	5 ~ 16	sabinene (17 ~ 28) α-pinene (14 ~ 22) β-pinene (9 ~ 15) terpinen-4-ol (6 ~ 9) myristicin (4 ~ 8) elemicin (2)	flavoring, carminative, aromatherapy, although the main constituents are terpenoids, most of the flavoring comes from the minor aromatic constituents, myristicin, elemicin, etc. myristicin is hallucinogenic
Wintergreen	Gaultheria procumbens (Ericaceae) or Betula lenta (Betulaceae)	leaves / bark	0.7 ~ 1.5 / 0.2 ~ 0.6	methyl salicylate (98)	flavoring, antiseptic, antirheumatic, prior to distillation, plant material is macerated with water to allow enzymic hydrolysis of glycosides, methyl salicylate is now produced synthetically

Aliphatic derivatives

A few aliphatic hydrocarbons have been discovered and isolated in essential oils, such as heptane extracted from pine needle oil. "Stearoptenes", which are composed of a blend of waxes, encompass higher members of the paraffin series. This wax-like substance, conceivably arising from the protective coatings of leaves, flowers, fruits, and seeds, can be found in significant quantities within rose and chamomile oils, leading these oils to solidify easily below room temperature. Waxes can also be found in rather high percentages in cold-pressed citrus oils. Within essential oils, only a few aliphatic alcohols and acids exist in free form, with the majority being esterified. The lower members of the saturated aliphatic alcohols and acids, including methyl and ethyl alcohol and formic, acetic, propionic, butyric, and valeric acid, which are water-soluble, can be found in oils obtained from "cohobation" by redistillation of the distillation water. These alcohols and acids are likely degradation products resulting from processes such as hydrolysis during steam distillation or fermentation before distillation. Saturated aliphatic alcohols, such as butyl, amyl, hexyl, octyl, nonyl, decyl, and undecyl, in the form of normal or branched isomers, have been extracted by fractional distillation from a range of essential oils. Additionally, the unsaturated aliphatic alcohol 3-hexen-1-ol, also called the "leaf alcohol", is found in many green leaves, herbs, and grasses, and has a grass-like aroma, being the primary component of tea leaf oil. Isomers of methyl heptenol are formed in lemon-grass oil.

The lower members of non-terpenic aliphatic aldehydes do not have a significant impact on essential oils. Examples of such aldehydes include formaldehyde and acetaldehyde, which are found in distillation water and are likely decomposition or degradation products created during steam distillation. Propyl, butyl, valeral, and caproaldehyde, on the other hand, can be seen in the lower fractions of oils such as eucalyptus and peppermint. Higher aliphatic aldehydes like octyl and nonyl, despite only being present in small quantities, have a more noticeable influence due to their strong characteristic odor and flavor. These aldehydes can be found in oils such as orris root, coriander seed, rose, lemon, and sweet orange. An unsaturated aliphatic aldehyde, beta hexanal, better known as "leaf aldehyde", contributes to the scent of green leaves. Violet leaf oil contains 2,6-nonadien-1-a1, or "violet leaf aldehyde".

Not many aliphatic ketones occur in essential oils. Again, the lowest members, such as acetone and di-acetyl, are formed mainly in distillation water and are probably decomposition products. Amyl methyl ketone has been found in the lower boiling fractions of clove oil, and methyl heptenone in many essential oils. The following fatty acids have been predominantly found in essential oils extracted from seeds and roots: alpha and beta-methyl butanoic, caproic, enanthic, caprylic, pelargonic, capric, nndecylic, lauric, myristic, hydroxy myristic, palmitic, stearic, methacrylic, isopropylidenacetic, angelic, tiglic, beta-propylacrylic, oleic, and succinic. Some essential oils contain substantial quantities of fatty acids. For instance, orris root oil may contain up to 85% myristic acid. Esters are among the most important constituents that greatly contribute to the odor and flavor characteristics of essential oils. Some oils are made up almost entirely of esters; wintergreen oil and sweet birch oil sometimes contain up to 99% methyl salicylate. Numerous alcohols and acids are present in the form of esters in essential oils, allowing for a great variety of possible combinations.

Miscellaneous

There are a few compounds that do not belong to the three categories previously discussed. These compounds are nitrogen-or sulfur-containing, including organic cyanides, indole, skatole, allyl, benzyl, phenylethyl cyanide, methyl anthranilate, and methyl-N-methyl-anthranilate, and they are found in specific citrus fruit peel, flower, and leaf oils. Additionally, organic disulfides and sulfides, such as allyl sulfides in garlic oil, and organic isothiocyanates, including allyl, butenyl, benzyl, and phenylethyl thiocyanate, are found in mustard oil. However, there are surprisingly few of these compounds found as essential oil components.

Section 2

Isolation and chemical analysis of essential oils

2.1 Isolation from Plants

Numerous volatile mono-and sesquiterpenoids are used in the flavor and fragrance industries due to their pleasant aroma and taste. Terpene-rich mixtures are extracted on a larger scale via steam distillation or extraction from plant parts such as fruits, flowers, leaves, roots, and stems, yielding essential oils that are strictly volatile. Given their alluring fragrances, certain essential oils are valuable components in the perfume industry. In addition, some essential oils are utilized in cooking as a seasoning, while others are classified as phytomedicines due to the pharmacological properties of their constituent compounds. Non-volatile higher terpenes isolated from plants play a crucial role in the pharmaceutical and nutrition industries, serving as emulsifying agents or pharmaceutical products.

Pure terpenes can be obtained from essential oils on a larger scale via distillation. Chromatographic techniques, such as gas chromatography (GC) or liquid chromatography (LC), allow for the isolation of small amounts with high purity. To isolate low-volatile sesqui-, di-, sester-, and triterpenes containing polar groups from organisms such as plants and fungi, the natural material is first dried, chopped, or ground. Inert solvents are then used for extraction at the lowest possible temperature to prevent the formation of artifacts. Petroleum ether is a suitable solvent for the extraction of less-polar terpenes, while polar terpenes, including saponins, are extracted with water, ethanol, or methanol. After extraction, the solvent is evaporated to dryness in a vacuum or via freeze-drying. The resulting substance is then fractionated by column chromatography, with well-separated spots in thin-layer chromatography (TLC) indicating separable constituents in the crude fractions. To purify the constituents, column chromatography is often utilized. Petroleum ether or cyclohexane, followed by increasing concentrations of more polar solvents like dichloromethane, chloroform, methanol, or ethanol, is used for elution of the constituents, based on preceding TLC analysis. For spectroscopic identification or structure elucidation and pharmacological screenings, the final purification of constituents is frequently accomplished using liquid chromatography with medium-or high-pressure (MPLC or HPLC).

2.2 Chemical analysis of essential oils by GC and GC-MS

The qualitative and quantitative analysis of essential oils is typically conducted using gas chromatography (GC). Identification is achieved through a combination of this separation method with mass spectrometry (GC-MS). As volatile metabolites, essential oils have low boiling points, allowing them to be easily converted into steam. For this reason, gas chromatography with GC-MS mass spectrometry is the ideal analysis method.

The use of capillary columns in essential oil analysis has enabled the separation of over 100 compounds with defined specificity. Typically, chromatographic separations are conducted on nonpolar columns containing 95% dimethylpolysiloxane, given that several components in essential oils contain polar groups such as hydroxyl (OH). Nonetheless, the use of intermediate polarity columns has enabled the separation of these components. Both assays together provide complete chemical inquiry of molecules and are complementary. Structural elucidation is achieved by combining various analyses, which include comparison with spectrum databases and the theoretical and experimental determination of the retention rates of compounds. There are currently databases available for this purpose, with the most widely used being the "Identification of Essential Oil Components by Gas Chromatography/Mass Spectrometry" database with approximately 4,000 compounds from essential oils.

The GC-MS technique is unable to effectively evaluate stereoisomers, therefore, it is necessary to use chiral columns or other techniques such as nuclear magnetic resonance imaging. Regardless of this limitation, GC-MS testing is widely regarded as the "gold standard" in the chemical analysis of essential oils. In essence, a GC-MS analysis identifies the various constituents in an essential oil and lists the percentage of each constituent present.

Indeed, the GC-MS test is indispensable in verifying the authenticity and purity of an essential oil. A report can be evaluated against published standards or international specifications, like those set by the ISO (International Organization for Standardization), to determine whether the oil meets the necessary criteria. Beyond purity testing, a fundamental understanding of GC-MS is advantageous to anyone who works with or enjoys using essential oils. This knowledge provides insight into the safe application of essential oils while also helping to enhance aromatic and therapeutic blending.

The analysis begins with the introduction of the essential oil into the GC element, which comprises of two phases — a stationary phase and a mobile phase. The stationary phase is a lengthy, coiled tube coated with a highly stable liquid that remains motionless while the mobile phase, an inert carrier gas such as Helium or Nitrogen, flows through it. The temperature of the set-up can be regulated as required by placing the coiled tube in an oven.

Once the essential oil sample is injected, it undergoes vaporization, and transforms into a gas. The molecules proceed to travel through the coiled tube. The molecules that have lower volatility are more attracted to the liquid stationary phase, thus requiring a longer duration to reach the end of the column. Conversely, molecules with higher volatility interact more readily with the inert gas and travel more swiftly through the column. As a result of the combined efforts of the mobile and stationary phases, the distinct constituents become effectively separated based on their volatility.

The gas chromatography method records the time taken for each constituent to reach the detector. Upon detection, it produces a peak in the gas chromatogram. However, it may become challenging to differentiate between two distinct constituents if they arrive at the detector simultaneously.

Subsequently, the partitioned compounds are channeled into the mass spectrometer, which bombards them with an electron beam. This causes the essential oil compounds to break up further into positively charged fragments. These fragments, also known as ions, are accelerated through an electric field and rerouted with a magnetic field to generate a mass spectrum. The identity of each original molecule can be deduced by matching the mass spectrum with a database consisting of established sample patterns (Figure 2-6).

In order to verify the spectra, the full m/z range TOF-MS data can be compared to the NIST library databases. Retention times of observed peaks can also be connected with the retention index via a known alkane standard for retention index matching with the NIST library databases, leading to greater confidence. A GC-MS chromatogram of a mint essential oil is depicted in Figure 2-7, while Table 2-3 presents the area quantification, aroma characteristics and identifications for the 30 most potent analytes in the sample.

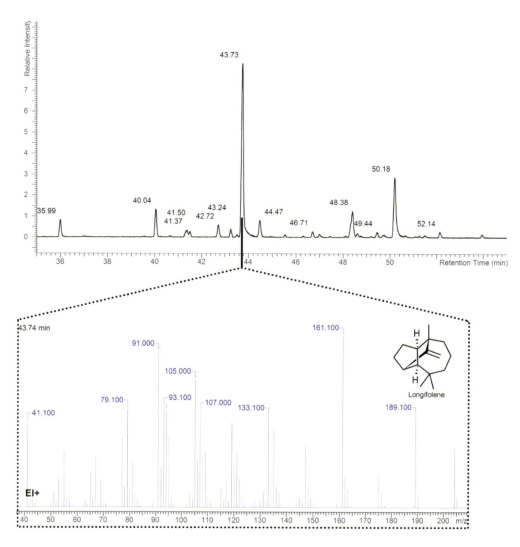

Figure 2-6 A mass spectrum extracted from a peak

A more in-depth analysis of the chemical composition of the molecules within an essential oil can be conducted with instrumentation that combines gas chromatography with spectrophotometric techniques, such as nuclear magnetic resonance imaging and infrared spectroscopy. Additionally, NMR or IR spectra can be examined in extracted molecules using column or thin layer chromatography.

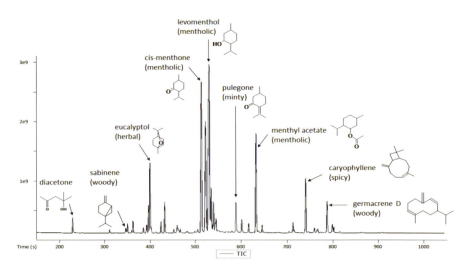

Figure 2-7 TIC Chromatogram for mint essential oil

Table 2-3 Identification information for top 30 analytes

	Name	R.T. (s)	Formula	RI	CAS	Area %
1	diacetone	228.7	$C_6H_{12}O_2$	839.8	123-42-2	1.102
2	sabinene	346.2	$C_{10}H_{16}$	976.5	3387-41-5	0.326
3	β-pinene	350.3	$C_{10}H_{16}$	981	127-91-3	0.656
4	3-octanol	362.7	$C_8H_{18}O$	994.6	589-98-0	0.835
5	α-terpinene	385.7	$C_{10}H_{16}$	1019.7	99-86-5	0.38
6	p-cymene	392.9	$C_{10}H_{14}$	1027.6	99-87-6	0.494
7	limonene	397	$C_{10}H_{16}$	1032.1	138-86-3	1.978
8	eucalyptol	400.1	$C_{10}H_{18}O$	1035.4	470-82-6	5.836
9	γ-terpinene	424.6	$C_{10}H_{16}$	1062.1	99-85-4	0.81
10	(Z)-sabinene	432.6	$C_{10}H_{18}O$	1070.7	15537-55-0	2.248

(continued)

	Name	R.T. (s)	Formula	RI	CAS	Area %
11	linalool	459.9	$C_{10}H_{18}O$	1100.4	78-70-6	0.538
12	*cis*-menthone	513.4	$C_{10}H_{18}O$	1160.4	491-07-6	16.828
13	menthofuran	521.7	$C_{10}H_{14}O$	1169.6	494-90-6	3.12
14	(±)-menthol	522.4	$C_{10}H_{20}O$	1170.4	1490-04-6	2.842
15	l-menthone	523.1	$C_{10}H_{18}O$	1171.3	14073-97-3	3.258
16	levomenthol	530.6	$C_{10}H_{20}O$	1179.7	2216-51-5	29.868
17	(-)-terpinen-4-ol	534.8	$C_{10}H_{18}O$	1184.3	20126-76-5	3.113
18	neoisomenthol	539.9	$C_{10}H_{20}O$	1190	491-02-1	2.075
19	(1*S*,2*R*,5*R*)-(+)-isomenthol	543.9	$C_{10}H_{20}O$	1194.5	23283-97-8	0.503
20	α-terpineol	545.4	$C_{10}H_{18}O$	1196.2	98-55-5	0.848
21	pulegone	588.7	$C_{10}H_{16}O$	1247.4	89-82-7	2.196
22	*p*-menth-1-en-3-one	601.4	$C_{10}H_{16}O$	1262.4	89-81-6	0.944
23	neomenthyl acetate	616.4	$C_{12}H_{22}O_2$	1280.3	2230-87-7	0.661
24	menthyl acetate	632.1	$C_{12}H_{22}O_2$	1298.8	89-48-5	9.468
25	isomenthyl acetate	645.8	$C_{12}H_{22}O_2$	1315.7	20777-45-1	0.599
26	(-)-β-bourbonene	712.4	$C_{15}H_{24}$	1398	5208-59-3	0.716
27	caryophyllene	740.6	$C_{15}H_{24}$	1434.9	87-44-5	4.379
28	germacrene D	787	$C_{15}H_{24}$	1496	23986-74-5	2.316
29	β-cyclogermacrane	798.7	$C_{15}H_{24}$	1512	24703-35-3	0.632
30	β-himachalene	801.9	$C_{15}H_{24}$	1516.6	1461-03-6	0.432

NMR

Nuclear Magnetic Resonance (NMR) was first experimentally detected at the end of 1945, nearly simultaneously with the work conducted by Felix Bloch's research team at Stanford University and Edward Purcell's team at Harvard University. The first NMR spectrum was published in the same issue of the Physical Review in January 1946. Bloch and Purcell were jointly awarded the 1952 Nobel Prize in Physics for their research on Nuclear Magnetic Resonance Spectroscopy. NMR is a critical analytical tool for organic chemists, contributing significantly to research in organic laboratories. It can not only provide information on the structure of the molecule, but also determine the content and purity of the sample. One of the most commonly used NMR methods by organic chemists is Proton (^{1}H) NMR. The protons present in the molecule exhibit different behavioral characteristics based on the surrounding chemical environment, rendering it possible to elucidate their structure.

NMR spectroscopy has proved to be a valuable method for quantifying individual components in crude extracts, essential oils, or dietary preparations without the need for fractionation or isolation procedures. This methodology is particularly useful when the studied compounds exhibit at least one well-resolved signal in the region from 3.0 to 10.0 ppm. Quantitative NMR spectroscopy provides a complete validation of reference compounds for natural products that can compete with or even surpass molecular analysis-based chromatographic validation. Specifically, ^{1}H NMR spectroscopy enables the precise determination of the sample content as well as the amount and type of non-active and/or marker compounds, producing useful fingerprints for identifying herbal drugs that can be used in reference compound certification and quality control. Additionally, contemporary NMR instrument designs allow for readily accessible calibrated and standardized analytical conditions.

Numerous studies have reported on the use of different NMR spectroscopic methods in the analysis of essential oils. A quick and simple method based on proton nuclear magnetic resonance spectroscopy was developed to determine the trans-anethole content in fennel essential oil (Figure 2-8 and Table 2-4). The spectrum showed two sets of two chemically equivalent aromatic protons, predicted from the integration values, resonating at δ 7.32 and 6.89 ppm. The two protons H-2 and H-6, which were adjacent to the inductive electron-withdrawing effect of the oxygen atom, were assigned to signal at δ 7.32 (2H, d, *J* = 8.8, H-2 & H-6). The other two aromatic protons were assigned to signal at δ 6.89 (2H, d, *J* = 8.8, H-3 & H-5). The two olefinic protons of the propenyl moiety resonated at δ 6.41 (1H, d, *J* = 15.6, H-1') and δ 6.15 (1H, qd, *J* = 6.8, 15.6, H-2'). Table 2-4 displays the chemical shift and coupling constants of ¹H NMR signals ascribed to their corresponding positions in the structure of trans-anethole. The signal from H-1' was selected as a target peak for quantitative analysis. This signal is in a region where there is no interference with other signals in the ¹H NMR spectrum of fennel essential oil. The signals of *trans*-anethole can be readily differentiated in the oil's spectrum because it represents the major component of the oil.

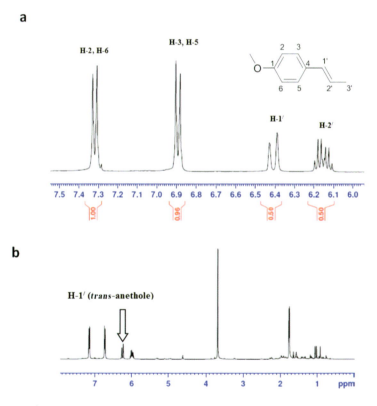

Figure 2-8 ¹H NMR spectra of *trans*-anethole (a) and of pure essential oil of fennel (b).

Table 2-4 1H NMR spectral data of trans-anethole in CDCl3 (δ ppm)

Position	δ (multiplicity, J in Hz)
2,6	7.32 (2H, d, $J = 8.8$)
3,5	6.89 (2H, d, $J = 8.8$)
1-OCH$_3$	3.84 (3H, s)
1'	6.41 (1H, d, $J = 15.6$)
2'	6.15 (1H, qd, $J = 6.8, 15.6$)
3'	1.92 (3H, dd, $J = 1.6, 6.8$)

Chapter II References

1. Baser, K. Husnu Can, Gerhard Buchbauer. Handbook of essential oils: science, technology, and applications [M]. London: CRC press, 2009.

2. Dewick, Paul M. Medicinal natural products: a biosynthetic approach [M]. New York: John Wiley & Sons, 2002.

3. Breitmaier, Eberhard. Terpenes: flavors, fragrances, pharmaca, pheromones [M]. New York: John Wiley & Sons, 2006.

4. Walsh, Christopher T, and Yi Tang. Natural product biosynthesis [M]. London: Royal Society of Chemistry, 2017.

Chapter III

Extraction Process and Equipment of Plant Essential Oils

Plant essential oil refers to the volatile substances or aromatic components found in various parts (organs) of the plants of aromatic plants. They are a type of secondary biomass. Different aromatic plant species exist in their roots, stems, leaves, bark, fruits, seeds, and the whole plant, for example, in the genus Cinnamomum, Cinnamomum camphora essential oils exist in stems, roots, and branches, respectively. The essential oil of C. burmannii in camphor is present in the branches and leaves; the essential oil in the flower of Jasminum sambac in the osmanthaceae family is present in the petals; the essential oil in the mint (Mentha canadensis) family is present in the whole plant; the essential oil of Rutaceae Citrus is mainly found in the peel and branches; the seed kernels of the ginger family *Amomum villosum* are rich in essential oils in addition to the essential oils contained in the whole plant.

The organs that form essential oils are the original secretions of their cells, and the cells of aromatic plant organs should have the ability to secrete essential oils. Such tissues are also called "oil cells". Such cells are analyzed from anatomical sections and can be divided into external essential oil secretion gland and the internal essential oil cell secretion gland. The external cells exist in the epidermal tissue, called the oil glands, while the internal cell secretion glands contain the fission secretory cells, secretory cavity or secretory tract, respectively. The essential oils contained in different species are present in glandular hairs, resin ducts, oil cells, and mesophytic secretory cells in various forms, such as white orchid leaves and citrus fruit epidermis; they also coexist with secreted resin, mucus quality, etc., such as liquidambar resin, agarwood; there are also a few essential oil components combined with sugar to form glycosides, such as amygdalin, wintergreen glycosides, etc., after their hydrolysis, bitter almond oil and wintergreen oil will be formed respectively. We extract essential oils contained in aromatic plants by destroying their glandular hairs, oil cells, schizonts, etc. For example, when we extract essential oils by steam distillation, we heat the steam to destroy their cell walls, and the method of extracting essential oils by solvents or pressing is the same principle.

Section 1

Pretreatment before extraction of plant essential oils

The raw materials for extracting plant essential oils, are normally various parts of aromatic plants, including flowers, fruits, branches and leaves, wood, roots, resins, etc. In general, the harvested raw materials should be processed as soon as possible, or delivered to the processing site in time, so that the harvested raw materials are not damaged during transportation. Different types of raw materials are treated differently.

1.1 Branches, leaves and herbs

Some of the branches, leaves and herbs should be put into processing as soon as possible after harvesting, such as rock rose, lavender, tea tree branches, white branch leaves and herbs, and other raw materials such as Bailan leaves, tree orchid leaves, hawksbill leaves, mint, Spearmint, etc., should be placed for a period of time, but must be evenly spread out and piled up to make it ventilated and ventilated to avoid heating. The essential oils extracted from some raw materials even need to dry for a period of time. For example, cassia leaves must be placed for 2~3 months before steam distillation. Some raw materials such as Agastache rugosa can meet the standard only if they were distilled after fermentation. If Bailan leaves are placed for several days, the oil output can be increased by 5% to 20% (according to the fresh weight after harvest). After drying, the cell pores on the leaf surface expand to make the oil spots easily to spread.

1.2 Flowers

The flowers after harvesting are easier to wither and lose their aroma. Different flowers have different treatment methods. For example, after the osmanthus is harvested, the flowers can be soaked in salt water, and the small flower jasmine is harvested when the flower buds are spread, spreading on the layers of flower stands, keep ventilation, and then put into extraction when the flower buds are about to open.

1.3 Underground stems

Underground stems, such as ginger and sage of ginger family, are generally processed into ginger oil. You need to slice the ginger and keep it in a cool place until it is dry before processing.

1.4 Roots and stem wood

Tree roots and stem wood are relatively solid. In order to extract the essential oil in the wood glands easier, the raw materials should be crushed, chopped or powdered such as maple, benzoin, sandalwood, cypress and agarwood.

1.5 Seeds

Some seed kernels contain essential oil but their seed shells are relatively hard, and it is not easy to obtain essential oils with higher yields by steam distillation or solvent extraction. Generally, crushing the seed shells can increase oil yield and shorten processing time.

1.6 Fermentation

Some aromatic plants need to ferment the raw materials to extract essential oils or concrete; such as iris root, vanilla bean, etc.

Section 2
Plant essential oils extraction method and equipment

The methods and equipment for extracting essential oils are different due to different objects and requirements. They can be divided into two types: one is purely for obtaining a small amount of essential oil samples for analysis and research, and the other is a large-scale equipment and process for production.

2.1 Method and equipments for sample extraction

2.1.1 The simplest sampling is to use an adsorbent-macroporous resin in an odorless PVC bag, and then use a portable atmospheric extractor to evacuate, and the air inlet is connected to a large pore resin tube for filtration and purification of the air into the sample bag. If the sample is flowers, the fragrance can be naturally released by the flowers and absorbed by the large-pore resin tube installed in the air outlet. If the sample is branches and leaves, you need to rub the branches and leaves of the sample bag with external force to accelerate the release of fragrance. The macroporous resin can be desorbed with refined ether or hexane to obtain a small amount of essential oil sample, which can be analyzed by preliminary GCMS (Figure 3-1). Refined mixed fats (mixed lard and butter) can also be used in the wild to absorb the fragrance released by flowers, and eventually obtain a small amount of balsam, but the analysis also needs to remove the fat. This process is cumbersome and only for the preliminary evaluations of fragrance, which is rarely used now.

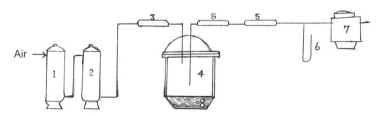

1.&2. Molecular sieve   3. Macroporous resin tube (for reference)   4.Container for adsorbed raw materials (unscented plastic bags can be used in the field)   5.Macroporous resin tube 6.Flow meter   7.Air extractor (field air sampler can be used in the field)   8. Water

Figure 3-1 Macroporous resin adsorption process

Use a portable complete steam distillation device to the vicinity of the sampling site, set up a complete set of steam distillation equipment, including the distillation pot, stove, condenser and the connection of condensate, and install the oil-water separator (Figure 3-2) . This method requires very heavy work such as firewood, which takes a long time to get, generally can only be distilled 2 to 3 times a day. This method is generally used for branches or herbs, and more samples are obtained. It can be used for further research in addition to component analysis.

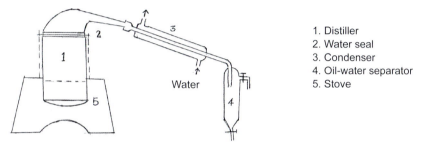

1. Distiller
2. Water seal
3. Condenser
4. Oil-water separator
5. Stove

Figure 3-2 Field steaming equipment

2.1.2 The extraction method of the essential oil used in laboratory analysis and research is relatively simple. The adsorption method is the same as that of field sampling. It may be necessary to add an additional glass tube with a molecular sieve inside the air inlet. The sampler can be directly plugged into the laboratory power supply. For steam distillation, glass equipment can be installed, and the steam generator uses an electric hot water steam generator to provide steam (Figure 3-3).

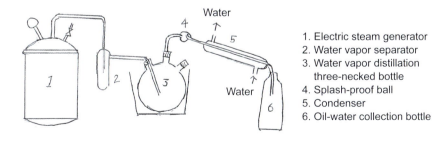

1. Electric steam generator
2. Water vapor separator
3. Water vapor distillation
   three-necked bottle
4. Splash-proof ball
5. Condenser
6. Oil-water collection bottle

Figure 3-3 Laboratory steam distillation device

2.2 Plant essential oil steam distillation production process and equipment

Due to the variety of aromatic plants that produce essential oils, the demand for essential oil product commodities is objectively different, so their production processes and equipments are also different.

2.2.1 Atmospheric pressure steam distillation process and equipment. This method is most commonly used for the production of plant essential oils. It also has a variety of similar methods and equipments: The principle is steam distillation. According to Dalton's Law, the definition of boiling is that the vapor pressure of the liquid itself reaches boiling when it is the same as the external atmospheric pressure. The temperature of the liquid itself when it reaches the external vapor pressure, called the boiling temperature of the liquid or the boiling point. The two-phase mixed liquid, when the sum of the vapor pressure of the two reaches the external atmospheric pressure, they can reach boiling at the same time. The mixed steam can be distilled out after cooling, and the mixed liquid can be separated by oil-water separator to obtain essential oil. The ratio of the two-phase liquid that is distilled out, named the oil-water ratio, depends on the ratio of their vapor pressures of the oil and water. Since the boiling point of water at normal pressure is 100°C, which is relatively constant, increasing the boiling point temperature of the oil phase will increase the ratio of the oil phase to the

(a)   (b)

Figure 3-4a, b Steam distillation equipment

water phase. Improving the distillation process and equipment in this way will change the ratio of oil to water. Therefore, higher temperatures will get the rate of steaming out essential oils higher (Figure 3-4a,b).

General steam distillation can be different according to the material stacking method. One is water steaming. The steamed material is mixed with water and heated to make the water boil out and steamed out. This method can be used for steaming flowers. The steam is separated into oil and water. The layer can be returned to the distillation kettle for recirculation and distillation. Another method is water-distillation, that is, the material is separated from the water in the distillation kettle by sieve. After heating the water to boiling, the steam rises directly from the bottom of the distillation kettle to the steamed material to achieve the purpose of distillation. This method is mostly used for small and medium-sized branches and leaves, herbs and so on. Similarly, the distilled water can be recycled back to the distillation kettle for distillation. The third method is direct steam distillation. The steam passes through the bottom of the pipeline to the bottom of the distillation kettle and then enters the steamed material from bottom to top for distillation. In this way, the distillation temperature can be higher, the oil-water distillation ratio is higher than that of the first two methods, and the distillation time will be shortened, but this method requires a steam generator boiler. This method has a short distillation cycle and is suitable for larger production scales. It can also use backwater (Figure 3-5a, b).

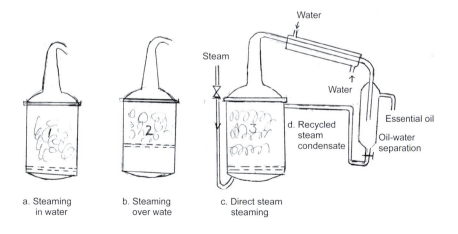

Figure 3-5a Three steam distillation methods

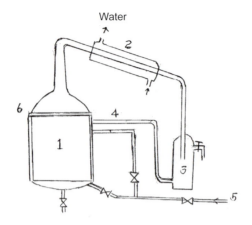

Water

1. Steam kettle
2. Condenser
3. Oil-water separator
4. The conduit allows the distilled water layer back to the steam kettle
5. Steam
6. Water seal

Figure 3-5b Steam distillation backwater schematic

2.2.2 Depressurized distillation and pressurized distillation. Based on the reduction of the external pressure during steam distillation, the boiling point is also reduced, and the steaming temperature is also reduced. This is beneficial for temperature-sensitive aromatic components, especially the extraction of essential oils from flowers, but the ratio of essential oils to water will be lower. Conversely, the external pressure increases, the boiling point naturally increases, which leads to a higher steaming temperature, and the ratio of the essential oils to water is higher. This is very suitable for steamed wood and tree roots, such as sandalwood, cypress, agarwood, etc. Regardless of the vacuum distillation and

Figure 3-6a, b High-pressure distillation equipment

pressurized distillation method, the equipment requires a complete set of special decompression and pressurization equipment, and the production environment also requires an explosion-proof environment and measures (Figure 3-6a,b).

2.2.3 Dry steaming method. This method is not widely used, mainly for some special raw materials and objective specific requirements. The basic principle of this method is to isolate the air and put the crushed materials into a steel dry steam kettle. When the temperature reaches to the boiling point of the essential oil in the material, it is distilled out. This method is mostly used in the distillation of Chinese fir and cypress. For example, cedrenol is dehydrated to cedrene during the dry steaming process. This component is a synthetic raw material of methyl cedrone. The dry steaming method is used for the dry distillation of Chinese fir, and a black sticky oily substance called "black oil" is obtained, and the crystalline cedrenol can be obtained after further treatment. Using the dry steaming method, the oil yield can reach 8%, and the cedrene content can reach 40% to 70%. If the general steam distillation method is used, the cedrene oil yield is only 1% to 2%. (Figure 3-7)

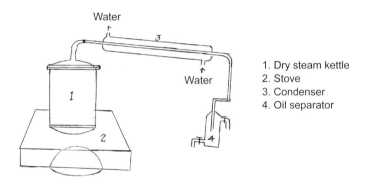

Figure 3-7 Diagram of dry steaming

2.2.4 Water osmotic distillation. This method is similar to the direct steam distillation method, the difference is that the steam is sprayed from the distillation kettle from top to bottom, the steam is re-heated before entering the material, and the steamed mixed liquid flows out from the bottom of the kettle. For this reason, the distillation temperature is higher, the oil-water ratio of the distillate is higher, and the hot steam is simultaneously diffused into the steamed material quickly, which accelerates the distillation, and the residence time of the distilled mixture in the distillation kettle is shortened, so that the aroma, color and oil yield of the essential oil are better than conventional steam distillation (Figure 3-8).

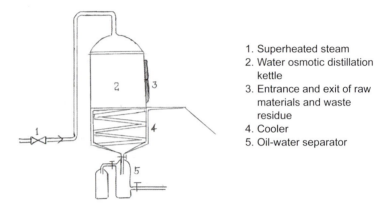

1. Superheated steam
2. Water osmotic distillation kettle
3. Entrance and exit of raw materials and waste residue
4. Cooler
5. Oil-water separator

Figure 3-8 Diagram of the process of water osmotic distillation

2.2.5 For the steam distillation process, the following process shall be noted:

1) For stacking, it should be noted that the steamed materials must be stacked evenly. If the materials become soft after heating, such as flowers and herbs, they need to be separated to the screen mesh, and the leaves need to be added with small branches to avoid blocking the diffusion of steam and affecting the distillation.

2) Remaining water must be removed by distillation to obtain crude essential oil. A small amount of water remaining in the essential oil will deteriorate the essential oil. Commonly used water-absorbing agents include anhydrous sodium sulfate, sintered dehydrated magnesium sulfate, heating, and high-speed centrifugation.

3) After the steam-distilled mixed liquid is separated into oil and water, there may still be a certain amount of polar aromatic components in the water layer, such as aldehydes, ketones, alcohols, phenols, etc., which can be re-distilled or directly extracted with a low-boiling solvent, Adding electrolytes such as table salt to the water layer is more effective. The extract is dehydrated with anhydrous sodium sulfate and then concentrated to remove the solvent to obtain polar aromatic components(Figures 3-9~3-10).

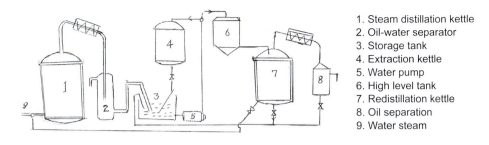

1. Steam distillation kettle
2. Oil-water separator
3. Storage tank
4. Extraction kettle
5. Water pump
6. High level tank
7. Redistillation kettle
8. Oil separation
9. Water steam

Figure 3-9 Diagram of the automatic recycling process of redistillation and extraction

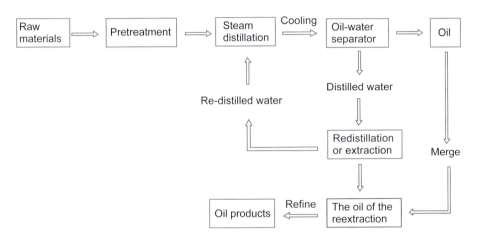

Figure 3-10 Schematic diagram of steam distillation oil extraction process

4) During the steam distillation process, cooling is a very important part. The cooling temperature depends on the nature of the essential oil being distilled. Generally, the final temperature of essential oils containing many monoterpenes needs to be close to or reach the outside temperature. If the essential oil contains solids at normal temperature, such as borneol and camphor, the cooling temperature is preferably 10°C~20°C higher than the outside temperature, to avoid plugging the cooling pipe. While steam distillation of cassia leaves, the cooling temperature needs to be higher than the outside temperature by 20°C~30°C, otherwise the distillate will be emulsified and difficult to separate oil and water. If it is depressurized distillation, the cooling temperature is below

10°C, and the cooling area needs to be increased.

5) Extracting essential oils from resins: Plant essential oils are mostly present in various organs in plants, but a few essential oils are in resins, such as liquidambar resin, agarwood, and pinaceae resins, masson pine, etc. Turpentine extracted from turpentine is the largest plant essential oil in China, with an annual output of 30,000~40,000 tons. It is one of the most important raw materials for light chemicals, medicine and food. The production process is as follows(Figures 3-11~3-15) :

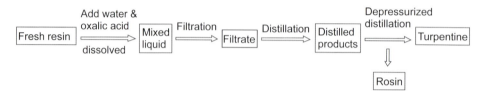

Figure 3-11 Resin production equipment and process

Remarks: Resin should be processed for the next step as soon as possible

Figure 3-12 Cut the resin manually

Figure 3-13 Cutting of the resin

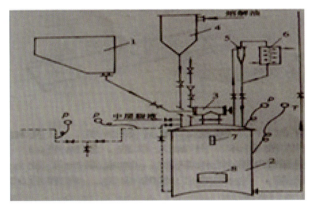

1.loadding hopper   2.dissolve tank   3.closed charging valve
4.solven tank 5.gas-liquid seperator   6.condenser   7.sight glass   8.outlet

Figure 3-14 Rosin dissolve process flow chart

Figure 3-15 Rosin extraction equipments

Section 3
Solvent Extraction Method

Using organic solvents with low volatility and low boiling point, the aromatic components in aromatic plants are extracted, and then the solvent is removed by distillation to obtain extracts containing essential oils and waxes. After treatment with ethanol, a pure essential oil is obtained, called pure oil.

3.1 The principle of extraction

Leaching is also known as extraction. The solvent leaching process means that the solvent completes the leaching or extraction process through three stages of penetration, dissolution and diffusion. In simple terms, the solvent and the aromatic component are distributed to achieve a balance. The amount of solute dissolved in the solvent, and the amount that remains in the solute, their ratio is called the partition coefficient.

$K=C1/C2$, C1 is the solvent concentration, C2 is the solute concentration, and their ratio K is the partition coefficient. This number K represents the solvent extraction capacity. The partition coefficient K of a solute differs different solvents, which is a strong basis for us to choose a solvent.

3.2 The choice of solvent

What solvent is used is very important for extracting aromatic components. The choice of solvent is based on the type of aromatic product and objective requirements. It is generally required that the solvent is colorless and odorless, does not affect the quality of the extracted aromatic components, that is, chemically inert, low volatility at room temperature, and requires that its boiling point is not high, it is easily distilled off, and its solubility in water is extremely low. It has a high selectivity, that is, it has a high solubility for aromatic components, and has a low solubility for impurities such as fat, wax, pigment, protein, starch, and sugar. The chemical inertness is high, and it basically does not react with aromatic components. Finally, the flash point of the solvent is higher, and it is not easy to explode and burn, and the toxicity is low. At present, the most commonly used solvent is petroleum ether (boiling point is around

65℃~75℃), which is the hexane range. In addition, there are alternatives such as ethanol, benzene, chloroform, dichloroethane, acetonitrile, etc., and mixed solvents are also used.

The choice of solvent also affects the determination of processes and equipments. Industrial solvents generally need to be refined or processed before extraction, especially the aromatic components that are extracted are more precious.Re-distillation is generally used for refining. Refined petroleum ether can be added with higher boiling point paraffin oil or refined with more efficient fractionation equipment. Adding potassium permanganate acidity solution, can remove aromatic impurities remaining in petroleum hydrocarbons.

3.3 Leaching processes and equipments

The processes and equipments are complementary to each other. You can base on the existing equipments to select the processes, or you can design the processes then to purchase equipments. The leaching processes include raw material processing, leaching solvent selection and aromatic raw material leaching, filtering out the solvent, evaporating the solvent back, concentration, etc. In the process of collection and concentration, the final products are extracts, balsams, oleoresins, tinctures, and pure oil obtained after removing impurities such as wax or pigments.

1. Classified by equipment and solute, it can be divided into static leaching and dynamic leaching. Static leaching refers to adding solvent into the leaching kettle. Generally, 4~5 liters of solvent are added per kilogram. After the solvent slowly penetrates into the material, the released solvent is after 4~5 hours before adding fresh or recycled solvent. The released solvent can be immediately distilled and concentrated to recover the solvent. The recovered solvent can be recycled. However, the recovered solvent generally contains more water than the fresh solvent. Some fresh solvent must be added to ensure that the solvent meets the extraction concentration requirements. This method has been improved to automatic circulation, the leaching solvent is directly transferred to the evaporating concentrator after leaching, and the recovered solvent is directly re-entered into the leaching kettle after adjusting the concentration and continuously circulated.

The equipments used in this method are simple, but the amount of solvent used is large, the material is still, and the penetration and diffusion speed are reduced to affect the leaching effect and time. The dynamic is that the material keeps turning, making the aromatic material more easily diffused and penetrated by the solvent, which helps accelerate the extraction of the aromatic material. Another dynamic leaching method is that the materials in the leaching kettle and the leaching solvent move in opposite directions to form countercurrent diffusion. Generally, the solvent or recovery solvent of the extraction kettle is added before the discharge valve, and the extraction liquid flows out from the charging valve. This operation is carried out continuously. The solvent is the first to contact the material near the inlet of the feed valve, so the concentration is the highest at this time, and the concentration of the extraction liquid is the lowest near the discharge valve. The situation is the opposite for the material. In continuous liquid-solid contact extraction and washing are carried out continuously, the extraction efficiency is high, the extraction rate can be as high as 90%, and the product quality and aroma are superior. This method has a large production capacity and great adaptability. It is suitable for the extraction of flowers such as roses, jasmine flowers, white orchids, tuberose, yellow orchids, etc. It is also suitable for stem wood, branches and leaves, roots, dry tubers, and granular materials. (Table 3-1)

2. Extraction temperature and extraction time. Temperature is related to the effect of the extraction speed and the quality of extraction products. The temperature is high and the leaching speed is fast, but the extract has many impurities and increases energy consumption. Generally, flowers are leached, especially the more expensive flowers are leached at room temperature. For woods such as sandalwood, agarwood, iris, etc., in order to get high-boiling point aromatic components, used to increase the penetration and diffusion of the solvent, then the leaching temperature is generally controlled below the boiling point of the solvent.

It is necessary to strictly control the leaching time in the leaching process. The leaching time of different raw materials is not the same, generally it is to control the leaching end point. The "end point" means that the concentration of the extract and the aroma components of the raw material show a dynamic balance (i.e. the concentration of the aroma components in the extract does not increase anymore). In practice, it will reach the balance when the leaching

Table 3-1 Comparison of different extraction methods

Equipment Methods and results	Fixed leaching	Stirred leaching	Rotary leaching	Countercurrent leaching
Extraction method	The raw material is immersed in the solvent to stand still, but it can be circulated and circulated	The raw material is immersed in the solvent. The scraper is used to agitate the raw material and the solvent to rotate at the same time	The raw material and the solvent move together in the drum	The raw material and the solvent move in the countercurrent direction to improve the extraction effectiveness
Raw material requirements	Suitable for delicate flowers such as jasmine, tuberose, violet	Suitable for small granular flowers such as osmanthus, Milan, etc.	Suitable for petal flowers such as white orchids, jasmine, roses, etc.	It has a wide adaptability and is mostly used for the extraction of large amounts of flowers
Production efficiency	lower	higher	high	highest
Leaching efficiency	60% ~ 70%	Around 80%	80% ~ 90%	Around 90%
Product quality	Because the raw material is still, it has certain advantages for delicate flowers. Less impurities are soaked	After the scraper is stirred, the raw materials are less damaged, so there are not many impurities	The raw materials are damaged more and the impurities are more	The leaching is more complete, but the impurities are also more

rate is 80%~95%, but it is not the same with different raw materials. For example, the extraction of 70%~75% of jasmine is basically balanced, and the length of extraction time of course has an impact on production cycle, energy consumption, solvent consumption, but also increases the impurities of the product and so on.

3. Evaporation and concentration of the extract. The extract becomes a concentrated solution after evaporation, but in order to protect the fragrance ingredients from damage, it is necessary to remove the remaining solvent as soon as possible. It is necessary to perform secondary concentration of depressurized distillation. In order to ensure the aroma of the pure oil, it is concentrated under reduced pressure during the process, the vacuum pressure, distillation temperature and cooling system must be controlled. To meet the requirements of fresh flower extract, the degree of vacuum is 80~84kpa, it is safer to decompress with a jet pump, and the heating temperature is higher than room temperature by 10°C~15°C, the cooling temperature of depressurized distillation is very important, the most suitable method is to use dry ice or ice water with salt or calcium chloride (below-10°C), and the collector must also have to keep the temperature between 5°C~10°C, the depressurized distillation kettle should be able to rotate. The obtained "crude product" will be formulated into concentrated solutions, tinctures, flower extracts, extracts and perfumes of different concentrations according to market and objective requirements (Figure 3-16).

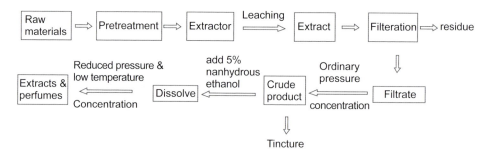

Figure 3-16 Schematic diagram of extract preparation process

3.4 Preparation of pure oil

Pure oil is the final product of the extraction of aromatic ingredients. Its aroma is basically a highly concentrated feature and embodiment of the aromatic ingredients of the aromatic plants. The extract, obtained through multiple steps, is a thick paste. This is the fragrance, wax (flower wax), pigment, etc., dissolved in the solvent during the extraction process. For this purpose, wax and other impurities must be removed. Because waxes are insoluble in ethanol at low temperature, but soluble in ethanol at higher temperature, essential oil can be dissolved in ethanol no matter how the temperature changes. According to this phenomenon, we add a certain amount (12~15 times) of 95% refined ethanol at room temperature or higher to completely dissolve the extract. Afterwards, slowly lower the temperature of the mixture, it can be frozen step by step or frozen at once (Figure 3-17).

For freezing step by step, the mixture is first frozen to -10°C and then filtered under reduced pressure. Then the filter is washed with 95% ethanol at -10°C, until the filter residue is unscented, and then frozen to -26°C. Repeat the process at -10°C. One-time dewaxing is to freeze the mixture to -25°C, keep 2~3 hours. Generally speaking, this operation can remove 85%~90% of waxes in one process. In order to obtain a higher quality pure oil, the wax removal must be repeated more than twice.

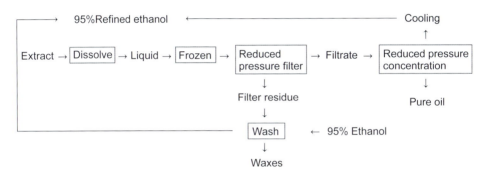

Figure 3-17 Diagram of the pure oil preparation process

The ethanol filtrate after dewaxing filtration and washing needs to be evaporated, concentrated and recovered. It can be done in two steps, firstly under reduced pressure (73.3~80kpa) at the temperature (water bath) 45~55°C. After the majority of the ethanol in the concentrated material is distilled, the distillation vacuum can be increased to 93.3kpa, so that the remaining ethanol can be basically distilled out. Generally, the residual amount of ethanol is below 0.5%. This method of obtaining pure oil is very cumbersome. At present, molecular distillation is used to obtain pure oil with extremely high purity.

3.5 Leaching equipments

The selection of the kind of extraction equipment is based on many aspects, including the scale of production, whether the type of extraction is a single type or multiple types, etc.

1. Fixed extraction equipment. This equipment is relatively simple, usually used for 1 to 4 sets, about 2,000~4,000 liters jacketed stainless steel kettle, prepared for the bottom outlet slag, the upper cover can be loaded into the material port and condensate reflux cooler, the bottom of the kettle is equipped with leachate outlet, the outlet pipe is connected to the evaporator to recover the solvent. If the materials need to be flipped, then install a stirring rod on the top of the extraction kettle and go straight into the kettle.

2. Scraper-type extraction equipment. Its working principle is to immerse the material in an organic solvent, and use a scraper-type stirrer to rotate the material and the solvent together, so that the material is always kept in dynamic contact with the solvent, and the rotation speed does not need to be too fast, about two revolutions per minute. After filtration, the leaching solution can be sent directly to the evaporator to recover the solvent. The recovered solvent can be reused after adjusting the concentration. (Figure 3-18)

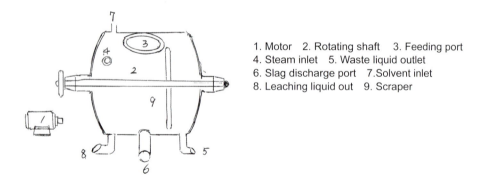

1. Motor   2. Rotating shaft   3. Feeding port
4. Steam inlet   5. Waste liquid outlet
6. Slag discharge port   7.Solvent inlet
8. Leaching liquid out   9. Scraper

Figure 3-18 Scraper-type extraction equipment

3. Floating filter leaching equipment. Its working principle is that the rotating impeller of the turbine agitator installed at the bottom of the extraction kettle, and the fixed guide wheel and baffle, its role is to make the material and the solvent can be quickly and fully mixed and contact, to achieve the purpose of

high-efficiency diffusion and penetration. The suction filter tube in the center of the float floats up and down to keep the filter disc on the surface of the liquid, so that the extraction liquid is separated from the material residue, and extracted faster. In this process, the clarified liquid and the filter residue are not easy to contact with the residue particles. It is especially suitable for the extraction of powdery and resinous materials. The leaching solution can be sent directly to the solvent in the evaporation recovery area. The equipment requires high processing precision for rotating the original seal. (Figure 3-19)

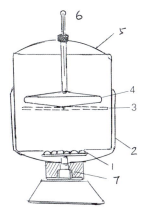

1. Turbo agitator   2. Jacket   3. Filter disc
4. Float   5. Quick-drying cover
6. Center suction tube   7.Power

Figure 3-19 Floating filter leaching equipment

4. Horizontal rotation countercurrent continuous extraction equipment. The working principle and leaching process of the equipment are continuous. According to the principle of percolation, the solvent is sprayed on the surface of the material, and the material is infiltrated from the top to the bottom. The aromatic components are dissolved and leached from the material and flow to the bottom layer, which is filtered into the reserve tank. It is then sent to the evaporation and recovery solvent equipment, which has a large processing capacity. Since the material hopper does not move, just spraying the solvent from top to bottom does not damage the materials (flowers) and ensures the quality of future products. Due to the higher solvents concent in the emission gas, the recover process should be done.

5. Swimming paddle blade continuous extraction equipment. The device is special, the equipment is a cylindrical stainless steel cylinder, it is about 11~12

meters long, about 1 meter in diameter, about 8 degrees inclination when installed, the bottom is equipped with a vertical high-level feed groove, the cylindrical leaching kettle is equipped with a belt-type spiral blade spindle that can rotate. Before the material enters the cylindrical type horizontal inclined extraction kettle through the high-level groove. It is first sprayed with high-concentration extraction solvent, and then enters the cylindrical horizontal extraction leaching kettle. The material is slowly pushed to the top by the belt spiral blade, and the solvent flows from the top to the bottom to form a reverse convection. Quickly penetrate and dissolve the aromatic components in the material. The most concentrated leaching solution is first sprayed with fresh material input from the feeding port of the high-level tank, and the concentrated leaching solution after spraying is sent to the evaporation recovery device. This device has a large feeding amount, can be continuously produced, and the production cycle is short, high efficiency. It is most suitable for extracting roses, jasmine, gardenia, ink red and other flowers. (Figure 3-20)

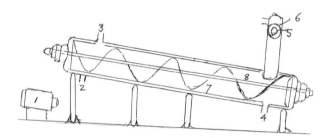

1. Motor
2. Waste outlet
3. Solvent inlet
4. Leachate outlet
5. Fresh material inlet
6. Feed funnel
7. Belt screwLiquid
8. Rotating shaft

Figure 3-20 Schematic diagram of swimming dip paddle continuous extractor

6. Hot reflux leaching equipment. The device is the same as the traditional general leaching device. The cylindrical stainless steel leaching kettle with interlayer, the bottom is an inclined cone, vertical, a feeding inlet is installed on the upper cover, the top center is equipped with a stirring rod, and the outlet is in the kettle at the bottom, the material is put in from the top feed port, the solvent can also be put in from the feed, and there is a reflux condenser. When heating, the jacket can be heated by steam from the jacket. At the same time, the material can be stirred with a stir bar. This device is most suitable for rock rose, iris, spirit vanilla, Yunmuxiang and some Chinese herbal medicines such as Tianqi and

Longan extraction. The extraction is carried out discontinuously, and the waste slag discharge port is a pneumatically opened bottom discharge cover. (Figure 3-21)

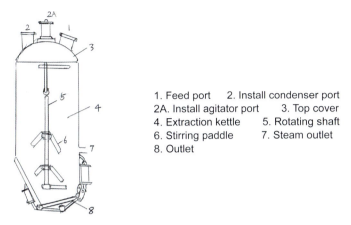

1. Feed port   2. Install condenser port
2A. Install agitator port   3. Top cover
4. Extraction kettle   5. Rotating shaft
6. Stirring paddle   7. Steam outlet
8. Outlet

Figure 3-21 Thermal reflux Extractor

Section 4
Grinding method

This method is mainly used for extracting citrus peel essential oil. This kind of essential oil produced by this method is mainly used in food, cosmetics and medicine and the output is very large. The extraction of essential oils from citrus peels can be divided into two categories, one is the squeezing method and the other is the whole fruit peeling method.

4.1 Pre-treatment

No matter what method is used, the purpose is to break the oil sacs stored in the outer skin layer to obtain the essential oil. Too soft and hard skins are not conducive to rupture the oil sacs. At the same time, there are thick, spongy inner skins in the pulp and fruit skin layers. Its presence tends to adsorb the essential oil of the peel and causes those pectins to be collected along with the essential oil, affecting the extraction rate. The most commonly used method is to soak the peel in water or lime water, which not only makes the peel suitable for processing, but also water can make the endothelial layer contain a large amount of water to reduce the adsorption of essential oils by the sponge-like peel. Lime water also makes the pectin of the endothelial layer calcified and makes it insoluble in water. In order to avoid mixing pectin with essential oils, this will affect the separation of essential oils. It is necessary to wash the water before feeding to remove the immersion agent. Because citrus essential oils are heat sensitive, so the general processing is carried out at room temperature.

4.2 The pressing method

Most of this method is for fruit-shaped, non-spherical and scattered citrus. The peel is processed and pressed. Multi-screw pressing is used. The upper and lower pressing plates are tapered, with spikes and holes. The turbid liquid that flows out after pressing has essential oils, water, a small amount of peel, pectin, you need to obtain citrus essential oil by centrifugation or standing, there is also squeezed out turbid liquid after absorption with a sponge, after the sponge is saturated, take out and squeeze out the oil and water, then the peel is broken, the pectin is left in the sponge, and the squeezed oil-water liquid is allowed to stand for oil-water separation. This method is mostly manual and batch methods to obtain essential oils.

4.3 Whole fruit grinding method

This is for the whole fruit. The shape of the fruit is spherical and the size is relatively consistent, such as sweet orange, lime, and wide orange. The grinding basin is a flat or funnel-shaped bottom basin with a slight cone shape. The surface of the basin has small holes and nails. The shape of the upper pressure rotating box is consistent with the shape of the bottom basin with nails. The distance between the bottom basin and the upper rotating box can be adjusted. Depending on the size of the material, it can be rotated in the upper basin and not in the lower basin, and the two can be reversed to accelerate the extraction speed. During the rotation process, water must be sprayed continuously. Before processing, the same pre-treatment as above is carried out according to the actual situation. After the general fruit can also be processed and used, such as squeezing juice, processing into candied fruit, etc., the grinding fluid flows out from the bottom of the grinding bowl, after filtration, it can stand still and layered, and then essential oils can be extracted. In order to improve the yield and obtain high-quality citrus peel essential oils, large-scale manufacturers now use continuous feed high-speed centrifugal processing to flow out the mixed liquid. After processing peel essential oils, the endothelial layer of the sponge can also produce very useful edible pectin. (Figures 3-22~3-24)

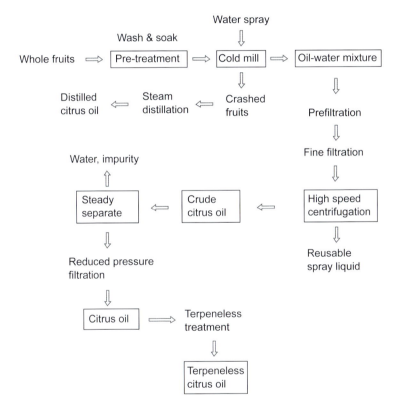

Figure 3-22 Diagram of the process of citrus whole fruit cold milled essential oil

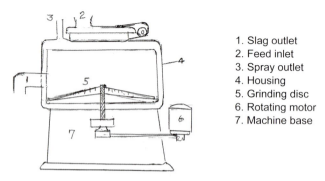

Figure 3-23 Diagram of a flat plate mill

1. Slag outlet
2. Feed inlet
3. Spray outlet
4. Housing
5. Grinding disc
6. Rotating motor
7. Machine base

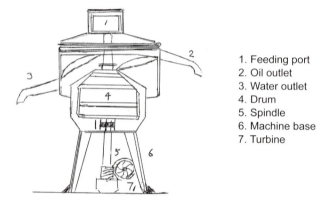

Figure 3-24 Diagram of a disc-type high-speed centrifugal separator

1. Feeding port
2. Oil outlet
3. Water outlet
4. Drum
5. Spindle
6. Machine base
7. Turbine

Adsorption

5.1 Resin adsorption method

Since plant essential oil is a volatile chemical component, it also emits a certain aroma naturally. Of course, some aromatic components such as branches and leaves can be broken up by mechanical methods to accelerate their release of aroma. A highly surface-active substance such as activated carbon or silica gel is used to adsorpt the released aromatic components naturally. Therefore, it is easy to desorb and recover aromatic components with low-boiling point solvents, and then remove the solvent to obtain high-quality flower essential oil. Since this adsorption process is a physical adsorption process, the so-called "Intermolecular force", the adsorbent does not have the adsorbed aromatic components to play a chemical role. At the same time, as long as the organic low-polarity solvent is added, it is very easy to desorb, which can be described as original. The previously introduced adsorbent is currently replaced by macroporous resin. It has a large adsorption capacity, strong adsorption capacity, and is more chemically inert than the above two types of adsorbents. It can also be regenerated repeatedly. It is an adsorbent with a large surface area formed by the accumulation of polypropylene particles. It has a variety of models, one of the trade names of macroporous resin is XAD series. For the adsorption of organic gas it is XAD-4, for the collection of trace organic matter from water is XAD-2, and the most used in environmental protection is XAD-7. If this type of adsorbent uses supercritical CO_2, the extraction effect is better, but the cost of the equipment will be higher.

5.2 Fat adsorption method

The adsorption method is to use refined fats such as lard, tallow and goat oil or both to mix and heat to make a fat base, and put multiple layers of glass plate, which is fat base coated on both sides, in a glass frame capable of convection semi-closed and transparent (50cm×40cm×5cm). The air entering the space must be filtered, and keep the humidity and temperature consistent with the outside condition. Then gently spread the material on the glass plate coated with fat base, absorb the material for 24 hours, remove the material, and then replace it with fresh material every 24 hours until the fat base is completely absorbed and saturated. This method is suitable for flowers that release the fragrance strongly, such as jasmine large flower, jasmine small flower, tuberose. For jasmine, a cycle takes about 30 days. The fat base scraped off after saturation is the product, collectively called balsam. Balsam is a precious product in natural fragrances and can be used directly in high-end cosmetics. (Figure 3-25)

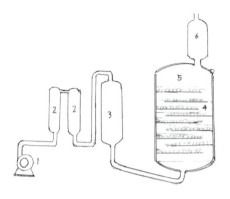

1. Blower
2. Air purification device
3. Warmer
4. Flowers
5. Adsorber
6. Adsorbent

Figure 3-25 Diagram of dynamic adsorption device

Section 6

Supercritical CO$_2$ fluid extraction method

Supercritical extraction is a relatively new extraction method, especially applied to the extraction of aromatic components. Its main principle is to make a fluid have characteristics such as close to liquid density, low gas viscosity and diffusion speed at a specific temperature and pressure. It has great dissolving and permeating ability, which is very good selectivity for a certain kind of fluid, and can reach the solute and solvent balance quickly.

Figure 3-26 Supercritical extraction equipment

Supercritical extraction of aromatic components (mainly fragrant flowers), because CO_2 is a gas state at normal temperature, non-toxic, chemically inert, not easy to burn and explode. If the temperature is lowered to -56°C and the pressure is 0.51MPa, carbon dioxide is in the state of coexistence of gas, liquid, and solid. At this time, its viscosity and density are already very close to the liquid state, and it is easy to penetrate and diffuse into the material, making the two achieve dynamic equilibrium, which basically completes the extraction process. After completion, the first step is to depressurize the pressure extraction kettle to the same pressure as the outside, which is equivalent to dismantling the carbon dioxide. Extract the kettle and slowly increase the temperature to 40~ 55°C at the same time. The carbon dioxide in the kettle is completely eliminated. The aromatic components of the extracted material mainly remain in the kettle. There may be a small amount of water in the waxy pigment. The water comes from the material. In addition, since supercritical carbon dioxide cannot be converted into a liquid state and then into a gaseous state, the carbon dioxide after evaporation and extraction is more easily eliminated. (Figure 3-27)

Generally, the crude extract obtained by carbon dioxide extraction can be used for molecular distillation to remove wax and other impurities to obtain pure oil.

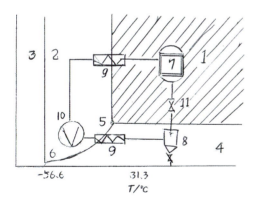

1. Supercritical state area
2. Liquid state area
3. Solid state area
4. Gaseous state area
5. Water bottle
6. Gas-liquid-solid three-phase Point
7. Extraction kettle
8. Separation
9. Cooler
10. Compressor
11. Valve

Figure 3-27 Diagram of the extraction process flow in the supercritical state of carbon dioxide

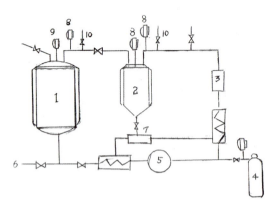

1. Extraction kettle
2. Separator (extract)
3. Carbohydrazine
4. Carbon dioxide cylinder
5. Compressor
6. Raw material or semi-
   finished product
7. Heat exchange
8. Pressure gauge
9. Temperature gauge
10. Safety valve

Figure 3-28 Diagram of the carbon dioxide extraction process

Section 7

Molecular distillation method

Molecular distillation is also called short-path distillation, which is completely different to the distillation principles we commonly use. The commonly used distillation is based on Dalton principle. When the distilled material is heated (accelerating the molecular movement) until its vapor pressure is equal to the outside atmosphere, it will boil, and its vapor escape process is the distillation process. Molecular distillation, also known as short-path distillation, works on the following principle: the equipment used is a short-path distillation kettle or short-path evaporator, the basic working condition is high vacuum (0.1~10Pa). The first thing is to pull out the residual air to reach the minimum in the evaporator to reduce the impact of the air on the escaped material molecules. Therefore, the material (Aromatic components) molecules can be kept at a low temperature, escape from evaporation without hindrance, and quickly reach the collection cooling surface. The distance between the evaporation surface of this evaporator and the cooling collection surface is shorter than the average degree of freedom of the material molecules, so it is called short-path evaporation. Due to the high degree of vacuum, the boiling point of the volatile components in the material can be greatly reduced, and the evaporation rate is also increased, so it can produce high-quality net oil. This method is mostly used in the production and processing of high-grade fragrant raw materials such as vetiver essential oil (Essen vetyer U.V.), lavender pure oil (Absolue lavandula U.V.), etc. The special term for such products is U.V. products.

The most fundamental of molecular distillation is higher vacuum. In the early stage, higher vacuum was carried out in stages. First, a mechanical pump was used to achieve a certain vacuum, and then a diffusion pump was used. In the early stage, mercury diffusion was used to change to oil diffusion pump. Recently, many factories have used it. More advanced molecular pumps have greatly reduced the internal and external repeated pollution of various diffusion pumps, and the front pumps have also been replaced with water ring pumps to reduce environmental pollution. (Figures 3-29~3-32)

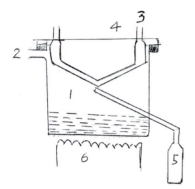

1. Molecular distillation kettle
2. Connect to vacuum pump
3. Cooling water
4. Condenser
5. Receiver
6. Heater

Figure 3-29 Static molecular distillation equipment

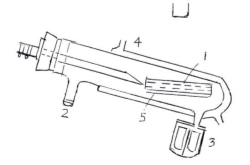

1. Evaporator
2. Connect to oil diffusion pump
3. Rotary collector
4. Cooling water
5. Heater

Figure 3-30 Static disk showing molecular steaming equipment

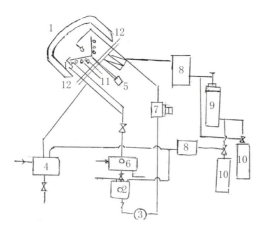

1. Molecular Distiller
2. Raw Material Tank
3. Feed Pump
4. Ride Out Tank
5. Evaporator Motor
6. Steam Residue Tank
7. Pre-Hairer
8. Lengfei
9. Oil diffusion pump (molecular pump)
10. Oil rotary pump (water ring pump)
11. Electric heater
12. Cooling water
13. Feed

Figure 3-31 Diagram of the industrialization process of centrifugal molecular distillation

Figure 3-32 Molecular distillation equipment

Section 8

Microwave Extraction Technology

Microwave is an electromagnetic wave whose wavelength is between infrared and radio waves. The wavelength of a microwave oven that is generally used as a heat source is 12.2cm (2.45GHz). The process of extracting plant essential oil in the microwave is as follows: Microwave heating causes plant cells (referred to as cells containing essential oil) such as oil cells, oil glands, resin channels and other polar substances (cell fluid) moisture to absorb microwave energy, thereby generating a lot of heat, promoting the rapid rise of the temperature in the cell, the vaporization and expansion of the liquid cell fluid (water) generates pressure to break the cell membrane and cell wall, and the continued heating accelerates the cracks and holes in the cell surface, so that the extracted liquid (solvent) enters the cell directly for penetration and dissolves, and dissolve the volatile matter (essential oil) in the cell, so that it is quickly heated to the boiling point under microwave irradiation. At this time, the temperature in the sealed container (extractor) should be higher than the normal boiling point of the solution, so that the essential oil components are extracted faster. The application of microwave technology to extract plant essential oils has the advantages of faster speed, reduced thermal damage, and is more beneficial to the extracted plant raw materials with more water content, because the larger the dipole of water, the faster the heat production. Higher moisture content of fresh plant materials is more beneficial. In addition, there are different reports on whether microwaves will rearrange or isomerize certain components of essential oils when extracting plant essential oils. This is mainly to master the time, solvent and temperature have little effect on the composition of essential oils. At present, this technology has more reports on the extraction of essential oils.

Chapter III References

1. An Xinnan. Forest Products Chemical Technology [M]. Beijing: China Forestry Press, 2002.

2. Jin Qi. Xiangke Production Technology [M]. Harbin: Northeast Forestry University Press, 1994.

3. Wang Afa. Natural perfume production technology [M]. Nanjing: Lecture Notes of Nanjing Forestry University, 1994.

4. He Jinke, Li Qiji. The Complete Book of Forest Products Chemical Industry (Volume Three) [M]. Beijing: China Forestry Publishing House, 1997.

5. Sun Lingfeng, Wang Hongwu, Yang Minyan. Natural source of safrole and its use in synthetic fragrances [M]. Perfume flavoring Makeup, 1992: 14-18.

Chapter IV

Uses of Plant Essential Oils and Their Components

Plant essential oils have always been closely tied to human life across the world. During our visit to southern Africa, we learned that most of the herbs used by indigenous people are aromatic plants prepared for oral administration, fumigation, or external application. Around 2000~3000 years ago, the Chinese were already using *Artemisia* spp. essential oil to fumigate the body for cleansing, disease prevention, and treatment. Jasmine flowers were also used to scent tea, creating the now well-known jasmine tea, a practice dating back about a thousand years. Since plant essential oils are used in food seasoning, perfumery, beauty, and healthcare, they have become a thriving industry closely linked to modern livelihoods.

Section 1

The active ingredients of plant essential oils and medical drugs

Many components of plant essential oils have biological activity and medicinal properties. Practice has shown that most of them have the functions of fragrant stomach, relieving appearance and divergence, eliminating phlegm, relieving cough and asthma, bacteriostasis, anti-inflammatory, analgesic, relieving itching and bleeding. All kinds of Fengyoujing directly prepared with essential oils are commonly produced and used in southern China, as well as overseas Southeast Asia, Europe and the United States, where many Chinese live. The biggest feature of this type of Fengyoujing is that it can be carried around and is convenient to use. Most of them only need one or two drops of external application. The main functions are anti-inflammatory and sterilization, muscle pain, snake and insect bites, cold discomfort, motion sickness, antipruritus, dizziness, headache, cold and nasal congestion, etc., have a certain relief effect, and most of them are one of the common medicines for home and travel. These products are commonly used in major cities in southern China, including Hong Kong Hexing White Flower Oil, Singapore's Axe Brand Medicated Oil, Malaysia's Tiger Balm Oil series, Guangzhou's Suicao Oil, Luofu Mountain's Baicao Oil, and Zuifeng Oil. And White tree Oil (from Melaleuca leucadeadron var. cajaputi) produced in Indonesia mainly contains 1,8-cineole, as well as various types of essential oils of medicinal plants. Generally speaking, these essential oils of medicinal plants are inseparable from peppermint oil, eucalyptus oil (Eucalyptus globulus), camphor, methyl salicylate (wintergreen oil), lavender oil, rosemary oil, clove oil, etc.

In addition, from the 2015 edition of the Pharmacopoeia of China, there are many Chinese patented medicines that add plant essential oils as main or auxiliary materials, such as Shixiang Pills series, Baoji Pills, Huoxiang Zhengqi Pills, Qiwei Xiangfu Pills, Fifteen Flavour Agarwood Pills, Eight Flavour Sandalwood Powder, Nine Flavour Jiang Huo Pills, Chuan Gong Cha Tiao Pills, Xiaoer Xiangju Granules, Xiaoer Jiebian Granules, Ruxiang Qufengzhitong Tincture, Muxiang Shunqi Pills, Cough Orange Red Pills, etc. Camphor and mint are also indispensable in tinctures of Western medicine.

Plant essential oil, on the other hand, is a newly developed industry that is suitable for home, fitness, beauty, and aromatherapy for sub-health treatment. This entire industry is carried out under the guidance of relevant professional medical personnel. The content includes sniffing, oral, massage, aromatherapy, bathing, etc. There are more than 70 kinds of plant essential oils commonly used, and more than 30 kinds of our common plant essential oils are now selected. See Table 4-1 for details.

Table 4-1 More than 30 common plant essential oils used in aromatherapy

Name	Mail component	Characteristic
1. *Melaleuca altemifolia*	terpineol-4, terpinene, 1,8-cineole	Inhibit bacteria, viruses, fungi and parasites
2. *Ocimum basilicum*	eugenol	Relieve pain, improve dysmenorrhea, treat intestinal diseases, abdominal distension, constipation, and eliminate irritability and depression
3. *Citrus medica* var. *sarcodactylis*	limonene, linalool, acetate	Cure vitiligo, sterilize, relax mood, aphrodisiac, healthy skin
4. *Aniba roseodora*	linalool, geraniol, linalyl acetate	Improve rough skin, anti-wrinkle, delay skin aging, eliminate irritability
5. *Fokienia hodginsii*	neroli, thujone	Improve prostate congestion
6. *Cinnamomum zeylanicum*	eugenol, cinnamaldehyde	Sterilizes, treats urethra, has curative effect on asthmatic bronchitis, kills human parasites and relieves fatigue
7. *Cistus ladaniferus*	α-pinene	Stop bleeding, improve detoxification for nosebleeds and hemorrhoids bleeding
8. *Citrus limon*	limonene, citral	Liver cleansing, purifying the digestive system, diuresis, and improving blood circulation
9. *Cymbopogon nardus*	citronellol, geraniol	Repelling insects, sterilizing and anti-inflammatory, fresh air, eliminating peculiar smell
10. *Boswellia carterii*	α-pinene, juniene, limonene, myrcene	Suppress depression and enhance immunity
11. *Artemisia dracunculus*	methyl wilt	Treat respiratory allergies and relieve rheumatic pain
12. *Eucalyptus citriodora*	citronellal, citronellol	Eliminate fungi, repel mosquitoes and flies, relieve spasms

(continued)

Name	Mail component	Characteristic
13. *Eucalyptus globulus*	1,8-cineole	Treat pneumonia, soothe the throat
14. *Eucalyptus robusta*	1,8-cineole, p-cymene	Treat colds, flu and tracheitis
15. *Gaultheria procumbens*	methyl salicylate	Improve pain caused by inflammation, anti-inflammatory, antipyretic, and diuretic
16.*Pelargonium graveolens*	citronellol, geraniol	Treat hypoglycemia and related diseases, improve body fatigue, treat hair loss, acne, improve skin aging, and tighten skin
17. *Zingiber officinale*	gingerene, camphene, 1,8-cineole	Relieve various symptoms of nausea, relieve physical fatigue, muscle tension, treat hair loss, and have obvious effects on arthritis and muscle pain
18. *Eugenia caryophyllata*	eugenol, syringyl acetate	Relieve all kinds of toothaches and oral ulcers, with strong sterilization, disinfection and fatigue
19. *Inula graveolens*	bornyl acetate	Treatment of excessive mucus secretion, effective for various inflammations, vaginitis, cystitis, etc.
20. *Laurus nobilis*	1,8-cineole, linalool, α-terpineol, terpineol, costanolactone	Sterilize, relieve pain, cure flu, treat rheumatism
21. *Lavandula latifolia*	1,8-cineole, linalool, camphor	Severe burns, snake bites and bee stings, scorpion, centipede bites
22. *L. angustifolia*	linalyl acetate, linalool, 1,8-cineole, lavender alcohol	Treatment of migraine, dizziness, insomnia, nervousness, palpitations, arrhythmia, vasculitis

(continued)

Name	Mail component	Characteristic
23. *Cymbopogon citratus*	α, β-Citral, linalool	Calm, soothe the emotions, reduce inflammation, sterilization, cellulite, and treat cellulitis
24. *Pistacia chinensis*	β,α-pinene, α,β-ocimene, limonene	Treat various congestion, varicose veins, have a certain effect on tinnitus
25. *Litsea cubeba*	α, β-citral, limonene	Suppress depression, reduce stress, sleep peacefully, and reduce inflammation
26. *Mentha piperita*	menthol, menthone, hot menthol	Relieve pain, improve breathlessness, cure pain and digestive diseases
27. *Nardostachys jatamansi*	β-gurusene, calamus terpenes	Respiratory tranquilizer, hair nourishing agent, stimulate hair growth
28. *Citrus aurantiaca*	linalool, α-terpineol, nerol	Resolve bad emotions and regulate nerves
29. *Origanum vulgare*	carvacrol, thymol, linalool	Disinfect and sterilize, cure viral inflammation, enhance immunity, treat ear, nose, and throat diseases
30. *Citrus paradisi*	limonene, linalool, nerol, α, β-citral	Sterilize, purify indoor air, tighten skin, face, body, thighs, buttocks, etc.
31. *Santalum album*	α-β-santalol, santalol isomers	Effective for various inflammations, firming the skin
32. *Abies fargesii*	β-pinene, bornyl, bornyl acetate	Cure colds, relieve symptoms of runny nose, treat tracheitis, and fight fatigue
33. *Citrus limon*	limonene, α, β-citral, geraniol, terpineol-4	

Section 2

Plant essential oils used in food

Plant essential oils are mainly used in food in the form of food additives. Among them, citrus peel essential oils are widely used in beverages, biscuits, cakes, and ice cream to enhance food flavor, with a large consumption volume. Another type of plant essential oils is directly used as seasonings, such as pepper oil, star anise oil, pepper oil, cumin, and other spicy seasonings. Some are also used in the form of oleoresins.

Due to their various potential properties such as antibacterial, antifungal, and antioxidant effects, plant essential oils have found wide applications in the food industry. The efficacy of plant essential oils as natural food preservatives has been tested in many foods, including baked goods, fruits and vegetables, and meat products. Essential oils extracted from oregano, cloves, thyme, and citrus fruits exhibit strong inhibitory activity against bacteria and yeast. Adding essential oils to dairy products not only enhances their aroma but also imparts antibacterial properties, making them more appealing to consumers. The use of plant essential oils in fruits and vegetables helps extend their shelf life. The primary concern with fruits and vegetables is fungal decay, and studies have shown that the application of plant essential oils in fruits can reduce microbial populations.

Despite their numerous beneficial properties, the volatility and hydrophobicity of plant essential oils make them difficult to use directly in food systems. Nanotechnology has been reported as a solution for plant essential oil delivery systems that can protect active compounds from degradation and enhance their efficacy in food systems. Nanoencapsulation protects plant essential oils from deteriorating factors such as light, moisture, and pH changes, both during processing and storage. It also aids in the solubilization of lipophilic compounds in aqueous media and their release to target sites.

Plant essential oils used in cosmetics, skin care products and detergents

Most plant essential oils are formulated into essences and added to various cosmetic and skin care products such as creams, washing shampoos, high-end perfumes, toilet water, etc. To be honest, without the plant essential oils, there would be no cosmetics, skin care products, shower gels and other products in the market. Another product of essential oil blending is cigarette flavor, which is also a very large amount of essential oil blending product, and it is a very important auxiliary material in the cigarette industry. At present, natural flavors for cigarettes used in China mainly include brandy leaf oil, patchouli oil, star anise oil, ginger oil, ling vanilla oil, cinnamon oil, cardamom oil, lavender oil, osmanthus absolute, jasmine absolute, geraniol oil, clove oil, agarwood oil, etc. It not only improves the aroma of cigarettes, but also plays a major role in reducing tar and nicotine and improving the aroma of tobacco. There is also a type of wine flavor added to the wine.

The effect of essential oils and plants on bacteriostasis and insecticide

Plant essential oils are generally considered to have a certain inhibitory or killing effect on microorganisms. However, due to the very complex chemical components of plant essential oils, the antibacterial effects of each chemical component are absolutely different. According to research reports, the main components of plant essential oils are terpenoids. Followed by aromatic compounds, their oxygen-free terpenes have poor antibacterial activity, and carbonyl compounds in oxygen-containing derivatives are better than hydroxy compounds. There are double bonds in front of the groups, especially unsaturated or conjugated double bonds. Bacterial efficiency is higher, and the carbon chain is shorter than the chain length and the bacteriostatic effect is higher. See attached Table 4-2 for reference.

Table 4-2 Bacteriostatic test of essential oil components

Bacteria species Samples	Aspergillus niger	A.Sydowi	A.terrous	Penicillium chrysogenum	Paecilomyces varioti	Chaetomium globosum	Cladosperium herbarum	Trichoderma sp.
Ylangene	--	--	--	--	--	--	--	--
Humulene	--	○	--	-	-	-	-	-
β-Elemene	--	○	--	-	-	○	○	--
Borneol	-	○△	-	○	○	++	○	○△
Dodecanal	-	+	-	○	+	++	○	○
Aldehyde	-	○△	-	○△	○	+△	+△△	-
Sylvestrene	○	○	○	○	+	++	+	+
β-Phellandrene	○	○	○	○	+	++	+	+
Octyl acetate	○△	++	○	○	○	++	+	○
α-Pinene	+	++	○	+	+	++	+	+
Octanal	○	++	+	+△	+△	+△△	+	++
α-Terpineol	++	+△	○	+△	+△	++	+	+
Linalool	++	++	+△	++	++	++	++	++
4-Methyl-6-acetoxyhexanal	+△	++△△	+	++	++	++	++	++
Citral	++	++	++	++	++	++	++	++
L-octanol	++	++	++	++	++	++	++	++

Note: + + means no mold growth, + means that the zone of inhibition is above 5mm, ○ means that the zone of inhibition is between 2 and 4mm,-means that the zone of inhibition is less than 2mm, but does not grow on the filter paper, --Means that mold has grown on the filter paper, △ means inhibiting growth, △△ means only growing individual colonies.

Recent research on plant essential oil components aldehydes such as cinnamaldehyde, citral, and anisaldehyde; phenols such as thymol, eugenol, carvacrol; ketones such as carvone, menthone, and peppermint; alcohols such as linalole Alcohol, citronellol, menthol (brain), etc. all have a strong inhibitory effect on the growth of fungi and the synthesis of mycotoxins. The other five essential oil components, terpineol-4, eugenol, 1,8-cineole, and thymol have obvious growth-inhibiting or killing effects against Fusarium, Aspergillus, and Penicillium. Most of them reach to 100% inhibition rate of mold.

Regarding the mechanism of plant essential oils and their active ingredients in inhibiting fungi, studies on citral, an essential oil component with strong antibacterial efficacy, have shown that it can damage the plasma membrane of Aspergillus flavus, causing it to lose selective permeability and enter the cell for other cells. Organs are affected, and the damaged fungal plasma membrane will change the spatial structure of macromolecules in the cell, and the metabolism will be disordered, thereby inhibiting the growth of molds.

Table 4-3 Broad-spectrum antimycotic activity of plant essential oils and their active ingredients

Name	Source	Inhibited Eumycetes
rosa multiflora	Rosaceae	12 kinds of fungi including Fusarium solani, Fusarium oxysporum, and Fusarium spp.
cuminum cyminum	Cuminum	17 kinds of fungi including Aspergillus fumigatus, Fusarium oxysporum, Penicillium, etc.
foniculum vulgare	Fennel	17 kinds of fungi including Aspergillus fumigatus, Fusarium oxysporum, Penicillium, etc.
ocimum sanctum	Ocimum Lamia	13 kinds of fungi including Aspergillus flavus, Aspergillus fumigatus, Penicillium citrinum
zanthoxylum armatum	Zanthoxylum	12 kinds of fungi including Aspergillus flavus, Aspergillus niger and Penicillium citrinum
thymus vulgaris	Thyme	9 kinds of fungi including Aspergillus flavus, Fusarium oxysporum, Aspergillus niger, etc.
menthol	Peppermint	Ringworm, Candida, Aspergillus flavus, Aspergillus niger, etc.

(continued)

Name	Source	Inhibited Eumycetes
thymol	Thyme essential oil	11 kinds of fungi including Aspergillus flavus, Aspergillus ochra and Penicillium citrinum
eugenol	Clove oil	10 kinds of fungi including Fusarium graminearum, Aspergillus charcoal and Penicillium
carvone	Parsley essential oil	Fusarium graminearum, Aspergillus, Aspergillus niger, etc.
citral	Litsea cubeba oil	5 kinds of fungi including Aspergillus flavus, Penicillium citrinum and Fusarium graminearum

Section 5

The effect of plant essential oils on controlling pests

Apply 66 kinds of plant essential oils, against the three main kinds of pests in the storage silos (the red pirate, the corn weevil, and the corn stupid) were tested for the reproduction inhibition of the pests, at a dose of 0.2% of the plant essential oils. Yellow bark, rue, artemisia scoparia, star anise, sandalwood incense, phoebe cinnamon, camphor, Cinnamomum angustifolia, and cinnamon oil can completely inhibit the occurrence of the F1 generation of the red piranha, indicating that 0.2% concentration of the above-listed plant essential oils can inhibit the reproduction rate of the F1 of the red worm to 94.36%, and 0.1% concentration of the essential oils listed above can inhibit the reproductive rate of the F1 of the worm completely, and the lethal rate of the corn weevil is as high as 85%. This indicates that the listed plant essential oils have a very effective inhibitory effect on the reproduction or poisoning of the three kinds of storage pests. Research reports have obtained a 93% estragole from the essential oil of the yellow bark branches and leaves of the genus Rutaceae. It has strong antibacterial activity against 7 kinds of molds, but it inhibits corn weevil (*Sitophilus zeamais*), *Tribolium castaneum*, *Tenebrio molitor*, and *Rhizopertha dominica* at a concentration of 0.2 mg/L. It can kill 100%. This is a kind of natural medicine that uses plant essential oils to prevent and suppress pests in grain storage. It is a natural medicine worthy of development. Since plant essential oils are all volatile natural products and their safety has been tested, there is basically no residual problem (See attached Table 4-4).

Table 4-4 Safety evaluation of plant essential oils and active ingredients

Common essential oil active ingredients	Evaluation method	Test subject	Concentration LD_{50} g/kg^{-1}	Evaluation (toxicity level)
Aegle marmelos	Acute oral toxicity	Mouse	23.660	Non-toxic
Thymus vulgaris	Acute oral toxicity	Mouse	17.500	Non-toxic
Syzygium aromaticum	Acute oral toxicity	Mouse	55.23	Non-toxic
Cinnamomum aromaticum	Acute oral toxicity	Mouse	5.038	Non-toxic
Litsea cubeba	Acute oral toxicity	Mouse	4.000	Low toxicity
Cymbopogon winterianus	Acute oral toxicity	Mouse	3.500	Low toxicity
Mentha canadensis	Acute oral toxicity	Mouse	2.000	Low toxicity
Cazadirachtin	Acute oral toxicity	Mouse	13.000	Non-toxic
Verbenone	Acute oral toxicity	Mouse	8.088	Non-toxic
1,8-Cineole	Acute oral toxicity	Mouse	7.220	Non-toxic
Eugenol	Acute oral toxicity	Mouse	3.000	Low toxicity
Cinnamaldehyde	Acute oral toxicity	Mouse	2.220	Low toxicity

Section 6

Plant essential oils as industrial raw materials

The largest output plant essential oils, and the widest range of application area is turpentine oil that mainly contains α-pinene and β-pinene. The production of turpentine in China is mainly called GT turpentine obtained from the distillation of masson pine resin. This is different from MPT turpentine that is recycled from paper pulp produced by conifer wood in North America (main countries). The annual production of turpentine in China is 45,000~50,000 tons, except for domestic use. For some exports, turpentine is mainly used for the synthesis of other chemical products, such as synthetic camphor, borneol, and other synthetic spices, binders, solvents, etc. except for a small amount of medicinal use.

The main components of turpentine are α-pinene, β-pinene, myrcene, terpinene, and α-phellandrene. They are all raw materials that are allowed to synthesize food flavors, such as synthetic terpineol, synthetic terpineol, synthetic linalool, synthetic geraniol and nerol, synthetic menthol, synthetic camphor, synthetic bornyl acetate, synthetic neryl acetate and geranyl acetate, synthetic geranyl acetone, synthetic isopropyl acetate, toluene and so on. Synthetic N-alkylimides are pesticide synergists, as well as synthetic insect juvenile hormones or insect growth regulators, insect attractants, insect repellents, and insect antifeedants.

In addition, external application of turpentine oil can improve local blood circulation and relieve swelling and mild pain relief. It is formulated into a cream emulsion, which is widely used to relieve exercise pain and fatigue. The external application of turpentine oil and cod liver oil can stop bleeding and accelerate wound healing.

Turpentine also replaces the toxic xylene used in cryosectioning agents. Although the output of other plant essential oil components is far less than that of turpentine, its structure is special and can be used as a synthetic raw material for some unique products such as safrole. There are 32 kinds of them with a content of more than 40% in plants. There are 20 species mainly found in rhizomes, including Lauraceae *Cinnamomum parthenoxylon*, *C. burmannii* f. *heyneanum*, *C. pauciflorum*, *C. camphora*, *C. parthenoxylon*, etc. There are 10 species that can exist in branches and leaves, and there are 4 species of herbs, just look at it. Laikeguo also has 14 kinds of safrole in continuous production. For example, the establishment of planting production bases does not require rooting or cutting trees to obtain safrole. It seems that this plant resource still has a lot of potential, and safrole is synthetic. It is the most ideal natural raw material for jasminaldehyde, anethole and vanilla.

Section 7

Antiseptic and fresh-keeping effect of plant essential oils

The role of plant essential oils with anti-fungal ingredients in the development of anti-corrosion and anti-fungal components is worth noting. First of all, the safety of this type of plant essential oil is easier to accept than chemically synthesized antiseptic preservatives. It uses citral, estragole, anisaldehyde, etc. to produce a kind of fresh-keeping film, which is used as a fruit fresh-keeping film. The effect of preventing the invasion of Penicillium digitatum and Penicillium italy is very good.

Section 8

Essential Oil Plants (Aromatic Plants) in Chinese Medicinal

Browse the first part of the Chinese Pharmacopoeia on Chinese medicinal materials. There are no less than 80~90 kinds of Chinese medicines containing essential oils. If you add folk herbal medicines, there are more than 100 kinds, but their medicinal properties are roughly similar. For example, lilac is pungent and moderate, good for spleen, stomach, lung, and kidney meridian; the nature and flavor of ginseng are sweet, slightly bitter, and lukewarm, good for spleen, lung, heart, and kidney meridian; Guri incense is pungent in nature, slightly bitter, warm, and slightly toxic, good for liver and stomach meridians. Most of their functions and indications are to promote Qi to relieve pain, rheumatism arthralgia, bruises, swelling and pain, insect bites, abdominal distension and pain, aroma and turbidity, release and relieve heat, regulate Qi, explode dampness and reduce phlegm, promoting blood circulation and dissipating blood stasis, detoxification, pain and swelling, malaria, dispelling blood stasis, detoxification and pain relief, traumatic bleeding, dispelling cold and relieving pain, menorrhagia, maintaining Qi and nourishing Yin, clearing heat and promoting body fluid, etc. See attached Table 4-5 for details.

Table 4-5 Chinese medicinal materials in aromatic plants

Name	Main essential oil ingredients	Characteristic
1. *Syzygium aromaticum*	eugenol	To invigorate the kidney and promote yang, deficiency and cold of the spleen and stomach, impotence due to deficiency of the kidney, vomiting and diarrhea after eating less food
2. *Panax ginseng*	α, β-elemene, β-caryophyllene, β-caryophyllene	To invigorate vitality, nourish body fluids, nourish blood, soothe the nerves and nourish the mind, blood-qi deficiency, lung deficiency, asthma and cough, palpitations and insomnia
3. *Murraya paniculata*	t-caryophyllene, α-curcumene, γ-cinderene, lupulene	Invigorating Qi and relieving pain, promoting blood circulation and removing blood stasis, rheumatic arthralgia, external treatment of dental diseases, insect bites, bruises, swelling and pain
4. *Inula helenium*	costus lactone, costene, eagle claw a	To invigorate the spleen and stomach, promote Qi to relieve pain, anti-fetus, vomiting and diarrhea, deworming
5. *Chenopodium ambrosioides*	terpineol-4 acetate, α-terpinene, ascarisin	Insecticide, treatment of ringworm, antipruritic (for external use)

(continued)

Name	Main essential oil ingredients	Characteristic
6. *Kaempferia galanga*	cis, t-p-methoxy ethyl cinnamate, ethyl cinnamate, anisin	In the middle of the temperature, digestion, pain relief, cold pain in the abdominal abdomen, poor diet
7. *Lonicera confusa*	cis-trans-linalool oxide, phenethyl alcohol, ocimene	To clear away heat and detoxify, evacuate wind-heat, heat toxin blood dysentery, wind-heat cold, for pain, swelling and boils
8. *Vladimiria souliei*	costus lactone, sesquiterpene	Invigorate Qi to relieve pain, abdominal distension and pain, bowel diarrhea, tenesmus
9. *Ligusticum sinense*	neocnidium, elemani, β-phellandrene, t-ocimene	Activating blood circulation, dispelling wind and relieving pain, used for chest pain, heartache, bruises, swelling and pain, irregular menstruation, headache, rheumatic arthralgia
10. *Zingiber officinale*	gingerone, 6-gingerol	Inhibition of bacteria and expel wind, asthma and cough with cold drink, warming and dispelling cold
11. *Citrus medica*	α, β-citral, limonene	Regulate Qi and resolve phlegm, promote blood circulation and relieve pain, sore throat, bruises and pain
12. *Pogostemon cablin*	patchouli azulene alcohol, patchouli alcohol, eugenol, α-patchoulene	Anti-vomiting agent, aromatic gastrointestinal, antipyretic, release heat relief, aromatize turbidity
13. *Foeniculum vulgare*	t- anethole, carvone, celery brain, limonene	Drive wind and invigorate the stomach, dispel blood stasis, reduce cold and pain in the abdomen, cold and abdominal pain in menstruation

(continued)

Name	Main essential oil ingredients	Characteristic
14. *Barenol*	d-borneol	Resuscitation, clearing away heat and relieving pain, stroke, phlegm and convulsions, red eyes and mouth sores, convulsions
15. *Aucklandia costus*	costuslactone, dehydrocostuslactone	Promoting Qi to relieve pain, invigorating the stomach and eliminating food, diarrhea after dysentery, treating diarrhea and abdominal pain
16. *Sliced turmeric* (contains turmeric)	aroma curcumene, curcumene, turmeric, curcumone	Promoting Qi to relieve pain, anti-bacterial and antihypertensive, easing menstruation and pain relief, dysmenorrhea and collateral closure, rheumatic shoulder and arm pain
17. *Nardostachys jatamansi*	glycinetoxin, aristolochne, glycineol	
18. *Conyza canadensis*	l-borneol	Resuscitation, clearing away heat and relieving pain, phlegm in stroke, red eyes and mouth sores
19. *Artemisia lavandulaefolia*	l-borneol, camphor, 1,8-cineole	
20. *Citrus grandis*'Juhong'	naringin, limonene, citral	Regulates Qi to be broad, heats irritability and resolves phlegm, eats and hurts wine, used for coughing and excessive phlegm
21. *Rosa chinensis*	limonene, citronellol, linalyl acetate, linalool	Activating blood to regulate menstruation, soothing liver and relieving depression, irregular menstruation, dysmenorrhea, amenorrhea, abdominal distension and pain

(continued)

Name	Main essential oil ingredients	Characteristic
22. *Acorus calamus*	β-asarone, mulberry, asarone, mulberry	Invigorate the spleen, refresh the mind and improve the mind, forgetfulness and insomnia, relieve dampness and appetite
23. *Angelica dahurica*	imperatorin, 1,8-cineole	Relieving the surface and dispelling cold, promoting the nasal orifice, reducing swelling and draining pus, nasal congestion and runny nose, toothache, sore swelling and pain
24. *Atractylodes macrocephala*	atractylone, atractylol, atractylenolide a, 1,8-cineole, caraway	Replenishing the middle energy and replenishing Qi, indigestion, dampness and diuresis, antiperspirant and anti-fetus, abdominal distension and diarrhea, edema and spontaneous perspiration
25. *Syzygium aromaticum*	eugenol	To reduce the adverse effects of warming, invigorate the kidney and assist yang, be used for spleen and stomach deficiency and cold, vomiting and diarrhea due to lack of food, and impotence due to kidney deficiency
26. *Angelica sinensis*	ferulic acid, safranine, limonene, 2,4,6-trimethylbenzaldehyde, bergamotene, limonene, artemene lactone, 2,4-xylene formaldehyde	Lowering blood pressure and diuresis, enriching blood and promoting blood circulation, regulating menstruation and relieving pain, dizziness and palpitations, amenorrhea and dysmenorrhea, abdominal pain due to coldness, rheumatic arthralgia, bruises
27. *Myristica fragrans*	dehydroisoeugenol, 1,8-cineole	Warming the middle and promoting Qi, astringent intestines to stop diarrhea, chronic diarrhea, less food and vomiting

(continued)

Name	Main essential oil ingredients	Characteristic
28. *Cinnamomum aromaticum*	cinnamic aldehyde, cinnamyl acetate, cinnamic acid	Replenishing fire and assisting yang, dispelling cold and relieving pain, cold waist and knee pain, kidney deficiency and asthma, dysmenorrhea, amenorrhea, vomiting cold, vomiting and diarrhea
29. *Daphne genkwa*	linalool, geranyl acetate, α-pine	To invigorate the spleen and stomach, drain water through drinking, pleural and abdominal effusion, anti-cough and asthma, topical insecticidal treatment of sores
30. *Liquidambar orientalis*	cinnamic acid, cinnamate acetate	Resuscitation, diversion, pain relief, used for stroke, phlegm, abdominal cold and pain
31. *Panax quinquefolius*	ginsenoside rg1, rb, re essential oil	To clear away heat and promote body fluid, used for Qi deficiency and yin deficiency, deficiency of heat and tiredness, coughing and wheezing, phlegm and blood, internal heat to reduce thirst
32. *Styrax tonkinensis*	benzyl esters, benzyl esters, benzoic acid and lipids	Promoting Qi and promoting blood circulation, resuscitating and awakening the mind, coma, coma, pain in the abdomen, apoplexy, phlegm
33. *Alpinia galanga*	1,8-cineole, camphor, α-pinene	Dispelling cold and dampness, soothing the spleen and eliminating food, fullness of food accumulation, excessive drinking
34. *Amomum kravanh*	1,8-cineole, terpineol-4, alpha-terpineol	To dissipate dampness and promote Qi, warm the middle to relieve pain, appetite and eliminate food, abdominal distension and pain, cold dampness and vomiting, the beginning of damp temperature

(continued)

Name	Main essential oil ingredients	Characteristic
35. *Vitex negundo* var. *cannabifolia*	1,8-cineole, terpineol-4, β-caryophyllene, β-sugarene	Expectorant, relieving cough and relieving asthma, used for coughing and phlegm
36. *Citrus medica* var. *sarcodactylis*	citral, nerol, hesperidin, α-bergerene, limonene	Soothing the liver and regulating Qi, relieving pain in the stomach, coughing with excessive sputum, distending and painful cavity and hypochondriac
37. *Notopterygium incisum*	cymenol, isoimperatorin, linalool, β-phellandrene, p-cymene, methyl cinnamate	Relieving the surface and dispelling cold, dispelling wind and dampness, relieving pain, used for cold and cold, shoulder and back pain, rheumatic arthralgia
38. *Rosa rugosa*	geraniol, linalool, farnesol, aromatene	Promoting Qi to relieve depression, harmonize blood, and relieve pain, used for liver and stomach Qi pain, poor appetite, nausea, irregular menstruation, bruises and pain
39. *Curcuma aromatica*	1.8-cineole, linalool, geranone, borneol, curcumone, camphor	Promoting blood circulation and relieving pain, promoting Qi and relieving depression, promoting choleretics and relieving jaundice, amenorrhea and dysmenorrhea, breast and fetal pain, jaundice and red urine
40. *Artemisia carvifolia*	artemisinin, artemisinone, linalool	Clearing deficiency and heat, removing bone steaming, relieving heat, reducing malaria, relieving jaundice, malaria cold and heat, damp-heat jaundice, yin deficiency and fever

(continued)

Name	Main essential oil ingredients	Characteristic
41. *Aguilaria sinensis*	agarwood tetraol, 5,6,7 trihydroxychromone, spirulinol, β-furan agarwood, furan ashwood alcohol	Invigorate Qi to relieve pain, warm up to stop vomiting, stomach-cold and vomiting, kidney deficiency and Qi to counteract asthma
42. *Selaginella tamariscina*	varieties of stilbine, terpineol-4, bornyl acetate	Promoting blood circulation, clearing menstruation, traumatic injuries, used for amenorrhea, clearing menstruation, and treating lumps
43. *Thuja orientalis*	pipin, morocene, cinderenol, carene-3	Cooling blood to stop bleeding, resolving phlegm and relieving cough, growing black hair, used for vomiting blood, hemoptysis, hematochezia, uterine bleeding, blood heat and hair loss
44. *Lonicera japonica*	lutein, oxalin, linalool, nerol, benzyl alcohol, isonerol, α-terpineol	Clear heat and detoxify, evacuate wind-heat, for pain, swelling, furuncle, throat numbness, erysipelas, wind-heat, cold, febrile disease and fever
45. *Schizonepeta tenuifolia*	pulegone	Relieving the surface, dispelling wind, removing scabies, and reducing sores, used for colds, headaches, measles, and sores at the beginning
46. *Alpinia hainanensis*	alpine, cardamom, alderone, farnesol, 1,8-cineole, cinnamaldehyde	To dry dampness and promote Qi, warm up to stop vomiting, used for internal resistance of cold and dampness, not thinking about eating, heating, vomiting, abdominal distension and cold pain

(continued)

Name	Main essential oil ingredients	Characteristic
47. *Amomum villosum*	bornyl acetate, borneol, neroliol, camphor	To dissolve dampness and appetite, warm the spleen to relieve diarrhea, regulate Qi to relieve the fetus, used for warming turbidity, preventing hunger, spleen and stomach deficiency and cold
48. *Dalbergia odorifera*	methyl benzoate, linalool, phenethyl alcohol, limonene	To remove blood stasis and stop bleeding, for vomiting blood, traumatic bleeding, liver depression, hypochondriac pain, bruises, vomiting, abdominal pain
49. *Amomum tsaoko*	1,8-cineole, 2-methylcinnamaldehyde, geranyl acetate, α, β-citral	Dry dampness and warm middle, cut malaria and remove phlegm, used for internal resistance of cold and dampness, fullness of vomiting, malaria, cold and fever, plague fever
50. *Citrus medica*	limonene, α, β-citral, citronellal, geranyl acetate	To soothe the liver and regulate Qi, broaden the middle and eliminate phlegm, it is used for stagnation of liver and stomach Qi, swelling and pain in the hypochondria of the cavity, coughing up phlegm, vomiting and gasping
51. *Mosla chinensis*	α, β-pinene, limonene, bornyl acetate, α-caryophyllene	Sweating to relieve the surface, neutralize the temperature, and be used for cold, aversion to cold, fever, headache, no sweat, edema, and difficulty in urination
52. *Pinus tabuliefomia*	α-pinene	To expel wind and dampness, relieve collaterals and relieve pain, used for wind-cold dampness, arthralgia, bruises
53. *Asarum sieboldii*	asarone, β-farnesene, camphor acetate	Eliminate phlegm and relieve cough, dispel blood stasis and swelling, warm the meridian and dispel cold

(continued)

Name	Main essential oil ingredients	Characteristic
54. *Curcuma longa*	curcumin, curcuminone, curcumene, 1,8-cineole, gingerene, geranone	To break the blood and promote Qi, relieve menstruation and relieve pain, used for chest and hypochondriac tingling, chest pain, dysmenorrhea, amenorrhea, swelling, bruises, swelling and pain
55. *Curcuma phaeocaulis*	gingerene, curcumone, gingerol, curcumone, camphor, borneolus	Promotes Qi to break blood, eliminates accumulation of pain and relieves pain. It is used to treat lumps, congestion, amenorrhea, chest pain, heartache, swelling and pain from food accumulation
56. *Alpinia officinarum*	greater galangal	Wet stomach to relieve vomiting, dispel cold and relieve pain, used for abdominal cold and pain, stomach cold and vomiting, heating sour
57. *Dendranthema morifolium*	luteolin, chlorogenic acid, borneol, camphor, 1,8-cineole	Dispelling wind and clearing heat, calming the liver and improving eyesight, clearing away heat and detoxification, used for wind-heat and cold, headache, dizziness, red eyes and swelling
58. *Perilla frutescens*	α, β-citral, geraniol, linalool, citronellol	Relieve the surface and dispel cold, promote Qi and stomach, for cold, cough and nausea, fish and crab poisoning, pregnancy vomiting
59. *Vitex trifolia* var. *simplicifolia*	vitexin, 1,8-cineole, β-caryophyllene	Evacuate wind-heat and clear up the boss, used for wind-heat, cold, headache, gum swelling and pain, red eyes and more tears

(continued)

Name	Main essential oil ingredients	Characteristic
60. *Mentha canadensis*	menthol, menthone, huthrone	Evacuate wind-heat, clear the leader, relieve sore rash, sooth the liver, for the first onset of wind and temperature, headache, red eyes and throat numbness, aphthous measles
61. *Ligusticum sinense*	ferulic acid, neoclonia lactone, γ-ylangolene, β-caryophyllene, t-ocimene, lavender alcohol	Dispel wind, dispel cold, dehumidify, and relieve pain, used for cold and cold, peak pain, rheumatic arthralgia
62. *Santalum album*	α-santalol, α-isosantalol, β-santalene, α-santalene	It is used for mid-temperature, appetizing and pain relief, used for cold coagulation, stagnation of Qi, uncomfortable chest and diaphragm, vomiting, lack of food, abdominal pain
63. *Artemisia capillaris*	cinchenone, eugenol, cynchenone, eugenol methyl ether	To clear away damp-heat, promote choleretics and reduce jaundice, used for jaundice, oliguria, dampness and heat, wet sores and itching
64. *Litsea cubeba*	α-citral, linalool, 6-methylheptenone	Warming the middle and dispelling cold, promoting Qi to relieve pain, used for stomach cold and vomiting, abdominal cold pain, cold colic, abdominal pain, turbid urine, cold dampness stagnation
65. *Boswellia carteri*	α-pinene, juniene, limonene, myrcene	—
66. *Artemisia dracunculus*	methylophylol	—

(continued)

Name	Main essential oil ingredients	Characteristic
67. *Zanthoxylum schinifolium*	β-ocimene-x, aigrain, limonene, 1,8-cineole	Dispelling cold, dispelling temperature, relieving itching, repelling insects, and helping Yang
68. *Artemisia waltonii*	eugenol methyl ether, elemane, saccharin, saccharin	Sterilization and anticorrosion
69. *Boswellia carteri*	α-pinene, juniene, myrcene, limonene	Invigorate blood and relieve pain, reduce swelling and build muscle, for chest pain, dysmenorrhea, amenorrhea, postpartum stasis, muscles and veins, bruises, pain, swelling and sore
70. *Asarum sieboldii*	asarone, aristolochic acid	Relieve the appearance and dispel cold, dispel wind and relieve pain, relieve the orifice, warm the lung and transform the drink, cold and cold, headache, toothache
71. *Zanthoxylum bungeanum*	1,8-cineole, linalool, α-terpineol, peppermint	Relieve pain, aid digestion, toothache, abdominal pain, relieve diarrhea, kill insects
72. *Liquidambar formosana*	α-pinene、β-pinene、camphene	Phlegm, wheezing and coughing, nasal congestion and runny nose

In addition, plants containing essential oils can be used in traditional Chinese medicine as well as seasonings, such as star anise, Sichuan pepper, pepper, cumin, Fennel seeds, cumin, cloves, etc. They are mainly used for spicy seasoning. There are also Lanke's Vanilla Beans, which are mainly used in liquor and cigarette flavors.

Section 9

Application of Terpenoids, Main Components of Plant Essential Oils

Plant essential oils, their chemical composition are basically monoterpenes and sesquiterpenes in terpenoids. Monoterpenes are mainly used in medicinal fragrances and light industry. The chemical structure of sesquiterpenes is more complex, so they have a wide range of biological activities. Many sesquiterpene oxygen-containing derivatives have strong and long-lasting aromas and are widely used in food, cosmetics and medical drugs. For example, artemisinin and its various derivatives have curative effects on various malaria, such as vivax malaria, falciparum malaria, cerebral malaria and critical malaria. Artemisinin derivatives, artemisinin methyl ether, artemisinin succinate esters have been officially used as clinical drugs.

1. There are the most natural sesquiterpenes with anti-tumor effects. Among them, are sesquilactones, and there are hundreds of cytotoxic activities with incomplete statistics. It has been reported that α-methylene-γ-butyrolactone or α-methylene cyclopentanone is an effective group that makes these sesquiterpenoids have anti-cancer biological activity. Except for lactone sesquiterpenes, for example, ginseng essential oil is rich in β-elemene. Studies have reported that it has inhibitory activity on a variety of tumors. It has been used to treat brain tumors, lung cancer, liver cancer, esophageal cancer, uterine cancer, leukemia, etc. The pharmacological effect is believed to interfere with the growth and metabolism of tumor cells and inhibit the proliferation of tumor cells. This compound has been used in clinical treatment.

2. Deworming and killing insects. A variety of sesquiterpenes have the ability to repel parasites in the human body. Santonin is a natural medicine for roundworms that has been used for a long time. Costunolide, Cremanthine and Inula helenium roots have insect repellent ingredients, eagle claw A. A sesquiterpene is isolated from the root, which has a strong inhibitory effect on the growth of Plasmodium murine. It is a sesquiterpene compound with peroxygen bond like Artemisia annua.

(Structural formula)

costunolide   eremanthine   santonin   artemisinin

3. Insect antifeeding, repellent and attracting effects. Insect antifeedants are derivatives of sesquiterpenes, such as dihydrofuran agarwood polyol derivatives, which are derived from the root bark of Celastrus angulatus. It has antifeedant or poisonous effect on many kinds of insects, and bitter leaf powder is also used as an antifeedant for vegetable insects. A sesquiterpene dihydroxy compound obtained from a kind of Allomyces sabviscula, is an attractant for certain insects and lower animals, and its physiological effect is very strong.

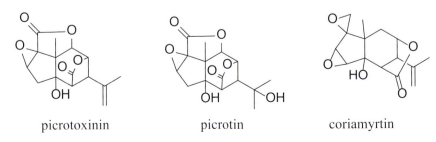

sirenin

4. Physiological effects on the nervous system. This type of physiological activity on the nervous system is mostly sesquiterpene lactones, such as Picrotoxinin and Picrotin obtained from Amamirta cocculus, as well as *Coriaria japonica* in Japan, *C. sinica* in China, both have Coriamyrtin and Tutin. The fruit, leaves, and bark of Illicium anisatum contain Anisatin, which has an excitatory effect on the central nervous system, and has a strong convulsive effect like the highly toxic paponyxine.

picrotoxinin   picrotin   coriamyrtin

anisatin   tutin

5. Regulate plant growth. As a regulator of plant growth, it has two interactions: inhibiting plant growth and promoting stimulation. (+)-abscisic acid was isolated from the shed cotton boll in the early stage. It is a hormone that promotes falling leaves. Control plant defoliation and inhibit the germination of seeds and bulb tips. Portulaca grandiflora contains Portulal, which is also a kind of auxin, that hinders the rooting of radish. In addition, the exudative secretion from cotton roots, strigol, a sesquiterpene derivative, can promote the germination of Scrophulariaceae seeds.

abscisic acid   Portulal

Strigol

6. Antibacterial activity. Antibacterial activity of plant essential oils In Chapter IV, Section 4, we mainly introduce the antimicrobial activities of monoterpenes in essential oils and someessential oils. This part mainly introduces the

antibacterial activity of sesquiterpenes in essential oils. Among them, Parthenium hysterophorus and Ambrosia maritma sesquiterpenes and their derivatives Parthenin and Ambrosin, have anti-Gram-positive bacteria effect. Oppositol, a sesquiterpene alcohol containing Br, is separated from Laurencia, which can well inhibit the activity of Staphylococcus aureus. They are the skeleton of azulene.

7. It can be used as a natural antioxidant in food and cosmetics-rosemary essential oil. Rosmarinus officinalis, a herb of the Lamiaceae family, has a strong antioxidant capacity and is completely non-toxic. It is native to the Mediterranean coast and is widely introduced all over the world. Rosemary essential oil also has strong biological activities such as sterilization, anti-inflammatory, anti-virus and tumor.

| Parthenin | Ambrosin | Oppositol |

Chapter IV References

1. Chinese Pharmacopoeia [M]. Part One. Medicinal materials and decoction pieces, vegetable oils and extracts. Prescriptions and single-flavor pharmaceuticals. 2015.

2. Daniel Festi. The Essential Oil Bible [M]. Liu Xi, Zhang Ting, Yu Chunhong, Song Meiyu, translated. Shanghai Science and Technology Literature Publishing House. 2011.

3. Ha Chengyong. Chemistry and Application of Natural Products [M]. Beijing Chemical Industry Press. 1996.

4. Introduction of safrole plant essential oil [M]. Chinese Wild Plant Resources. 1994(4): 21-23.

5. Li Duan, Zhou Ligang, Jiang Weiwei, et al. Research progress on antibacterial components of Umbelliferae plants [J]. Journal of Northwest A&F University (Natural Science Edition). 2005(33):161-166.

6. Liu Buming, Lai Maoxiang, Wei Xiuzhi. Essential oil resources and medicinal use in Guangxi [M]. Guangxi Forestry. 2004(6).

7. Liu Buming, Lai Maoyang. Resources and Application of Guangxi Medicinal Essential Oil Plants [J]. Journal of Guangxi University of Traditional Chinese Medicine. 2005, 8(2): 83- 87.

8. Chang Hongfei. Development and research of medicinal plant resources[M]. Chinese Herbal Medicine. 2000, 31(9): 711-715.

9. Miao Qing, Zhao Xiangsheng, Yang Meihua, et al. Research progress on the chemical constituents and harmful substances of aromatic plants[M]. Chinese Herbal Medicine. 2013, 44(8): 1062-1068.

10. Tian Yuhong, Zhang Xiangmin, Huang Taisong, et al. Plant essential oil resources and their application in tobacco industry[M]. Guangdong Chemical Industry. 2007, 34(10): 73-75.

11. Turpentine use [M]. Northwest Agriculture and Forestry University.

12. Liu Donglian, Ma Songtao, Zeng Renyong, et al. Determination of the 50% lethal dose of cinnamon volatile oil in mice [J]. Medical Journal of National Defense of Southwest China. 2010, 20(5): 481-482.

13. Ma Songtao, Liu Donglian, Lan Xiaoping, et al. Determination of the volatile oil of clove in mice [J]. Journal of Liaoning University of Traditional Chinese Medicine. 2010. 12(5): 67-68.

14. Sun Lingfeng, Wang Hongwu, Tang Minyan. Safrole and its application in synthetic fragrance [M]. Fragrance, Flavor and Cosmetics. 1998 (2): 14-18.

15. Tian Yuhong, Zhang Xiangmin, Huang Taisong, et al. Plant essential oil resources and their application in tobacco industry[M]. Guangdong Chemical Industry. 2007, 34(10): 73-75.

16. Li Yanjun, Kong Weijun, Li Menghua, et al. Research progress of plant essential oils in inhibiting fungi and mycomycin synthesis. Chinese Herbal

Medicine. 2011-2018, 47(11).

17. Yu Boliang, Luo Huibo. Experimental report of citral antifungal and inhibition of Aspergillus flavus toxin production [J]. Food Science and Technology. 2002 (4): 47- 49.

18. Cheng Shifa, Zhu Liangfeng, Lu Baiyao, et al. Study on the chemical constituents and antibacterial activity of the essential oil of Leziguo [J]. Bulletin of Botany. 1990, 32(1): 49-53.

19. Zhu Liangfeng, Lu Baiyao, Yu Yaoxin. Research on natural fruit preservation agent [M]. National Food Additives Newsletter. 1991: 28-29.

20. Xu Hanhong, Zhao Shanhuan, Zhu Liangfeng, et al. Study on the insecticidal effect and active ingredients of the essential oil of yellow bark [J]. Journal of South China Agricultural University. 1994, 15(2): 36-40.

21. Xu Hanhong, Zhao Shanhuan, Zhu Liangfeng. Study on the Reproductive Inhibition Effect of Essential Oils on Stored Grain Pests [J], Journal of the Chinese Cereals and Oils Association. 1993, 8(2): 290-296.

22. Jia Jinlian. Current Situation and Market of Turpentine Industry [M]. Forest Products Chemistry and Chemical Industry. 21(4): 60-64.19.

23. Liu Zhiqiu, Chen Jin, Xu Yong. The status quo of development and utilization of safrole plant resources [M]. Fragrance, fragrance and cosmetics. 2001 (4).

24. Shanghai Chemical Reagent Supply Station of China National Pharmaceutical Corporation. Reagent Manual [M]. Shanghai: Shanghai Science and Technology Press. 1985: 1074.

25. Cheng Biqiang, Yu Xuejian, Ding Jingkai. Resources of Cinnamomum in China and its aromatic components [M]. Kunming: Yunnan Science and Technology Press. 1997.

26. Cheng Biqiang, Xu Yong, Zeng Xianxian. Development and Utilization of Cinnamomum in Southern Yunnan[C]. Collection of Reports on Tropical Plant Research Papers. Yunnan University Press. 1994:38-45.

27. Pang Jianguang, Zhang Mingxia, Han Junjie. Research and application of plant essential oils [J]. Journal of Handan Agricultural College. 2003, 20(1): 26-29.

28. Yu Boliang, Luo Huibo. Experimental report of citral antifungal and

inhibition of Aspergillus flavus production [J]. Food Science and Technology. 2002, (4): 47-49.

29. Zhang Youlin, Zhang Runguang, Zhong Yu. The chemical composition, bacteriostasis, antioxidant activity and toxicological properties of thyme essential oil[J]. China Agricultural Sciences. 2011, 44(9): 1888-1897.

30. Luo M, Jiang L.K, Zou GL. Acute and genelie foxicity of essential oil extracted from Litsea cubeba(Lour)Pers [J]. J Food Prot. 2005. 68(3):581-588.

31. Morcin C.Malnati M, Teregi V. In vitro antidivngal avtivity of terpinen-4-d. engenol carvono.1,8-cineole(eucalyptol)and thymol against mycotoxigenie plant pathogen s [J]. Food Addit contam A. 2012, 29(3): 415-422.

32. Singh P, Kumar R, Dubey NK, et al. Essential oil of Aegle marmelos as a safe plant-based antimicrobial against postharvest microbial infestations and aflatoxin contamination of food commodities [J]. J Food Sci. 2009; 74(6): 302-307.

Appendix: Comparison of Chinese and English Compound Names

A

abscisic acid	脱落酸
acetophenone	苯乙酮
2-acetylpyrrole	2-乙酰基吡咯
acoradiene	菖蒲二烯
acyclic monoterpeneoid	无环单萜
acyclic sesquiterpenoid	无环倍半萜
ageratochromene	胜红蓟素
alantic acid	土木香酸
alantol	土木香醇
alantolactone	土木香内酯
alloaromadendrene	别香树烯
allo-ocimene	别罗勒烯
allyl methyl disulfide	烯丙基甲基二硫醚
allylcyclohexane	烯丙基环己烷
ambrettolide	黄葵内酯
ambrosin	豚草素
anethole	大茴香醚
angelicin	当归素
anisaldehyde	大茴香醛
anisatin	莽草毒素
anisole	茴香醚
anthracene	蒽
apiole	芹菜脑
aplotaxene	单紫杉烯
abietatriene	枞三烯
α-curcumene	姜黄烯

aristolene	土青木香烯
aristolene	马兜铃烯
aristolene-1,9-diene	马兜铃烯-1,9-二烯
9-aristolene-l-ol	9-马兜铃烯-1-醇
aromadendrene	香树烯
artemisia ketone	蒿酮
artemisin	蒿素
artemisinin	青蒿素
ar-zingberone	芳姜酮
asaricin	细辛醚
asarone	细辛脑
asarylaldehyde	细辛醛
asatone	细辛酮
atractylocin	苍术素
atractylone	苍术酮
azulene	薁

B

benyzl pentanoate	戊酸苯甲酯
benzaldehyde	苯甲醛
benzene	苯
benzofuran	苯并呋喃
benzothiazole	苯并噻唑
benzyl 2-hydroxybenzoate	2-羟基苯甲酸苯甲酯
benzyl 3-methylbutanoate	2-甲基丁酸苯甲酯
benzyl acetate	乙酸苯甲酯
benzyl acetone	苄基丙酮
benzyl alcohol	苯甲醇
benzyl benzoate	苯甲酸苄酯, 苯甲酸苯甲酯
benzyl butyrate	丁酸苄酯, 丁酸苯甲酯
benzyl cyanide	苯乙腈
benzyl formate	甲酸苯甲酯
benzyl hexanoate	己酸苄酯
benzyl isothiocyanate	异硫氰酸苯甲酯
benzyl salicylate	水杨酸苯甲酯

benzyl tiglate	惕各酸苯甲酯
bergamotene	佛手烯
bergamotene	香柠檬烯
betulene	香桦烯
betulenol	香桦烯醇
bicyclicterpenoid	双环单萜
bicyclo[6.1.0]non-5,8-diene-4-one	二环 [6.1.0] 壬 -5,8-二烯-4-酮
bisabolene	比沙泊烯
bisabolene	甜没药烯
α-bisabolol	α-甜没药醇
α-bisabolol oxide β	α-甜没药醇氧化物 β
2,10-bornediol	2,10-龙脑二醇
bornene	龙烯
bornene	菠烯
borneol	龙脑
bornyl 2-methylbutanoate	2-甲基丁酸龙脑酯
bornyl acetate	乙酸龙脑酯
bornyl butyrate	丁酸龙脑酯
bornyl formate	甲酸龙脑酯
bornyl isovalerate	异戊酸龙脑酯
bornyl propionate	丙酸龙脑酯
bornylene	龙脑烯
β-bourbonene	β-波旁烯
1,3-butanediol	1,3-丁二醇
bulnesol	布黎醇
butanedione	丁二酮
butanol	丁醇
butenolide A	白术内酯 A
butyl 3-methylbutanoate	3-甲基丁酸丁酯
butyl acetate	乙酸丁酯
butyl angetate	当归酸正丁酯
butyl isobutyrate	异丁酸丁酯
butyl salicylate	水杨酸丁酯
butyl tetradecanoate	十四酸丁酯
t-butyl 3,3-dimethylacrylate	3,3-二甲基丙烯酸叔丁酯
t-butyl-m-cresol	叔丁基间甲酚

2-butyltetrahydrofuran	2-丁基四氢呋喃
3-butylphthalide	3-丁基苯酞
butylidenephthalide	亚丁基 -2-苯并 [C] 呋喃酮
butyric acid	丁酸
3,5-di-t-butyl-salicylaldehyde	3,5-二叔丁基水杨醛

C

cadinene	杜松烯
cadinenol	杜松烯醇
cadinol	杜松醇
calacorene	甜旗烯
calamenene	去氢白菖烯
calameneol	白菖醇
calarene	水菖蒲烯
calarene	白菖烯
camphene	莰烯
camphene hydrate	水合莰烯
camphor	樟脑
α-camphorenal, α -campholenic aldehyde	α-龙脑烯醛
α-camphorene aldehyde	α-樟脑烯醛
capillene	茵陈炔
capillone	茵陈炔酮
γ-caprylolactone	γ-辛内酯
(+)-caran-cis-4-ol	(+)-蒈烷-顺-4-醇
carbamyl benzoate	氨基甲酰苯甲酸酯
carene	蒈烯
carotol	胡萝卜醇
carvacrol	香芹酚 , 香荆芥酚
carvacryl acetate	乙酸香芹酯
carvendione	香芹二烯酮
carvendione	葛缕二烯酮
carveol	葛缕醇
carveol methyl ether	葛缕醇甲基醚
carvone	葛缕酮
carvotanacetone	别二氢香芹酮

carvyl acetate	乙酸葛缕酯
caryophyllene	石竹烯
caryophyllene oxide	氧化石竹烯
caryophyllenol	石竹烯醇
cedrene	柏木烯
cedrenol	柏木烯醇
cedrol	柏木脑
cedrone	柏木酮
cedryl acetate	乙酸柏木酯
chamanzulene	母菊薁
β-chamigrene	β-恰米烯
chavibetol	蒌叶酚
chavicol	胡椒酚
chavicol	黑椒酚
chrysanthenol	菊醇
chrysanthenyl acetate	乙酸菊酯
cinnamic acid	桂酸
cinnamic alcohol	桂醇
cinnamic aldehyde	桂醛
cinnamyl acetate	乙酸桂酯
1,8-cineole	1,8-桉叶油素
cis-1,3-dimethyl-8-isopropyl-3-decene	顺式-1,3-二甲基-8-异丙基-3-癸烯
cis-3-hexenyl isovalerate	异戊酸顺式-3-己烯酯
cis-3-hexenyl propionate	丙酸顺式3-己烯酯
cis-3-hexenyl tiglate	惕各酸顺式-3-己烯酯
cis-4,11,11-trimethyl-8-methylenebicyclo [7.2.0] undec-4-ene	顺式-4,11,11-三甲基-8-亚甲基二环 [7.2.0] 十一碳-4-烯
cis-8-menthene	顺式-8-蓋烯
cis-caraway formate	甲酸顺式-葛缕酯
cis-p-menthan-9-ol	顺式-对蓋-9-醇
cis-terpin hydrate	顺式-水合萜二醇
citral	柠檬醛
citronellal	香茅醛
citronellene	香茅烯
citronellol	香茅醇
citronellyl acetate	乙酸香茅酯

citronellyl formate	甲酸香茅酯
citronellyl propionate	丙酸香茅酯
clovene	丁子香烯
cnidilide	蛇麻酞内酯
copaene	古巴烯
copaenol	古巴烯醇
coriamyrtin	马桑毒素
costene	广木香烯
costene	木香烯
costic acid	木香酸
costol	木香醇
costunolide	木香（烯）内酯
coumarane	香豆烷
coumarin	香豆素
cresol	甲酚
cubibene	荜澄茄烯
cumaldehyde, cuminic aldehyde	枯茗醛
cumic alcohol, (cuminol)	枯茗醇
cuparene	花柏烯
cuparenone	花柏酮
curcumene	姜黄烯
curcumenol	姜黄烯醇
curzerenone	莪术酮
cyanobenzene	苯甲腈
cyclodecanol	环癸醇
cyclofenchene	环小茴香烯
cycloheptane	环庚烷
cyclohexanol	环己醇
cyclohexanone	环己酮
cyclohexenyl formate	甲酸环己烯酯
1,3,5,7-cyclooctatetraene	1,3,5,7-环辛四烯
1,3,5-cycloheptatriene	1,3,5-环庚三烯
1-(1-cyclohexenyl)-2-propanone	1-(1-环己烯基)-2-丙酮
cyclohexyl benzoate	苯甲酸环己酯
3-cyclohexenyimethano	3-环己烯基甲醇
5-cyclohexenylethanone	5-环己烯基乙酮

cymene	伞花烃
cymol	伞花醇
cyperene	香附烯
cyperone	香附酮

D

d-[2(12)]thujopsene-3-α-ol	d-[2(12)] 罗汉柏烯-3-α-醇
dammaradienyl acetate	乙酸达玛二烯酯
decadienal	癸二烯醛
decahydro-1,1,4,7-tetramethyl-4aHcyclopro[e] azulen-4a-ol	十氢-1,1,4,7-四甲基-4aH-环丙 [e] 薁-4a-醇
decahydro-1,5,6,8a-tetramethyl-1,2,4-methyleneazulene	十氢 1,5,5,8a-四甲基-1,2,4-亚甲基薁
γ -decalactone	γ-癸内酯
decanal	癸醛
decane	癸烷
decanoic acid	癸酸
decanol	癸醇
2-decanone	2-癸酮
2-decenal	2-癸烯醛
2-dedecanone	2-十二酮
1-decene	1-癸烯
9-decenoic acid	9-癸烯酸
decyl acetate	乙酸癸酯
dehydrcostuslactone	去氢木香内酯
dehydroabietane	去氢松香烷
β-dehydroelsholtzione	β-去氢香薷酮
dehydrocostunolide	脱氢木香内酯
dehydroledol	去氢喇叭茶醇
6-demethyoxyageratochromen	6-去甲氧基胜红蓟素
1,6-diacetoxyhexane	1,6-二乙酰氧基己烷
diallyl disulfide	二烯丙基二硫醚
diallyl sulfide	二烯丙基硫醚
diallyl tetrasulfide	二烯丙基四硫醚

diallyl trisulfide	二烯丙基三硫醚
dibutanone	二丁酮
dibutyl phthalate	邻苯二甲酸二丁酯
1,1-diethoxyethane	1,1-二乙氧基乙烷
diethyl P-benzdioate	对苯二甲酸二乙酯
diethyl phthalate	邻苯二甲酸二乙酯
1,1-diethylpropylbenzene	1,1-二乙基丙基苯
3,3-diethylpentane	3,3-二乙基戊烷
dihydroactinidiolide	二氢猕猴桃（醇酸）内酯
dihydroalantolactone	二氢土木香内酯
dihydrocarvone	二氢葛缕酮
dihydrocarvyl acetate	乙酸二氢葛缕酯
dihydrocostuslactone	二氢木香内酯
dihydrodehydrocostuslactone, Mokko lactone	二氢脱氢广木香内酯
dihydrofarnesol	二氢金合欢醇
dihydroionone	二氢紫罗兰酮
dihydrolinalool	二氢芳樟醇
2,3-dihydro-3,5-dihydroxy-6-methylpyran-4-one	2,3-二氢-3,5-二羟基-6-甲基吡喃-4-酮
2,3-dihydro-4-methylfuran	2,3-二氢-4-甲基呋喃
2,3-dihydrobenzofuran	2,3-二氢苯并呋喃
2,5-dihydro-3,4-dimethylfuran	2,5-二氢-3,4-二甲基呋喃
3,4-dihydro-1(2H)-naphthalenone	3,4-二氢-1(2H)-萘酮
3,4-dihydro-2,5-dimethyl-2H-pyran-2-carboxaldehyde	3,4-二氢-2,5-二甲基-2H-吡喃-2-甲醛
3,4-dihydro-4,6,8-trimethyl-1(2H)-naphthalenone	3,4-二氢-4,6,8-三甲基-1(2H)-萘酮
3,4-dihydydro-8-hydroxy-3-methyl-1H-2-benzopyran-l-one	3,4-二氢-8-羟基-3-甲基-1H-2-苯并吡喃-1-酮
3a,7a-dihydro-5-methylindene-1,7(4H)-dione	3a,7a-二氢-5-甲基茚-1,7(4H)-二酮
dillapiolal	莳萝醛
dillapiole	莳萝脑
o-dimethoxybenzene	邻二甲氧基苯
β-dimethylstyrene	β-二甲基苏合香烯
o-dimethylbenzene	邻二甲苯
1,2-dimethoxybenzene	1,2-二甲氧基苯
1,3-dimethylcyclopentane	1,3-二甲基环戊烷
1,2-dimethyl-4-ethenyibenzene	1,2-二甲基-4-乙烯基苯
1,5-dimethyl-2-isoproyl-2,5-eporycyclopentanol	1,5-二甲基-2-异丙基-2,5-桥氧环戊醇

1,6-dimethyl-4-isopropylnaphthalene	1,6-二甲基-4-异丙基萘
2,2-dimethyl-2,4-heptadienal	2,2-二甲基-2,4-庚二烯醛
2,2-dimethyl-3-(2-methyl-1-propenyl)-cyclopropane carboxlic ethyl ester	2,2-二甲基-3-(2-甲基-1-丙烯基)环丙羧酸乙酯
2,2-dimethylhexanal	2,2-二甲基己醛
2,2-dimethyl-l-methylene-3-butene-2-one	2,2-二甲基-1-亚甲基-3-丁烯-2-酮
2,3-dimethyl-2-butanol	2,3-二甲基-2-丁醇
2,3-dimethyl-5-methoxyphenol	2,3-二甲基-5-甲氧基苯酚
2,3-dimethylbutyric acid	2,3-二甲基丁酸
2,4-dimethyl-2-decene	2,4-二甲基-2-癸烯
2,4-dimethyl-2-pentanone	2,4-二甲基-2-戊酮
2,4-dimethyl-2-pentene	2,4-二甲基-2-戊烯
2,4-dimethylhexane	2,4-二甲基己烷
2,4-dimethylhexanone	2,4-二甲基己酮
2,4-dimethylphenylethanone	2,4-二甲基苯乙酮
2-(3,3-dimethylcyclohexylidene)-ethanol	2-(3,3-二甲基亚环己基)乙醇
3-(4,8-dimethyl-3,7-nonadienly)furan	3-(4,8-二甲基-3,7-壬二烯基)呋喃
2,5-dimethylnonane	2,5-二甲基壬烷
2,5-dimethylpyrazine	2,5-二甲基吡嗪
2,6-dimethyl-2,4,6-octatriene	2,6-甲基-2,4,6-辛三烯
2,6-dimethyl-5-decanal	2,6-二甲基-5-癸醛
2,6-dimethyl-6-(4-isohexenyl)-cylo[3.1.1]hept-2-ene	2,6-二甲基-6-(4-异己烯基)环[3.1.1]庚-2-烯
2,6-dimethylheptane	2,6-二甲基庚烷
2,6-dimethylheptenal	2,6-二甲基庚烯醛
2,6-dimethylstyrene	2,6-二甲基苯乙烯
2,6-di-t-butyl-1,4-benzoquinone	2,6-二叔丁基对苯醌
2,6-di-t-butyl-p-cresol	2,6-二叔丁基对甲酚
2,7-dimethyloctane	2,7-二甲基辛烷
2,8-dimethyltridecane	2,8-二甲基十三烷
3,4-dimethoxybenzoic acid	3,4-二甲氧基苯甲酸
3,4-dimethoxyphenylaldehyde	3,4-二甲氧基苯甲醛
3,4-dimethyl-3-hexene-2-one	3,4-二甲基-3-己烯-2-酮
3,5-dimethoxytoluene	3,5-二甲氧基甲苯
3,5-dimethyldihydr o-2(3H)-furanone	3,5-二甲基二氢-2(3H)-呋喃酮
3,6-dimethyl-1,6-octadiene-3-ol	3,6-二甲基-1,6-辛二烯-3-醇

4,10-dimethyl-7-isopropyclo[4.4.0] deca-1,4-diene	4,10-二甲基-7-异丙基二环 [4.4.0] 癸-1,4-二烯
5,5-dimethyl-2-furanone	5,5-二甲基-2-呋喃酮
5,5-dimethyl-2-propyl-1,3-cyclohexanedione	5,5-二甲基-2-丙基-1,3-环己二酮
6,10-dimethyl-2-undecanone	6,10-二甲基-2-十一酮
6,10-dimethyl-5,9-undecadien-2-one	6,10-二甲基-5,9-十一碳二烯-2-酮
7,11-dimethyl-3-methylene-1,6,10-dodecatriene(Z)	7,11-二甲基-3-亚甲基-1,6,10-十二碳三烯 (Z)
8,8-dimethy1-4-methylene-1-Oxaspiro[2.5] oct-5-ene	8,8-二甲基-4-亚甲基-1-氧杂螺 [2.5] 辛-5-烯
4,7-dimethylundecane	4,7-二甲基十一烷
4,4-dimethyl-2-pentanol	4,4-二甲基-2-戊醇
dineopentyl carbonate	碳酸二新戊酯
1,4-dinethoxy-3,3,5,6-tetramethyl-benzene	1,4-二甲氧基-3,3,5,6-四甲基苯
1,4-dinethyl-3-cyclohexenol	1,4 甲基-3-环己烯醇
3,4-dinethylpentanol	3,4-二甲基戊醇
dioctyl phthalate	邻苯二甲酸二辛酯
dipentadiene oxide	二聚戊二烯氧化物
dipentene	二戊烯
1,4-dipropylbenzene	1,4-二丙基苯
dithiacyclopentene	二硫杂环戊烯
1-docosene	1-廿二烯
dodecanal	十二醛
dodecane	十二烷
dodecanol	十二醇
1-dodecenyl acetate	乙酸-1-十二烯酯
1,11-dodecadiene	1,11-十二碳二烯
2-dodecenal	2-十二烯醛
duryl aldehyde	杜基醛
dysoxylonene	樫木烯

E

eicosadiene	廿碳二烯
eicosanal	廿醛
eicosanol	廿醇

1-eicosene	1-廿烯
elemene	榄香烯
elemicin	榄香脂素
elemol	榄香醇
elsholzione	香薷酮
epicamphor	表樟脑
epiguaipyridine	表愈创吡啶
epoxycaryophyllene	环氧石竹烯（石竹素）
epoxyhumulene	环氧蛇麻烯
eremophiladienone	雅槛蓝二烯酮
eremophilene	雅槛蓝烯
eremophilone	雅槛蓝酮
estragole	爱草脑
ethenoxybenzene	乙烯氧基苯
ethenylbenzaldehyde	乙烯基苯甲醛
ethyl 2-hexenoate	2-己烯酸乙酯
ethyl 2-hydroxybenzoate	2-羟基苯甲酸乙酯
ethyl 2-methyl-2-phenylpropanoate	2-甲基-2-苯基丙酸乙酯
ethyl 2-methylbutanoate	2-甲基丁酸乙酯
ethyl 2-methylpropionate	2-甲基丙酸乙酯
ethyl 3-hydroxyhexanoate	3-羟基己酸乙酯
ethyl 3-hydroxyobutyrate	3-基丁酸乙酯
ethyl 3-methyl-2-butanoate	3-甲基-2-丁烯酸乙酯
ethyl 3-methylthiopropionate	3-甲基硫丙酸乙酯
ethyl 3-phenyl-2-propenoate	3-苯基-2-丙烯酸乙酯
ethyl 5-hydroxy-cis-7-decenoate	5-羟基-顺式-7-癸烯酸乙酯
ethyl acetate	乙酸乙酯
ethyl benzoate	苯甲酸乙酯
ethyl butyl ether	乙基丁基醚
ethyl butyrate	丁酸乙酯
ethyl cinnamate	桂酸乙酯
ethyl cis-4-decenoate	顺式-4-癸烯酸乙酯
ethyl crotonate	巴豆酸乙酯
ethyl decanoate	癸酸乙酯
ethyl dodecanoate	十二酸乙酯
ethyl formate	甲酸乙酯

ethyl furoate	糠酸乙酯
ethyl hexanoate	己酸甲酯
ethyl laurate	月桂酸乙酯
ethyl linoleate	亚油酸乙酯
ethyl linolenate	亚麻酸乙酯
ethyl myristate	肉豆蔻酸乙酯
ethyl octadecanoate	十八酸乙酯
ethyl octadecenoate	十八烯酸乙酯
ethyl octanoate	辛酸乙酯
ethyl oleate	油酸乙酯
ethyl palmitate	棕榈酸乙酯
ethyl pentadecanoate	十五酸乙酯
ethyl pentanoate	戊酸乙酯
ethyl phenylpropionate	苯丙酸乙酯
ethyl P-methoxycinnamate	对甲氧基桂酸乙酯
ethyl propionate	丙酸乙酯
ethyl propyl ether	乙基丙基醚
ethyl salicylate	水杨酸乙酯
ethyl stearate	硬脂酸乙酯
ethyl t-butyl ether	乙基叔丁基醚
ethyl tetradecanoate	十四酸乙酯
ethylbenzaldehyde	乙基苯甲醛
ethylbenzene	乙基苯
ethylphenol	乙基苯酚
ethylpyrazine	乙基吡嗪
1-ethenyl-4-methoxy benzene	1-乙烯基-4-甲氧基苯
1-ethoxyethyl acetate	乙酸-1-乙氧基乙酯
1-ethyl-2,3-dimethylbenzene	1-乙基-2,3-二甲基苯
1-ethyl-2-hexenylcyclopropane	1-乙基-2-己烯基环丙烷
1-ethylpropylbenzene	1-乙基丙基苯
7-ethyl-4a,5,6,7,8,8a-hexahydro-1,4-dimethyl-2(1*H*)-naphthalenone	7-乙烯基-4a,5,6,7,8,8a-六氢-1,4-二甲基-2(1*H*)-萘酮
2-(5-ethenyltetrahydro-5-methyl 2-furanyl)-6-methyl-5-hepten-3-one	2-(5-乙烯基四氢-5-甲基-2-呋喃基)-6-甲基-5-庚烯-3-酮
2-ethenyl-2,5-dimethyl-4-hexenol	2-乙烯基-2,5-二甲基-4-己烯醇
2-ethylcyclobutanol	2-乙基环丁醇

2-ethylfuran	2-乙基呋喃
2-ethylheptanoic acid	2-乙基庚酸
3-ethyl-1, 2-dithi-4-cycloherene	3-乙基-1,2-二硫杂-4-环己烯
3-ethyl-2,5-dimethylpyrazine	3-乙基-2,5-二甲基吡嗪
4-ethenylcyclohexymethanol	4-乙烯基环己基甲醇
4-ethyl-2,6-di-t-butylphenol	4-乙基-2,6-二叔丁基苯酚
4-ethyl-2-methoxyphenol	4-乙基-2-甲氧基苯酚
6-ethyldihydro-2,2,6-trimethyl-2*H*-pyran-3(4*H*)-one	6-乙基二氢-2,2,6-三甲基-2*H*-吡喃-3(4*H*)-酮
6-ethyltetrahydro-2,2,6-trimethyl-pyran-3-ol	6-乙基四氢-2,2,6-三甲基-吡喃-3-醇
eucarvone	优藏茴香酮
β-eudesmol	β-桉叶醇
eugenol	丁香酚
eugenol 2,2-dimethylpropionate	2,2-二甲基丙酸丁香酯
eugenyl 2-methylbutanoate	2-甲基丁酸丁香酯
eugenyl 2-methylpropionate	2-甲基丙酸丁香酯
eugenyl pentanoate	戊酸丁香酯
eugenyl propionate	丙酸丁香酯
exo-isocamphenone	外-异莰烷酮

F

farneseal	金合欢醛
farnesene	金合欢烯
farneseol	金合欢醇
farnesyl acetate	乙酸金合欢酯
fenchene	葑烯
α-fenchene	α-小茴香烯
fenchol	小茴香醇
fenchol	葑醇
fencholic acid	小茴香酸
fenchone	小茴香酮
fenchone	葑酮
fenchyl acetate	乙酸小茴香酯
2-furaldehyde	2-呋喃甲醛 ,2-呋喃醛

frufuryl alcohol	糠醇
furfural, 2-furaldehyde	糠醛

G

geranene	香叶烯
geranial	香叶醛
geraniol	香叶醇
geranyl 2,2-dimethylbutyrate	2,2-二甲基丁酸香叶酯
geranyl 2,2-dimethylpropionate	2,2-二甲基丙酸香叶酯
geranyl 2-methylpropionate	2-甲基丙酸香叶酯
geranyl acetate	乙酸香叶酯
geranyl acetone	香叶基丙酮
geranyl formate	甲酸香叶酯
geranyl isobutyrate	异丁酸香叶酯
geranyl propionate	丙酸香叶酯
germacrene-D	吉玛烯-D
germacrone	大根香叶酮
germacrone	吉玛酮
germacrone	牻牛儿酮
guaiazulene	愈创木薁
guaiene	愈创木烯
guaiol	愈创木醇
gurjnene	古芸烯

H

henicosane	廿一烷
heptadecanal	十七醛
heptadecane	十七烷
heptadecanoic acid	十七酸
2-heptadecanone	2-十七酮
heptanal	庚醛
2-heptanone	2-庚酮
2-heptenal	2-庚烯醛
heptanoic acid	庚酸
2-heptenoic acid	2-庚烯酸

heptanol	庚醇
heptenol	庚烯醇
heptyl acetate	乙酸庚酯
heptyl isobutyrate	异丁酸庚酯
hexadecahydropyrene	十六氢芘
hexadecanal	十六醛
hexadecane	十六烷
hexadecanoic acid	十六酸
hexadecanol	十六醇
hexadecenal	十六烯醛
hexadecyl isovalerate	异戊酸十六酯
9-hexadecenoic acid	9-十六烯酸
hexahydrofarmesyl acetone	六氢金合欢基丙酮
hexanal	己醛
hexanoic acid	己酸
hexanol	己醇
2,4-hexadienal	2,4-己二烯醛
2,4-hexadienol	2,4-己二烯醇
1-hexene	1-己烯
5-hexen-2-one	5-己烯-2-酮
1-hexenyl formate	甲酸1-己烯酯
2-hexenal	2-己烯醛
2-hexenyl acetate	乙酸-2-己烯酯
2-hexylthiophene	2-己基塞吩
3-hexenoic acid	3-己烯酸
3-hexenol	3-己烯醇
3-hexenyl benzoate	苯甲酸-3-己烯酯
3-hexenyl butyrate	丁酸3-己烯酯
3-hexenyl hexanoate	己酸3-己烯酯
hexyl acetate	乙酸己酯
hexyl butyrate	丁酸己酯
hexyl formate	甲酸己酯
hexyl hexanoate	己酸己酯
hexyl tiglate	惕各酸己酯
hexylcyclohexane	己基环己烷
5-hexyldihydro-2(3H)-furanone	5-己基二氢-2(3H)-呋喃酮

himachanlene	雪松烯
hinesol	茅术醇
humuladienone	蛇麻二烯酮
humulene	蛇麻烯
humulene oxide	氧化蛇麻烯
β-humnlenyl-7 acetate	乙酸 β-蛇麻-7-烯酯
hydroxyacetophenone	羟基苯乙酮
1-(2-hydroxy-4-methoxyphenyl)ethanone	1-(2-羟基-4-甲氧基苯基) 乙酮
2-hydroxy-5-methoxyacetophenone	2-羟基-5-甲氧基苯乙酮
2-hydroxybenzhydrazide	2-羟基苯甲酰肼
2-hydroxymyristic acid	2-羟基肉豆蔻酸
3-hydroxy-2-butanone	3-羟基-2-丁酮
2-hydroxyphenylaldehyde	2-羟基苯甲醛
4-hydroxy-1,1,4,7-tetramethyldecahydro-cyclopropylazulene	4-羟基-1,1,4,7-四甲基十氢环
4-hydroxy-3-methoxyphenylaldehyde	4-羟基-3-甲氧基苯甲醛
4-hydroxyphenylacetonitrile	羟基苯乙腈
4-hydroxyy-5-methylacetophenone	4-羟基-5-甲基苯乙酮
α-hydroxy-aristolene-9	α-羟基-马兜铃烯 -9

I

indole	吲哚
ionic aldehyde	紫罗兰醛
ionol	紫罗兰醇
ionone	紫罗兰酮
irone	鸢尾酮
isobutyl butyrate	丁酸异丁酯
isobutyl phenylacetate	苯乙酸异丁酯
isocalamendid	异白菖二醇
iso-calamendiol	异水菖蒲二醇
iso-calamendiol	异菖蒲二醇
isooctanal	异辛醛
isooctanol	异辛醇
isopenoid	异戊二烯
isopentanol	异戊醇

isopentyl benzoate	苯甲酸异戊酯
isopentyl undecanoate	十一酸异戊酯
isopinocamphorone	异蒎樟脑酮
isopiperitenone	异胡椒烯酮
isopropyl decanoate	癸酸异丙酯
isopropyl-2-cyclohexanone	异丙基-2-环己酮
2-isopropyl-5-methylanisole	2-异丙基-5-甲基茴香醚
isopulegol	异胡薄荷醇
3-isothiocyanate-1-propene	3-异硫氰基-1-丙烯
4-isopropyl-3-cyclohexylmethanol	4-异丙基-3-环己基甲醇
4-isopropyltropolone	4-异丙基草酚酮
4-isothiocyanate-1-butene	4-异硫氯基-1-丁烯

J

jasmine lactone	茉莉内酯
jasmone	茉莉酮
jatamansin	甘松素
juniper camphor	刺柏脑
juniper camphor	桧脑

K

(-)-kauran-16 α -ol	(-)-16 α-贝壳杉醇
4-ketone-β-ionone, 4-oxo-β-ionone	4-酮基-β-紫罗兰酮

L

larandulol	熏衣草醇
lauric acid	月桂酸
laurye alcohol	月桂醇
ledol	喇叭茶醇
l-heptadecene	1-十七烯
ligustilide, ligusticum lactone	藁本内酯
limettin	白柠檬亭
limonene	柠檬烯
limonene oxide	氧化柠檬烯
linalool	芳樟醇

linalool oxide (pyran type)	氧化芳樟醇（吡喃型）
linalool oxide(furan type)	氧化芳樟醇（呋喃型）
linalyl acetate	乙酸芳樟酯
linalyl propionate	丙酸芳樟酯
linoleic acid	亚油酸
l-methylbutyl formate	甲酸2- 甲基丁酯
longicyclene	长叶环烯
longifolene	长叶烯
luparenol	卢杷烯醇
luparol	卢杷醇

M

β-maaliene	β- 马榄烯
maglianol	马榄烯醇
mayurone	麦由酮
memthol	薄荷醇 , 薄荷脑
1,8-menthadiene	1,8- 孟二烯
2,4-*p*-menthadiene	2,4- 对蓋二烯
menthene	蓋烯
menthene	薄荷烯
menthone	薄荷酮
1-*p*-menthen-8-ol	1- 对蓋烯 -8- 醇
1-*p*-menthen-8-yl acetate	乙酸1- 乙酸松油酯-8- 酯
1-*p*-menthen-9-al	1- 对蓋烯 -9- 醛
1-*p*-menthen-9-ol	1- 对薄荷烯-9- 醇
menthyl acetate	乙酸薄荷酯
messoialactone	玛索依内酯
1,4-*p*-menthadienol	1,4- 对蓋二烯醇
methoxy propyl bengene	对甲氧基丙基苯
methoxy-4-(2-propenyl)phenol	2- 甲氧基-4-(2- 丙烯基）苯酚
2-methoxy-4-(1-propenyl)phenol	2- 甲氧基-4-(1- 丙烯基）苯酚
2-methoxy-4-isoallyphenol	2- 甲氧基-4- 异烯丙基苯酚
2-methoxycinnamic aldehyde	2- 甲氧基桂醛
2-methoxyphenol	2- 甲氧基苯酚
2-methoxyphenylaldehyde	2- 甲氧基苯甲醛
1-(4-methoxyphenyl)ethanone	1-(4- 甲氧基苯基）乙酮

methyl 1,2-dimethyltridecanoate	1,2-二甲基十三酸甲酯
methyl 12-methyltridecanoate	12-甲基十三酸甲酯
methyl 14-methylpentadecancate	14-甲基十五酸甲酯
methyl 16-methylheptadecanoate	16-甲基十七酸甲酯
methyl 2- butenoate	2-丁烯酸甲酯
methyl 2,4-hexadienoate	2,4-己二烯酸甲酯
methyl 2,5-octadecadienoate	2,5-十八碳二烯酸甲酯
methyl 2,8-dimethylundecanoate	2,8-二甲基十一酸甲酯
methyl 2-formylaminobenzoate	2-甲酰氨基苯甲酸甲酯
methyl 2-hexenoate	2-己烯酸甲酯
methyl 2-hydroxy-3-methylpentanoate	2-羟基-3-甲基戊酸甲酯
methyl 2-hydroxybenzoate	2-羟基苯甲酸甲酯
methyl 2-methoxybenzoate	2-甲氧基苯甲酸甲酯
methyl 2-methylbutanoate	2-甲基丁酸甲酯
methyl 2-methylenebutyrate	2-亚甲基丁酸甲酯
methyl 2-octenoate	2-辛烯酸甲酯
methyl 3-methyl-2-butanoate	3-甲基-2-丁烯酸甲酯
methyl 3-phenyl-2-propenoate	3-苯基-2-丙烯酸甲酯
methyl 3-propylbenzoate	3-丙基苯甲酸里酯
methyl 4-decenoate	4-癸烯酸甲酯
methyl 4-methoxybutyrate	4-甲氧基丁酸甲酯
methyl 4-methyl-5-phenyl-3-carbonylpentanoate	4-甲基-5-苯基-3-羰基戊酸甲酯
methyl 7-methylnonanoate	7-甲基壬酸甲酯
methyl allyl pentasulfide	甲基烯丙基五硫醚
methyl allyl tetrasulfide	甲基烯丙基四硫醚
methyl allyl trisulfide	甲基烯丙基三硫醚
methyl anthranilate	邻氨基苯甲酸甲酯
methyl benyl ether	甲基苯甲醚
methyl benzoate	苯甲酸甲酯
methyl butyl ether	甲基丁基醚
methyl butyrate	丁酸甲酯
methyl chavicol	黑椒酚甲醚
methyl cinnamate	桂酸甲酯
methyl citronellate	香茅酸甲酯
methyl cyclopropanenonanoate	环丙烷壬酸甲酯
methyl decanoate	癸酸甲酯

methyl dodecanoate	十二酸甲酯
methyl eugenol	丁香酚甲酯
methyl furoate	糠酸甲酯
methyl geranate	香叶酸甲酯
methyl heptadienone	甲基庚二烯酮
methyl heptanoate	庚酸甲酯
methyl hexadecanoate	十六酸甲酯
methyl hexanoate	己酸乙酯
methyl isohexanoate	异己酸甲酯
methyl isopropionate	异丙酸甲酯
methyl jasmonate	茉莉酮酸甲酯
methyl linoleate	亚油酸甲酯
methyl linolenate	亚麻酸甲酯
methyl methylthiolacetate	甲基硫烃乙酸甲酯
methyl myristate	肉豆蔻酸甲酯
methyl nonanoate	壬酸甲酯
methyl octadecynoate	十八炔酸甲酯
methyl octanoate	辛酸甲酯
methyl oleate	油酸甲酯
methyl palmitate	棕榈酸甲酯
methyl pentanoate	戊酸甲酯
methyl pentyl ketone	甲基戊基酮
methyl phenethy alcohol	甲基苯乙醇
methyl phenylethyl ether	甲基苯乙基醚
α-methyl-α-(2-propenyl)-phenylmethanol	α-甲基-α-(2-丙烯基)-苯甲醇
methyl salicylate	水杨酸甲酯
methyl stearate	硬脂酸甲酯
methyl tetradecanoate	十甲酸甲酯
methyl tiglate	惕各酸甲酯
methyl tridecanoate	十三酸甲酯
methyl α-acetoxyphenylacetate	α-乙酰氧基苯乙酸甲酯
2-(1-methyl-2-isopropenylcyclobutyl)-ethanol	2-(1-甲基-2-异丙烯基环丁基)乙醇
12-methyldihydrocostunolide	12-甲基二氢木香内酯
methylerythritol phosphate (MEP)	磷酸甲基赤藓糖
methylheptenone	甲基庚烯酮
methylnonanone	甲基壬酮

methylododecanone	甲基十二酮
methylpentanone	甲基戊酮
methylpentene	甲基戊烯
methylpyrazine	甲基吡嗪
1-methyl-2-(2-propyl)cyelohexane	1-甲基-2-(2-丙基) 环戊烷
1-methyl-2-ethylbenzene	1-甲基-2-乙基苯
1-methyl-2-isopropylbenzene	1-甲基-2- 异丙基苯
1-methyl-2-pentylcyclopropane	1-甲基-2-戊基环丙烷
1-methyl-3-phenylpropanol	1-甲基-3-苯基丙醇
1-methyl-4-(1-methylethyl)cyclohexanol	1-甲基-4-(1-甲基乙基) 环己醇
1-methyl-4-isopropyl-1,3-cyclohexadiene	1-甲基-4-异丙基-1,3- 环己二烯
1-methyl-5, 6-diethylcyclohexanone	1-甲基-5,6-二乙基环己酮
1-methyl-l,2-diethyl-5-cyclohexene	1-甲基-1,2-二乙烯基-5-环己烯
1-methylnaphthalene	1-甲基萘
1-methylpropyl butyrate	丁酸1-甲基丙酯
1-methylene-2-methyl-4-isopropylcyclohexane	1-亚甲基-2- 甲基-4- 异丙基环己烷
1-methyl-l,2-diethylene-5-cyclohexene	1-甲基-1,2- 二亚乙基-5-环己烯
2-methyl benzofuran	2-甲基苯并呋喃
2-methyl butanoic acid	2-甲基丁酸
2-methyl decahydronaphthalene	2-甲基十氢萘
2-methyl-2-butenal	2-甲基-2-丁烯醛
2-methyl-2-butenol	2-甲基-2-丁烯醇
2-methyl-2-propenyl acetate	乙酸2-甲基-2-丙烯酯
2-methyl-3-(1-methylethyl)oxirane	2-甲基-3-(1-甲基乙基) 环氧乙烷
2-methyl-3-buten-2-ol	2-甲基-3-丁烯-2-酮
2-methyl-3-pentanol	2-甲基-3-戊醇
2-methyl-5-(1-methylethenyl)cyclohexanone	2-甲基-5-(1-甲基乙烯基) 环己酮
2-methyl-5-(1-methylethy)-2,5-cyclohexadiene-1,4-dione	2-甲基-5-(1-甲基乙基)-2,5-环己二烯-1,4-二酮
2-methyl-6-methylene-1,7-octadiene-3 -one	2-甲基-6-亚甲基-1,7-辛二烯-3-酮
2-methyl-6-propylundecane	2-甲基-6-丙基十一烷
2-methyl-8-propyldecane	2-甲基-8-丙基癸烷
2-methylbutanal	2-甲基丁醛
2-methylbutane	2-甲基丁烷
2-methylbutyl 2-methylpropionate	2-甲基丙酸2-甲基丁酯
2-methylbutyl 3-methylbutanoate	3-甲基丁酸2-甲基丁酯

2-methylbutyl acetate	乙酸2-甲基丁酯
2-methylbutyl propionate	丙酸2-甲基丁酯
2-methylcyclopentanol	2-甲基环戊醇
2-methylcyclopentenol	2-甲基环戊烯醇
2-methylene-6-methyl-5,7-octadiene	2-亚甲基-6-甲基-5,7-辛二烯
2-methylformamide	2-甲基甲酰胺
2-methylhexadecane	2-甲基十六烷
2-methylhexane	2-甲基己烷
2-methylhexanoic acid	2-甲基己酸
2-methylpropyl 2-methyl-2-propenoate	2-甲基-2-丙烯酸2-甲基丙酯
2-methylpropyl 2-methylbutanoate	2-甲基丁酸2-甲基丙酯
2-methylpropyl 2-methylonate	2-甲基丙酸2-甲基丙酯
2-methyltetradecane	2-甲基十四烷
2-methypropyl hexanoate	己酸2-甲基丙酯
3-methoxy-3-methyl-2-butanone	3-甲氧基-3-甲基-2-丁酮
3-methyl-1H-pyrazole	3-甲基-1H-吡唑
3-methyl-2-(1,8-pentadienyl)-2-cyclopentenone	3-甲基-2-(1,3-戊二烯基)-2-环戊烯酮
3-methyl-2-butenyl acetate	乙酸3-甲基-2-丁烯酯
3-methyl-2-pentyl-2-cyclohexenone	3-甲基-2-戊基-2-环己烯酮
3-methyl-3-butanone	3-甲基-3-丁酮
3-methyl-6-hydroxyacetophenone	3-甲基-6-烃基苯乙酮
3-methylbutanol	3-甲基丁醇
3-methylbutyl 2-phenydlacetate	2-苯基乙酸3-甲基丁酯
3-methylbutyl butyrate	丁酸3-甲基丁酯
3-methylbutyl isovalerate	异戊酸3-甲基丁酯
3-methylcyclopentene	3-甲基环戊烯
3-methylene-2,2-dimethylbicyclo[2.2.1] heptane	3-亚甲基-2,2-二甲基二环 [2.2.1] 庚烷
4-methoxy-1-t-butoxybenzene	4-甲氧基-1-叔丁氧基苯
4-methoxydiphenylacetylene	4-甲氧基二苯基乙炔
4-methoxytoluene	4-甲氧基甲苯
4-methyl-1,2-dithi-3-cyclopentene	4-甲基-1,2-二硫杂-3-环戊烯
4-methyl-2,6-di-t-butylphenol	4-甲基-2,6-二叔丁基苯酚
4-methyl-2-heptanone	4-甲基-2-庚酮
4-methyl-3-pentenal	4-甲基-3-戊烯醛
4-methyl-4-ethyl-2-cyclohexenone	4-甲基-4-乙基-2-环己烯酮
4-methyl-4-hydroxy-2-pentanone	4-甲基-4-烃基-2-戊酮

4-methyl-4-pentene-2-one	4-甲基-4-戊烯-2-酮
4-methyl-6-acetylhexanal	4-甲基-6-乙酰基己醛
4-(5-methyl-2-furanyl)-2-butanone	4-(5-甲基-2-呋喃基)-2-丁酮
5-methyl-2-furanaldehyde	5-甲基-2-呋喃醛
5-methyl-2-hexanol	5-甲基-2-己醇
5-methyl-2-isopropyl-2-cyclohexenene	5-甲基-2-异丙基-2-环己烯
6-methyl-5-hepten-2-ol	6-甲基-5-庚烯-2-醇
6-methylbicyclo[3.2.0]hept-6-en-2-one	6-甲基二环 [3.2.0] 庚-6- 烯-2-酮
6-methylheptanol	6-甲基庚醇
7-methyl-4-decene	7-甲基-4-癸烯
5-methylhexanal	5-甲基己醛
8-methylheptadecane	8-甲基十七烷
5-(1-methylethyl)bicyclo[3. 1. 0]hexan-2-one	5-(1-甲基乙基) 二环 [3.1.0] 己 -2-酮
1-methoxy-3,7-dimethyl-2,6-octadiene	1-甲氧基-3,7-二甲基-2,6-辛二烯
mevalonate (MVA)	甲羟戊酸
monocyclic sesquiterpene	单环倍半萜
monocyclic monoterpene	单环单萜
myrcene	月桂烯
myrcenol	月桂烯醇
myrcenyl acetate	乙酸月桂烯酯
myricyl alcobol	蜂花醇
myristic acid	肉豆蔻酸
myristicin	肉豆蔻醚
myrtan acetate	乙酸桃金娘酯
myrtanal	桃金娘醛
myrtanol	桃金娘醇
myrtenal	桃金娘烯醛
myrtenol	桃金娘烯醇

N

naphthalene	萘
nardol	甘松奠醇
nardosinone	甘松酮
nardostachnol	甘松醇
n-butylisobutyl phthalate	邻苯二甲酸丁基异丁基酯

n-eicosanol	正廿醇
neocnidilide	新蛇床内酯
neocurzerene	杜鹃次烯
neofuranodinene	杜鹃烯
neomenthol	新薄荷醇, 新薄荷脑
neral	橙花醛
nerol	橙花醇
nerolidol	橙花叔醇
neryl acetate	乙酸橙花酯
neryl propionate	丙酸橙花酯
n-heptadecanol	正十七醇
1,2,3,4,4a,7-hexahydro-1,6-dimethyl-4- 　(1-methylethyl)naphthalene	1,2,3,4,4a,7-六氢-1,6-二甲基-4- 　(1-甲基乙基) 萘
n-hexyl isobutyrate	异丁酸正己酯
nonadecane	十九烷
nonadecanoic acid	十九酸
1-nonadecene	1-十九烯
2,4-nonadienal	2,4-壬二烯醛
nonanal	壬醛
nonane	壬烷
2-nonanone	2-壬酮
nonanoic acid	壬酸
nonanol	壬醇
nonanolide	壬内酯
nonene	壬烯
1-nonen-3-ol	1-壬烯-3-醇
8-nonen-2-one	8-壬烯-2-酮
3-nonenal	3-壬烯醛
2-nonenoic acid	2-壬烯酸
nonyl acetate	乙酸壬酯
nootkatene	诺卡烯
nootkatene	黄柏烯
nootkatone	诺卡酮
nootkatone	黄柏酮
norlapachol	降拉帕醇
n-pentylbenzene	正戊基苯

N-phenyl-1-naphthalenamine	N-苯基-1-萘胺
N-phenylanilline	N-苯基苯胺
N-phenylformamide	N-苯基甲酰胺

O

occidentalol	金钟柏醇
ocimene	罗勒烯
octadecanal	十八醛
octadecane	十八烷
octadecene	十八烯
9-octadecenal	9-十八烯醛
9,12-octadecadienal	9,12-十八碳二烯醛
9,12-octadialdehyde	9,12-辛二烯醛
1,2,3,4,4a,7,8,8a-octahydro-l,6-dimethyl-4-(1-methylethyl) naphtha lenol	1,2,3,4,4a,7,8,8a-八氢-1,6-二甲基-4-(1-甲基乙基) 萘醇
octanal	辛醛
octane	辛烷
octanoic acid	辛酸
octanol	辛醇
2-octanone	2-辛酮
2-octene	2-辛烯
octenol	辛烯醇
octenyl acetate	乙酸辛烯酯
octyl acetate	乙酸辛酯
octyl isobutyrate	异丁酸辛酯
octylcyclopropane	辛基环丙烷
γ -onantholactone	γ-庚内酯
o-hydroxylamine	邻癸基羟胺
o-methylacetophenone	邻甲基苯乙酮
o-methylanethole	邻甲基茴香醚
oppositol	对凹顶藻醇
o-propenyltoluene	邻丙烯基甲苯
osthole	欧芹酚甲醚
9-oxabicyclo[6.1.0]nonane	9-氧杂二环 [6.1.0] 壬烷

8-oxabicyclo[5.1.0]octane   8-氧杂二环 [5.1.0] 辛烷

9-oxonerolidol   9-氧化橙花叔醇

P

palmitic acid   棕榈酸

palmitoleic acid   棕榈油酸

parthenin   银胶菊素

patchoulane   广藿香烷

patchoulene   广藿香烯

patchoulipyridine   广藿香吡啶

patchoulol   广藿香醇

p-benzenediol   对苯二酚

p-dimethylbenzene   对二甲苯

pentacosane   廿五烷

pentadecane   十五烷

2-pentadecenol   2-十五烯醇

pentadecanoic acid   十五酸

7,10-pentadecadiynoic acid   7,10-十五碳二炔酸

pentadienylcyclopentane   戊二烯基环戊烷

pentanal   戊醛

pentanoic acid   戊酸

3-pentanol   3-戊醇

3-pentanone   3-戊酮

1-penta cosene   1-廿五烯

1-pentadccene   1-十五烯

5-pentoxy-2-pentene   5-戊氧基-2-戊烯

2-pentylfuran   2-戊基呋喃

pentyl acetate   乙酸戊酯

pentyl butyrate   丁酸戊酯

pentylcyclopropane   戊基环丙烷

perillaldehyde   紫苏醛

perillene   紫苏烯

petadecanal   十五醛

petadecenal   十五烯醛

p-ethylbenzaldehyde   对乙基苯甲醛

phellandrene	水芹烯
phellandrol	水芹醇
phenol	苯酚
phenyl benzoate	苯甲酸苯酯
phenyl isovalerate	异戊酸苯酯
phenyl methyl ether	苯甲醚
phenyl *o*-hydroxybenzoate	邻羟基苯甲酸苯酯
phenyl-2-propanone	苯基-2-丙醇
3-phenyl-2-propenal	3-苯基-2-丙烯醛
3-phenyl-2-propenol	3-苯基-2-丙烯醇
phenylacetaldehyde	苯乙醛
phenylbenzene	联苯
phenylethanol	苯乙醇
phenylethyl acetate	乙酸苯乙酯
phenylethyl isobutyrate	异丁酸苯乙酯
2-phenyl-2-propanol	2-苯基-2-丙醇
4-phenyl-2-butanone	4-苯基-2-丁酮
1,1-(1,4-phenylene)biethanone	1,1-(1,4-亚苯基)双乙酮
2-phenylenthanol	2-苯基乙醇
2-phenylethy benzoate	苯甲酸-2-苯乙酯
2-phenylethyl 2-methylpropionate	2-甲基丙酸2-苯基乙酯
β-phenylethyl isovalerate	异戊酸β-苯乙酯
β-phenylethyl propionate	丙酸β-苯乙酯
phenylpropanal	苯丙醛
phenylpropanol	苯丙醇
phenylpropyl acetate	乙酸苯丙酯
5-phenylmethoxypentanol	5-苯甲氧基戊醇
1,4-phthalic aldehyde	1,4-苯二甲醛
phytane	植烷
phytol	植物醇, 植醇
phytol ketone	植物酮
picrotin	印防己苦内酯
picrotoxinin	印防己毒内酯
pimara-8, 15-diene	海松-8,15-二烯
pinene	蒎烯
pinocamphone	松樟酮

pinocarveol	蒎葛缕醇
piperfenol methylether	胡椒酚甲醚
piperitenone	胡椒烯酮
piperitol	胡椒醇
piperitol	辣薄荷醇
piperitone	胡椒酮
piperitone	辣薄荷酮
piperitone oxide	氧化辣薄荷酮
piperonal	胡椒醛
piperonal	洋茉莉醛
p-isobutyltoluene	对异丁基甲苯
p-isopropyl benzoic acid	对异丙基苯甲酸
p-methylethyl-benzaldehyde	对甲基乙基苯甲醛
p-octylmethoxybenzene	对辛基甲氧基苯
podocarprene	罗汉松烯
pogostol	广藿香奠醇
portulal	马齿苋醛
p-propylphenol	对丙基苯酚
preisocalamendiol	前异白菖二醇
pronylcyelpropane	丙基环丙烷
propenyl benzoate	苯甲酸丙烯酯
propenyl pyrocatechol	丙烯基焦儿茶酚
propoxyanisole	丙氧基茴香醚
propyl 2-methylpropionate	2-甲基丙酸丙酯
propyl cinnamate	桂酸丙酯
propyl isothiocyanate	异硫氰酸丙酯
propyl salicylate	水杨酸丙酯
propylphenylaldehyde	丙基苯甲醛
6-propylbicyclo[3.2.0]hept-3,6-dien-2-one	6-丙基二环 [1.2.0] 庚-3,6-二烯-2-酮
6-propylbicyclo[3.2.0]hept-6-en-2-one	6-丙基二环 [3.2.0] 庚-6-烯-2-酮
protoplasma	细胞原生体
pseudo-limonene	伪柠檬烯
p-t-butylpheny methanol	对叔丁基苯甲醇
pulegone	胡薄荷酮
pyridine	吡啶

2-pyridinecarbonitrile	2-吡啶腈
4-pyridinecarboxylic acid hydrazide	4-吡啶羧酸酰肼
1*H*-pyrrole-2-formaldehyde	1*H*-吡咯-2-甲醛
1-(1*H*-pyrrole-2-ly)ethanone	1-(1*H*-吡咯-2-基) 乙酮

R

rimuene	芮木烯
robe oxide (pyran type)	氧化玫瑰(吡喃型)
rose oxide	玫瑰醚

S

sabinene	桧烯
sabinene hydrate	水合桧烯
sabinol	桧醇
sabinyl acetate	乙酸桧酯
safranal	藏红花醛
safrole	黄樟油素
salicylic acid	水杨酸
santalene	檀香烯
santalol	檀香醇
santene	檀烯
saussurea lactone	风毛菊内酯
scaridole	驱蛔素
sedanenolide	瑟丹烯内酯
selina-3,7(11)-diene	芹子-3,7(11)- 二烯
selinene	芹子烯
selinenol	芹子醇
sesquicitronellene	倍半萜香茅烯
β-sesquiphellandrene	β-倍半水芹烯
α-sinensal	α-甜橙醛
sirenin	雌诱素
somtonin	蛔蒿素
spathulenol	斯潘连醇
spiro[4.5]dec-1-ene	螺 [4.5] 癸-1-烯
strigol	独角金醇

styrene	苏合香烯
sylevestrene	枞油烯

T

tegatenone	万寿菊烯酮
1,2,9,10-tetradehydroaristolane	1,2,9,10-四脱氢马兜铃烷
11-tetradecenyl acetate	乙酸-11-十四烯酯
teresantalol	对檀香醇
terpene	萜
terpin-4-yl acetate	乙酸松油醇-4-酯
terpinene	松油烯
terpinol	松油醇
terpinolene	异松油烯
1,8-terpin hydrate	水合-1,8-松油二醇
terpinyl acetate	乙酸松油酯
tetradecanal	十四醛
tetradecane	十四烷
tetradecanoic acid	十四酸
tetradecanone	十四酮
tetrasantalol	四檀香醇
2,4,6,14-tetramethyl pentadecane	2,4,6,14-四甲基十五烷
3,7,11,15-tetramethyl-2-hexadecenol	3,7,11,15-四甲基-2-十六烯醇
thujanol	侧柏醇
thujene	侧柏烯
thujenol	侧柏烯醇
thujone	侧柏酮
thujopsene	斧柏烯
thujopsene, widdrene	罗汉柏烯
β-thujyl acetate	乙酸 β-侧柏酯
thymol	麝香草酚
thymol acetate	乙酸麝香草酯
thymol methyl ether	麝香草酚甲醚
thymyl hydroquinone dimethylether	麝香草氢醌二甲醚
thymyl isobutyrate	异丁酸麝香草酯
thymyl pentanoate	戊酸麝香草酯

2-tolyl acetate	乙酸2-甲苯酯
torreyol	香榧醇
trans-2-octenal	反式-2-辛烯醛
trans-4-hydroxy-3-methyl-6-isopropyl-2-cyclohexenone	反式-4-羟基-3-甲基-6-异丙基-2-环己烯酮
trans-marmelolactone	反式-紫花前胡内酯
trans-methadiene	反式-蓋二烯
trans-*p*-methan-9-ol	反式-对蓋-9-醇
tricosane	廿三烷
1-tricosene	1-廿三烯
1-tridecene	1-十三烯
tricyclene	三环烯，三环萜
tricyclo[3.2.1.01.5]octane	三环[3.2.1.01.5]辛烷
tricyclocaryophyllene	三环石竹烯
2-tridecanone	2-十三酮
tridecanal	十三醛
trigonelline	胡芦巴碱
1,2,4-triethylbenzene	1,2,4-三乙基苯
1,3,3-trimethyl-2-oxabicyclo[2.2.2]octane	1,3,3-三甲基-2-氧杂二环[2.2.2]辛烷
1,2,4-trimethylbenzene	1,2,4-三甲苯
2,2,3-timethyl-3-cyclopentenylacetaldehyde	2,2,3-三甲基-3-环戊烯基乙醛
trimethoxy-allylbenzene	三甲氧基烯丙基苯
trimethoxyphenylpropene	三甲氧基苯丙烯
1,1,5-trimethyl-2-acetyl-2,5-cyclohexadiene-4-one	1,1,5-三甲基-2-乙酰基-2,5-环己二烯-4-酮
2,2,3-trimethyl bicyclo[2.2.1]hept-2-ol	2,3,3-三甲基二环[2.2.1]庚-2-醇
2,2,6-trimethyl-1,4-cyclohexanedione	2,2,6-三甲基-1,4-环己二酮
3,3,5-trimethy cyclohexene	3,3,5-三甲基环己烯
3,3,5-trimethyl-4-(3-hydroxy-1-butenyl)-2-cyclohexen-1-one	3,3,5-三甲基-4-(3-羟基-1-丁烯基)-2-环己烯-1-酮
(-)-*cis*-2,6,6-trimethyl-2-ethenyl-5-hydroxytetrahydropyran	双花醇
1-(2,6,6-trimethyl-1,3-cyclohexadienyl)-2-utenone	1-(2,6,6-三甲基-1,3-环己二烯基)-2-丁烯酮
2,4,6-trimethylphenylaldehyde	2,4,6-三甲基苯甲醛

2,5,5-trimethyl-1,3,6-heptatriene	2,5,5-三甲基-1,3,6-庚三烯
2,5,5-trimethyl-1,6-heptadiene	2,5,5-三甲基-1,6-庚二烯
2,5,6-trimethyl-2,6-heptadien-4-0ne	2,5,5-三甲基-2,6-庚二烯-4-酮
2,6,6-trimethyl-1-cyclobexenylcarbaldehyde	2,6,6-三甲基-1-环己烯基甲醛
2,6,6-trimethyl-2-vinyl-5-hydroxy-pyran	2,6,6-三甲基-2-乙烯基-5-羟基吡喃
3,4,5-trimethyl-2-cyclopentenone	3,4,5-三甲基-2-环戊烯酮
3,4,6-trimethoxyphenylaldehyde	3,4,5-三甲氧基苯甲醛
3,5,5-trimethylhexanol	3,5,5- 三甲基己醇
3,6,5-trimethyl-2-cyclohexene-1,4-dione	3,5,5-三甲基-2-环己烯-1,4-二酮
3,6,6-trimethyl bicyclo[3.1.1]hept-2-ene	3,6,6-甲基二环 [3.1.1] 庚-2-烯
3,6,6-trimethyl-2-norpinaneol	8,6,6-三甲基-2- 降蒎醇
4,6,6-trimethyl bicyclo[3.1.1]hept-3-ene-2-one	4,6,6-三甲基二环 [3.1.1] 庚-3-烯-2-酮

4-(2,6,6-trimethyl-2-cyclohexen-1-ylidene)-
    2-butanone
	4-(2,6,6-三甲基-2-环己烯亚基)-
	    2-丁酮

4,4,6-timethylcyclohexadienone	4,4,6-三甲基环己二烯酮
4,4,6-trinethyl-2-cyclohexenone	4,4,5-三甲基-2-环己烯酮

3,7,11-trimethyl-14-(1-methylethy1)-
    1,3,6,10-cyclotetradecatetraene
	3,7,11-三甲基-14-(1-甲基乙基)-
	    1,3,6,10-环十四碳四烯

6,10,14-trimethyl pentadecane	6,10,14-三甲基十五烷
6,10,14-trimethyl-2-pentadecanone	6,10,14- 三甲基-2-十五酮
turmerone	姜黄酮
tutin	羟基马桑毒素

U

umbellulone	加州月桂酮
1,4-undecadiene	1,4-十一碳二烯
undecanal	十一醛
undecane	十一烷
2-undecanone	2-十一酮
2-undecenal	2-十一烯醛
10-undecenol	10-十一烯醇
undecanoic acid	十一酸
undecanol	十一醇
undecene	十一烯

V

valencene	瓦伦烯
valeranone	缬草酮
valerene	缬草烯
valeri acid	缬草酸
vanillin	香兰素
vanillyl acetate	乙酸香兰酯
verbanone	马鞭草酮
verbenol	马鞭草烯醇
verbenone	马鞭草烯酮
vetivene	岩兰草烯
vetivenic acid	岩兰草酸
vetiver ketone	岩兰草酮
vetiverol	岩兰草醇

Y

ylangene	依兰烯
ylangoilalcohol	依兰油醇
ylangoilene	依兰油烯

Z

zingiberene	姜烯

后　记

　　《植物精油（中英文版）》的书稿经过构思、筹划、组织、编辑、出版，终于要与广大读者见面了。感谢所有参编者的辛苦努力，同时还要感谢以往长期共事的同行们，是他们的研究成果给了我们很多的启发和数据支持，为图书的出版提供宝贵的资源。感谢广州市浩立生物科技有限公司朱宝璋教授、冯志豪总经理的大力支持，阳江市有故事的香科技有限公司高志恳先生，茂名市淘婷香农业科技有限公司杨桂娣总经理、广东德盛沉香发展有限公司提供有关沉香结香的照片。在此我以作者身份再次向陆碧瑶、李宝灵、刘驰、李用华、罗友娇、陈振焕、徐金富、程世法、李毓敬、朱宝璋、冯志豪、曾幻添、高志恳、杨桂娣等表示衷心的感谢。最后我还要特别感谢广州柏桐文化传播有限公司吴文静、王颢颖女士以及团队成员的鼎力相助！